# Pitman Research Notes in Mathematics Series



## Submission of proposals for consideration

Suggestions for publication, in the form of outlines and representative samples, are invited by the Editorial Board for assessment. Intending authors should approach one of the main editors or another member of the Editorial Board, citing the relevant AMS subject classifications. Alternatively, outlines may be sent directly to the publisher's offices. Refereeing is by members of the board and other mathematical authorities in the topic concerned, throughout the world.

## Preparation of accepted manuscripts

On acceptance of a proposal, the publisher will supply full instructions for the preparation of manuscripts in a form suitable for direct photo-lithographic reproduction. Specially printed grid sheets can be provided and a contribution is offered by the publisher towards the cost of typing. Word processor output, subject to the publisher's approval, is also acceptable.

Illustrations should be prepared by the authors, ready for direct reproduction without further improvement. The use of hand-drawn symbols should be avoided wherever possible, in order to maintain maximum clarity of the text.

The publisher will be pleased to give any guidance necessary during the preparation of a typescript, and will be happy to answer any queries.

## Important note

In order to avoid later retyping, intending authors are strongly urged not to begin final preparation of a typescript before receiving the publisher's guidelines. In this way it is hoped to preserve the uniform appearance of the series.


**Longman Scientific & Technical**
**Longman House**
**Burnt Mill**
**Harlow, Essex, CM20 2JE**
**UK**
**(Telephone (0279) 426721)**

## Hiroshi Kunita

Kyushu University, Japan

and

## Hui-Hsiung Kuo

Louisiana State University, USA

(Editors)

---

# Stochastic analysis on infinite dimensional spaces

## Proceedings of the U.S.–Japan Bilateral Seminar, January 4–8 1994, Baton Rouge, Louisiana

Longman Scientific & Technical

Copublished in the United States with
John Wiley & Sons, Inc., New York

**Longman Scientific & Technical**
Longman Group Limited
Longman House, Burnt Mill, Harlow
Essex CM20 2JE, England
*and Associated companies throughout the world.*

*Copublished in the United States with*
*John Wiley & Sons Inc., 605 Third Avenue, New York, NY 10158*



First published 1994

AMS Subject Classifications: (Main) 46F25, 58G32, 60F10
(Subsidiary) 60E07, 60H15, 60H20

ISSN 0269-3674

ISBN 0 582 24490 0

**British Library Cataloguing in Publication Data**

A catalogue record for this book is
available from the British Library

**Library of Congress Cataloging-in-Publication Data**

Kunita, H.
Stochastic analysis on infinite dimensional spaces / H. Kunita, H.
-H. Kuo.
p. cm. -- (Pitman research notes in mathematics ; )
1. Stochastic analysis. 2. Function spaces. I. Kuo, H.-H.
II. Title. III. Series.
QA274.2.K85  1994
519.2--dc20                                    94-10257
                                                   CIP

Printed and bound in Great Britain
by Biddles Ltd, Guildford and King's Lynn

# Contents

# Preface

The U.S.-Japan Bilateral Seminar *"Stochastic Analysis on Infinite Dimensional Spaces"* was held at Louisiana State University, January 4-8, 1994. The seminar covered the following topics:

(1) Stochastic analysis related to Lie groups.

(2) Stochastic partial differential equations.

(3) Stochastic flows and analysis on Wiener functionals.

(4) Large deviations.

(5) White noise calculus.

(6) Stable laws.

This volume is the collection of all lectures delivered during this seminar. We would like to thank all contributors for their participation in this seminar and for their effort to prepare the manuscript in a timely manner.

This seminar was supported by the National Science Foundation and the Japan Society for the Promotion of Science. It is our pleasure to thank both organizations. We thank also the Mathematics Department and the College of Arts and Sciences of Louisiana State University for their assistance to the seminar. Our special thanks go to student worker Cam Nguyen who did a superb job in handling the secretarial work for the meeting and to Fukuko Kuo who wrote beautiful calligraphic name tags for the participants. Finally, we would like to thank the members of the Local Committee, W. G. Cochran, A. Sengupta, and P. Sundar, for their contribution to the success of the meeting.

May 25, 1994

Hiroshi Kunita
Kyushu University

Hui-Hsiung Kuo
Louisiana State University

A. ALEMAN, B. S. RAJPUT, & S. RICHTER

# On an extremal problem in $\mathbf{H}^p$ and prediction of p-stable processes $0 < p < 1$

## 1. Introduction

Let $0 < p < \infty$, and let $\mathbf{H}^p$ denote the Hardy space of the unit disc $\mathbf{D}$. Let $\varphi(z) = \sum_{k=0}^{N-1} a_k z^k$, $a_0 \neq 0$. The (dual) extremal problem of finding all possible function(s) in $\mathbf{H}^p$ closest to the rational kernel $z^{-N}\varphi$ is of interest in analysis [3, p.136] as well as in other fields [2, 9]. If $1 \leq p$, then it is well known that $z^{-N}\varphi$ has a unique best approximation ($\equiv$ closest element) in $\mathbf{H}^p$, and it can be described "more or less explicitly" [3, p.136]. Making use of this, it is pointed out in [2] (see also [9]) that under an additional condition this best approximation is $\mathbf{P}_N(\varphi) \equiv z^{-N}[\varphi - \{P_N(\varphi^{p/2})\}^{2/p}]$, where $P_N(\varphi^{p/2})$ is the polynomial of the first N terms of $\varphi^{p/2}$ (see Remark 2 for more on this point). Cambanis and Soltani [2] used this representation to derive the N-step linear predictor for a discrete time harmonizable regular p-stable process $1 \leq p < 2$ (for the case p=1, one must supplement the noted formula of [2] with the Remark 2.2.4(a) of [8]). For the case $1 \leq p \leq 2$, the corresponding (dual) extremal problem of the upper half plane was recently solved by Rama-Murthy et al [8]. Further, using this solution they derived the linear predictor for a continuous time harmonizable regular p-stable process, $1 \leq p < 2$. In a more recent paper [9], a simple and unified method is used to prove the fact "$\mathbf{P}_N(\varphi)$ is the best approximation of $z^{-N}\varphi$" and its counterpart in the upper half plane, for all $1 \leq p < \infty$. In this paper the noted discrete and continuous time linear predictors have also been obtained in a unified manner.

In one of the results of the present paper (Theorem 1), we provide a solution to the above extremal problem for all $0 < p < \infty$. Specifically, we show that a best approximation of $z^{-N}\varphi$ always exists in $\mathbf{H}^p$ and that every best approximation admits a particular form which coincides with the known formula when $p \geq 1$ [3, p.138]. The classical proof of this result for the case $p \geq 1$ is based on duality methods [3, pp.133-138]. These methods are not applicable in the case when $0 < p < 1$. Our proof relies upon a modification of the usual $L^p$-variational argument, and, of course, we also use additional standard tools available in all $\mathbf{H}^p$ spaces. Making use of Theorem 1, we derive explicit formulae for the best approximations of $z^{-N}\varphi$ in $\mathbf{H}^p$,

$0 < p < 1$, for $N = 1, 2$ . As a consequence of these formulae, we show that, unlike the case $1 \leq p$, $z^{-2}\varphi$ may have more than one best approximations in $\mathbf{H}^p$ and that $\mathbf{P}_2(\varphi)$ may not be one of the best approximations (Theorem 2, Remark 2). Using Theorem 2, we provide precise formulae for one- and two-step linear predictors and the corresponding prediction errors for a discrete parameter harmonizable regular p-stable process, $0 < p < 1$ (Theorem 3). Here also, unlike the case $1 \leq p < 2$, there can be two two-step linear predictors, and, moreover, some of these predictors can have a different form than the form of their counterparts for the case $1 \leq p < 2$. These facts are included in Remark 3 and Theorem 3. This Remark and Theorem complement the results concerning linear predictors for p-stable processes, $1 \leq p$, obtained in [2, 4, 8, 9].

## 2. Notation and Preliminaries

Throughout, for simplicity of notation, we will abbreviate the integral of a function $f$ over $[0, 2\pi]$ relative to the normalized Lebesgue measure $\frac{d\theta}{2\pi}$ with $\int f$. Let $0 < p < \infty$ and $f \in L^p \equiv L^p([0, 2\pi], \frac{d\theta}{2\pi})$, we shall use the norm notation $\|f\| \equiv \|f\|_p$ for $(\int |f|^p)^{1/p}$ (even though, for $p < 1$, it is not a genuine norm). Let $\varphi \in \mathbf{H}^p$, and $N \in \mathbf{N}$, the set of positive integers. Recall that if there exists a function $\tilde{\varphi}_N \in \mathbf{H}^p$ satisfying $\|z^{-N}\varphi - \tilde{\varphi}_N\|_p \leq \|z^{-N}\varphi - f\|_p$, for all $f \in \mathbf{H}^p$, then it is called *a best approximation* of $z^{-N}\varphi$ (in $\mathbf{H}^p$). We shall denote the set of all such possible functions by $\mathbf{B}_N(\varphi)$. The number $\|z^{-N}\varphi - \tilde{\varphi}_N\|_p^{p\wedge 1}$ is referred to as the *error* (in the approximation); we shall denote it by $\mathbf{e}_N(\varphi)$. Further, if for some $\varphi$, there is a unique best approximation of $z^{-N}\varphi$, then we shall denote it by $\mathbf{b}_N(\varphi)$. The following facts are straightforward consequences of the definitions, these are recorded here for ready reference. ( Here and in the following $\mathbf{C}$ denotes the set of complex numbers): Let $\varphi(z) = \sum_{k=0}^{\infty} a_k z^k \in \mathbf{H}^p$, and set $P_N(\varphi)(z) = \sum_{k=0}^{N-1} a_k z^k$ and $R_N(\varphi) = \varphi - P_N(\varphi)$, then we have

$$\mathbf{B}_N(w\varphi) = w\mathbf{B}_N(\varphi), \quad w \in \mathbf{C}, \qquad \mathbf{B}_N(\varphi) = \mathbf{B}_N(P_N(\varphi)) + z^{-N}R_N(\varphi), \quad (2.1)$$

$$\mathbf{e}_N(w\varphi) = |w|^{p\wedge 1}\mathbf{e}_N(\varphi), \quad w \in \mathbf{C}, \qquad \mathbf{e}_N(\varphi) = \mathbf{e}_N(P_N(\varphi)). \qquad (2.2)$$

Now we succinctly introduce a few necessary definitions, notation and facts needed to state and discuss the prediction result for stable processes. For additional information on these and related materials, we refer the reader to the papers [2, 6, 8, 9]. Let $0 < p < 2$ and let $M$ be a (complex) rotationally invariant p-stable random measure on $[0, 2\pi]$ with the (finite) control measure $\mu$.

Then, for any $0 < q < p$, the stochastic integral $\int f dM \equiv \int_{[0,2\pi]} f dM$ is well defined for all $f \in L^p(\mu) \equiv L^p([0,2\pi],\mu)$; and the following crucial equality holds $(E|\int f dM|^q)^{1/q} = C(p,q)(\int |f|^p)^{1/p}$, $f \in L^p(\mu)$, where $C(p,q)$ is a known constant which depends only on $p$ and $q$ (and not on $f$, $M$ or $\mu$) [1] . Set $\mathcal{L}(M) = \{\int f dM : f \in L^p(\mu)\}$ and if $Y = \int f dM \in \mathcal{L}(M)$, define $\|Y\| = \|f\|$. Then the map

$$L^p(\mu) \ni f \longmapsto \int f dM \in \mathcal{L}(M) \tag{2.3}$$

establishes an "isometric isomorphism" between $(\mathcal{L}(M),\|.\|)$ and $L^p(\mu)$. Now let $X \equiv \{X(m) = \int_0^{2\pi} e^{im\theta} dM(\theta) : m \in \mathbf{Z}, \text{the set of all integers}\}$ be a p-stable harmonizable stochastic process. In view of the above noted crucial equality, the metric induced by $\|.\|$ on $sp(X)$, the linear span of $X$, is the natural distance for defining the notions needed for linear prediction of $X$. As we will point out later, this observation and the isomorphism defined by (2.3) reduce the problem of linear prediction of $X$ to an analytic problem of finding a best approximation (in $\mathbf{H}^p$ ) of a certain $L^p-$ function. Now we recall a few definitions: Let $n \in \mathbf{Z}$ and set $\mathcal{L}_n(X) = \|.\|-$closure of $sp\{X(m) : m \le n\}$. Let $N \in \mathbf{N}$. If there exists a $\tilde{Y}(n,N) \in \mathcal{L}_n(X)$ satisfying $\|X(n+N) - \tilde{Y}(n,N)\| \le \|X(n+N) - Y\|$, for all $Y \in \mathcal{L}_n(X)$, then $\tilde{Y}(n,N)$ is called *a best linear predictor* or just a *linear predictor* of $X(n+N)$ (in $\{X(m) : m \le n\}$ or in $\mathcal{L}_n(X)$). We also say that $\tilde{Y}(n,N)$ is an N-step (linear) predictor (of X(n+N) ) in $\mathcal{L}_n(X)$. The number $e_N(X) = \|X(n+N)-\tilde{Y}(n,N)\|^{p\wedge 1}$ is called the *prediction error*. Whenever there exists a unique linear predictor , we shall denote it by $\tilde{X}(n,N)$. Finally , we recall that the process $X$ is called *regular* if its remote past contains only the zero random variable; i.e, if $\bigcap\{\mathcal{L}_n(X) : n \in \mathbf{Z}\} = \{0\}$. It is only the regular processes for which the problem of determining linear predictors is of interest (see [2, 6, 8]).

## 3. Statements and Proofs of Results

We now state and prove our main result concerning the extremal problem in $\mathbf{H}^p$.

**Theorem 1** *Let $0 < p < \infty$, $N \in \mathbf{N}$ and $\varphi \in \mathbf{H}^p$. Then there always exists a best approximation of $z^{-N}\varphi$ in $\mathbf{H}^p$. If $\varphi(0) \neq 0$ and if $\tilde{\varphi}_N$ is a best approximation of $z^{-N}\varphi$ , then $\tilde{\varphi}_N$ must satisfy*

$$\varphi(z) - z^N\tilde{\varphi}_N(z) = c \prod_{k=1}^{\tilde{N}-1}(1 - \bar{\alpha}_k z)^{2/p} \prod_{k=1}^{M-1}\left(\frac{z - \alpha_k}{1 - \bar{\alpha}_k z}\right), \quad z \in \mathbf{D}, \tag{3.1}$$

*where $M, \tilde{N} \in \mathbf{N}$, $M \leq \tilde{N} \leq N, c \in \mathbf{C}$, $c \neq 0$, $|\alpha_k| \leq 1$, and $0 < |\alpha_k| < 1$, for $1 \leq k \leq M - 1$. The analytic roots $(1 - \bar{\alpha}_k z)^{2/p}$ can be and are chosen so that their value at $0$ is $1$.*

**Proof** Let $d \equiv \inf\{\|\varphi - z^N h\|_p : h \in \mathbf{H}^p\}$, and let $\{h_k\}$ be a sequence from $\mathbf{H}^p$ such that $\|\varphi - z^N h_k\|_p \to d$. Then, using $\|h_k\|_p^{p \wedge 1} = \|z^N h_k\|_p^{p \wedge 1} \leq \|\varphi - z^N h_k\|_p^{p \wedge 1} + \|\varphi\|_p^{p \wedge 1}$, we have that $\{\|h_k\|_p\}$ is bounded. Hence, using a well known fact [3, p.35], it follows that $\{h_k\}$ is uniformly bounded on compact subsets of $\mathbf{D}$. Therefore, there exists a subsequence which we denote also by $\{h_k\}$ and an analytic function $h_0$ on $\mathbf{D}$ such that this subsequence converges to $h_0$ uniformly on compact subsets of $\mathbf{D}$. Let $0 < r < 1$. For a complex valued function $h$ defined on $\mathbf{D}$, set $(h)_r(e^{i\theta}) = h(re^{i\theta}), 0 \leq \theta \leq 2\pi$. Then, clearly $(\varphi - z^N h_0)_r \in L^P$; and, by Fatou's Lemma and the fact that $\sup_{0 < r < 1} \|(\varphi - z^N h_k)_r\|_p = \|\varphi - z^N h_k\|_p$, we have that

$$\|(\varphi - z^N h_0)_r\|_p \leq \liminf_{k \to \infty} \|(\varphi - z^N h_k)_r\|_p \leq \lim_{k \to \infty} \|\varphi - z^N h_k\|_p = d,$$

for any fixed $r$. Therefore $\varphi - z^N h_0$ and $h_0$ belong to $\mathbf{H}^p$, and $d \leq \|\varphi - z^N h_0\|_p \leq d$. This completes the proof of the existence of a best approximation.

Now we begin to establish the representation (3.1). Write $\varphi - z^N \tilde{\varphi}_N = Bh$, where B and h are, respectively, the inner and the outer factors of $\varphi - z^N \tilde{\varphi}_N$ (note that $\varphi - z^N \tilde{\varphi}_N \not\equiv 0$). Let $\alpha \in \mathbf{C}$, $k \in \mathbf{N}$, $k \geq N$. Since $Bh + \alpha z^k h = \varphi - z^N(\tilde{\varphi}_N - \alpha z^{k-N} h)$ and $\tilde{\varphi}_N - \alpha z^{k-N} h \in \mathbf{H}^p$, it follows that $\|Bh + \alpha z^k h\|_p^p \geq \|Bh\|_p^p$, equivalently, $\int |h + \alpha z^k h \bar{B}|^p - \int |h|^p \geq 0$. Adding and subtracting $p \int |h|^p \Re(\alpha z^k \bar{B})$ to the left side of this inequality, one gets

$$\int |h|^p[\{1 + |\alpha|^2 + 2\Re(\alpha z^k \bar{B})\}^{p/2} - \{1 + p\Re(\alpha z^k \bar{B})\}] + p \int |h|^p \Re(\alpha z^k \bar{B}) \geq 0. \quad (3.2)$$

Recalling that for a positive constant $C$ one has $|(1 + w)^{p/2} - (1 + \frac{p}{2}w)| \leq C|w|^2$, if $|w| \leq 1/2$, it follows that there exists a $C' > 0$ such that the absolute value of the first integral in (3.2) is bounded by $C'|\alpha|^2 \int |h|^p$, for all small $\alpha$. This fact and (3.2) then implies that $\int |h|^p \bar{B} z^k = 0$ for all $k \geq N$. Therefore, by a well known Theorem of F. and M. Riesz [3, p.41], $\psi \equiv |h|^p \bar{B} z^{N-1}$ belongs to $\mathbf{H}^1$. Observing that $z^{-N+1} B \psi = |h|^p \geq 0$ a.e. on the circle $S \equiv \{|z| = 1\}$, the Fourier coefficients of $z^{-N+1} B \psi$ are symmetric about zero. Thus, since the negative Fourier coefficients of the $\mathbf{H}^1$-function $B\psi$ are zero, any non-zero Fourier coefficient of $z^{-N+1} B \psi$ will correspond to a $k$ satisfying $|k| \leq N - 1$. Then it follows by a Theorem of Fejér and Riesz [7, p.11] that there is a polynomial $P(z) = A\prod_{k=1}^{\tilde{N}-1}(z - \alpha_k)$ of degree $\tilde{N} - 1, \tilde{N} \leq N$ with $|\alpha_k| \leq 1$ such that

$$|h(z)|^p = z^{-N+1}B(z)\psi(z) = |P(z)|^2 = |A|^2\left|\prod_{k=1}^{\tilde{N}-1}(1 - z\bar{\alpha}_k)^2\right| \ a.e. \text{ on } S. \qquad (3.3)$$

Let $(1 - z\bar{\alpha}_k)^{2/p}$ be that analytic root of $1 - z\bar{\alpha}_k$ on $\mathbf{D}$ which is 1 at 0. Then, clearly $A^{2/p}\prod_{k=1}^{\tilde{N}-1}(1 - z\bar{\alpha}_k)^{2/p}$ is an outer function; and by (3.3) the absolute value of its boundary function coincides a.e. on $S$ with $|h|$. Therefore, $h(z) = c\prod_{k=1}^{\tilde{N}-1}(1 - z\bar{\alpha}_k)^{2/p}$, $c \in \mathbf{C}, c \neq 0$. Denote the polynomial $|A|^2 z^{N-\tilde{N}}\prod_{k=1}^{\tilde{N}-1}(1 - z\bar{\alpha}_k)(z - \alpha_k)$ by $Q$. Then, by (3.3), $B\psi = Q$ a.e. on $S$. Therefore, since both $B\psi$ and $Q$ are $\mathbf{H}^1$ functions, $B\psi = Q$ on $\mathbf{D}$. Thus, $B\psi$ is the restriction (to $\mathbf{D}$) of an entire function. Therefore, $B\psi$ and, hence, $B$ does not have any singular factor. Thus, from $B\psi = Q$ on $\mathbf{D}$ and from $\varphi(0) \neq 0$, it follows that $B$ is a finite Blaschke product, and that the zeros of $B$ are some of the $\alpha_r$'s. After relabeling, if necessary, let $\alpha_1, ...\alpha_{M-1}$, $M \leq \tilde{N}$ denote the zeros of B; then $B(z) = \prod_{k=1}^{M-1}\left(\frac{z-\alpha_k}{1-\bar{\alpha}_k z}\right)$, $0 < |\alpha_k| < 1$. This completes the proof.

**Remark 1** From Theorem 1, one also gets easily the analog of the representation (3.1) for a non-zero function $\varphi \in \mathbf{H}^p$ when $\varphi(0) = 0$ (if $\varphi \equiv 0$ then trivially 0 is the unique best approximation of $z^{-N}\varphi$ for every $N$): Let $\varphi = z^{N_0}\varphi_0$, $\varphi_0(0) \neq 0$; and let $N \in \mathbf{N}$. If $N \leq N_0$, then $z^{-N}\varphi = z^{-(N-N_0)}\varphi_0 \in \mathbf{H}^p$. So the only interesting case is when $N > N_0$. Let $\tilde{\varphi}_N \in \mathbf{B}_N(\varphi)$, then, since $\mathbf{B}_N(\varphi) = \mathbf{B}_{(N-N_0)}(\varphi_0)$, $\varphi_0 - z^{(N-N_0)}\tilde{\varphi}_N$ admits the representation on the right side of (3.1). Then clearly $\varphi - z^N\tilde{\varphi}_N$ is equal to the product of this representation with $z^{N_0}$. Before we conclude this remark, we would like to mention here that we have also obtained the complete analog of Theorem 1 in the Hardy space of the upper half plane. This and the related research will be reported elsewhere.

For the proof of the Theorem 2, we shall need the following two Lemmas which we shall state and prove first.

**Lemma 1** *Let $0 < p < 1$ be fixed and let $q = p/(2 - p)$. Define the functions $f_j(p, .) \equiv f_j(.), j = 0, 1, 2$ on the interval $(0, \infty)$ by :*

$$f_0(u) = 1 + 1/4\left((2-p)\sqrt{u} + p/\sqrt{u}\right)^2; \ f_1(u) = q^{-p}u^{p/2}(1 + q^2/u); \ f_2(u) = u^{-p/2}(1+u).$$

*Then $f_1(q) = f_2(q) = 2q^{-p/2}(2-p)^{-1} > f_0(q)$, $f_0(1) = f_2(1) = 2 > f_1(1)$, $f_2(u) > f_j(u), j = 0, 1$, for all $q < u < 1$; and there exists a unique number $u_0(p) \equiv u_0, q <$*

$u_0 < 1$, such that $f_0(u_0) = f_1(u_0)$. Thus, $f_0(u) < f_j(u)$, $j = 1, 2$, for every $q \le u < u_0$, $f_1(u) < f_j(u)$, $j = 0, 2$, for every $u_0 < u \le 1$, and $f_0(u_0) = f_1(u_0) < f_2(u_o)$. Further, all three functions are strictly increasing on $(q, \infty)$ and strictly decreasing on $(0, q)$.

**Proof** Before we give the proof, we note that the conclusions stated in the last line of the Lemma are not needed here. These are included for completeness and to help the reader to visualize the graphs of the three functions. It is suggested that the reader draw rough graphs of these functions while reading the following proof.

A direct computation shows $f_1(q) = f_2(q) = 2q^{-p/2}(2 - p)^{-1}$ and $f_0(1) = f_2(1) = 2$. It can be verified that

$$\frac{d}{du} \ln\big(f_0/f_2\big)(u) = \frac{p(2 - p)^2(u - q)(u - 1)^2}{2u(1 + u)\big(u^2(2 - p)^2 + 2u(2p - p^2 + 2) + p^2\big)} \tag{3.4}$$

$$\frac{d}{du} \ln\big(f_2/f_1\big)(u) = \frac{(1 - p)(u - q)^2}{u(1 + u)(u + q^2)} , \tag{3.5}$$

for all positive $u$. Clearly, the derivative in (3.5) is positive for all $u > 0, u \ne q$. Hence $f_2/f_1$ is strictly increasing for $u > q$ (and also for $0 < u < q$). Thus, since $f_1(q) = f_2(q)$, $f_2(u)/f_1(u)$ is larger than 1 for all $u > q$; i.e, $f_1 < f_2$ on this interval. One checks that the sum of the roots of the quadratic term in the denominator of the right side of (3.4) is negative and their product is positive; so both roots are negative. Thus, since this term is positive at 0, it is positive for all $u > 0$. Using this it follows from (3.4) that the derivative in (3.4) is positive for $q < u < 1$ (and negative for $0 < u < q$). Hence $f_0/f_2$ is strictly increasing for $q < u < 1$. Thus, since $f_0(1) = f_2(1)$, $f_0(u)/f_2(u)$ is less than 1 for all $q < u < 1$; i.e, $f_0 < f_2$ on this interval. Using continuity of $f_0/f_2$, we also have $f_0(q) < f_2(q)(= f_1(q))$. This fact along with $f_0(1) > f_1(1)$ and the Intermediate Value Theorem assures the existence of at least of one point in $(q, 1)$ where $f_0$ and $f_1$ must agree. However, since $f_0/f_1$, being equal to the product $(f_0/f_2)(f_2/f_1)$, is also strictly increasing on $q < u < 1$, there can be at most one point in $(q, 1)$ where $f_0$ and $f_1$ agree. This completes the proof of the first part of the Lemma. The last part follows by noting that $\frac{df_0}{du}(u) = \frac{(2-p)^2(u^2 - q^2)}{4u^2}$; $\frac{df_1}{du}(u) = \frac{p(u-q)}{2q^p u^{2-p/2}}$ and $\frac{df_2}{du}(u) = \frac{(2-p)(u-q)}{2u^{1+p/2}}$ on the interval $(0, \infty)$.

**Lemma 2** Let $0 < p < 1$ and set $\kappa(p) \equiv \kappa = p(2 - p)$. Define the functions $g_1$ and $g_2$ on $[\sqrt{\kappa}, \infty)$ by

$$g_1(v) = v(2 - p)^{-1}\big[1 - \sqrt{1 - \kappa v^{-2}}\big] = (2 - p)^{-1}\big[v - \sqrt{v^2 - \kappa}\big],$$

$$g_2(v) = v(2-p)^{-1}\left[1 + \sqrt{1 - \kappa v^{-2}}\right] = (2-p)^{-1}\left[v + \sqrt{v^2 - \kappa}\right].$$

*Then $g_1 < 1$ on $[\sqrt{\kappa}, \infty)$; and $g_2 < 1$ on $[\sqrt{\kappa}, 1)$, $g_2 > 1$ on $(1, \infty)$ and $g_2(1) = 1$.*

**Proof** Since $g_1(\sqrt{\kappa}) = \sqrt{p/(2-p)}$ is less than 1, and since $\frac{d}{dv}g_1(v) = (2-p)^{-1}[1 - v(v^2 - \kappa)^{-1/2}]$ is negative $\Leftrightarrow v > \sqrt{\kappa}$, the assertion about $g_1$ follows. Trivially $g_2(1) = 1$; the other two assertions about $g_2$ now follow since clearly $g_2$ is strictly increasing on $[\sqrt{\kappa}, \infty)$.

Before we state Theorem 2, we introduce some more notation. Let $0 < p < 1$; and let $g_1, g_2, q$ and $\kappa$ be as above. Set $v(t) = pt/2$, $t > 0$; and let $u(t) = [g_2(v(t))]^2$, $v(t) \geq \sqrt{\kappa}$ ($\Leftrightarrow t \geq 2/\sqrt{q}$). Note that u is a strictly increasing function of t; and as t varies from $2/\sqrt{q}$ to $2/p$, u varies from $q$ to 1. We shall denote the number $u^{-1}(u_0(p))$ by $\gamma^*(p) \equiv \gamma^*$, where $u_0 = u_0(p)$ is as in Lemma 1. Note that $2/\sqrt{q} < \gamma^* < 2/p$. Define the functions $\tilde{\beta}_j, \tilde{c}_j$, and $\tilde{\alpha}$ as follows: $\tilde{\beta}_j(b) = g_j(v(|b|)) \exp i(\pi - Arg(b)), \tilde{c}_j(b) = -\tilde{\beta}_j(b)^{-1}$, $|b| \geq 2/\sqrt{q}$ ($\Leftrightarrow v(|b|) \geq \sqrt{\kappa}$), $,j = 1, 2$; and $\tilde{\alpha}(b) = -(p\bar{b})/2$, $b \in \mathbf{C}$. We shall need the following facts:

$$\left\{\begin{array}{l} |\tilde{\alpha}(b)| \leq 1 \Leftrightarrow |b| \leq 2/p \\ |\tilde{\beta}_1(b)| < 1 \Leftrightarrow |b| \geq 2/\sqrt{q} \end{array}\right\} \quad \left\{\begin{array}{l} |\tilde{\beta}_2(b)| < 1 \Leftrightarrow 2/\sqrt{q} \leq |b| < 2/p \\ |\tilde{\beta}_2(b)| > 1 \Leftrightarrow |b| > 2/p; \ \tilde{\beta}_2(2/p) = 1 \end{array}\right\}. \quad (3.6)$$

The first is of course trivial, for the rest one uses Lemma 2. For $|\alpha| \leq 1$ and $|\beta| < 1$, define $Q_0(\alpha)(z) = (1 - \bar{\alpha}z)^{2/p}$ and $Q_1(\beta)(z) = (1 - \bar{\beta}z)^{2/p}(\frac{z-\beta}{1-\bar{\beta}z})$, $z \in \mathbf{D}$, where the roots are chosen as in Theorem 1. Finally, set

$$\mathcal{P}_0(b) = Q_0(\tilde{\alpha}(b)), |\tilde{\alpha}(b)| \leq 1 \text{ and } \mathcal{P}_j(b) = \tilde{c}_j(b)Q_1(\beta_j(b)), |\tilde{\beta}_j(b)| < 1, j = 1, 2. \quad (3.7)$$

**Theorem 2** *Let $0 < p < 1$ and let $\varphi = \sum_{n=0}^{\infty} a_n z^n$, $a_0 \neq 0$, be in $\mathbf{H}^p$. Then we have*

(a) *The function $z^{-1}\varphi$ has a unique best approximation in $\mathbf{H}^p$. The approximation and the error are given by*

$$\mathbf{b}_1(\varphi) = z^{-1}(\varphi - a_0), \qquad \mathbf{e}_1(\varphi) = |a_0|^p. \qquad (3.8)$$

(b) *If either (i) $a_1 = 0$ or (ii) $a_1 \neq 0$ and $|\frac{a_1}{a_0}| \neq \gamma^*$, then the function $z^{-2}\varphi$ has a unique best approximation in $\mathbf{H}^p$. The approximation and the error are given by*

$$\mathbf{b}_2(\varphi) = \left\{\begin{array}{ll} z^{-2}(\varphi - a_0), & \text{if } a_1 = 0, \\ z^{-2}(\varphi - a_0\mathcal{P}_0(\frac{a_1}{a_0})), & \text{if } a_1 \neq 0 \text{ and } |\frac{a_1}{a_0}| < \gamma^*, \\ z^{-2}(\varphi - a_0\mathcal{P}_1(\frac{a_1}{a_0})), & \text{if } a_1 \neq 0 \text{ and } |\frac{a_1}{a_0}| > \gamma^*; \end{array}\right.$$

$$\mathbf{e}_2(\varphi) = \begin{cases} |a_0|^p, & \text{if } a_1 = 0, \\ |a_0|^p(1 + |\tilde{\alpha}(\frac{a_1}{a_0})|^2), & \text{if } a_1 \neq 0, \ |\frac{a_1}{a_0}| < \gamma^*, \\ |a_0\tilde{c}_1(\frac{a_1}{a_0})|^p(1 + |\tilde{\beta}_1(\frac{a_1}{a_0})|^2), & \text{if } a_1 \neq 0, \ |\frac{a_1}{a_0}| > \gamma^*. \end{cases}$$

(c) *If $a_1 \neq 0$ and $|\frac{a_1}{a_0}| = \gamma^*$, then $z^{-2}\varphi$ has two best approximations: $z^{-2}(\varphi - a_0\mathcal{P}_0(\frac{a_1}{a_0}))$ and $z^{-2}(\varphi - a_0\mathcal{P}_1(\frac{a_1}{a_0}))$ with the error of approximation equal to $|a_0|^p(1 + |p\gamma^*|^2/4)$.*

**Proof** If $N = 1$ the right side of (3.1) is $a_0$. Then Theorem 1 immediately shows that $z^{-1}\varphi$ has a unique best approximation and that the assertions of (3.8) are valid. This completes the proof of (a). The proofs of (b) and (c) are given below:

In view of (2.1) and (2.2), it is sufficient to prove the stated assertions in (b) and (c) when $\varphi = \varphi_b$ where $\varphi_b(z) = 1 + bz$, $b \neq 0$ (the case b=0, follows easily and is therefore omitted). Let $\tilde{\varphi}_j$, $j = 0, 1$ be any functions in $\mathbf{H}^p$ which satisfy the following equations on $\mathbf{D}$

$$(i) \ \varphi_b - z^2\tilde{\varphi}_0 = c\,Q_0(\alpha), \qquad (ii) \ \varphi_b - z^2\tilde{\varphi}_1 = \tilde{c}\,Q_1(\beta), \qquad (3.9)$$

where $c, \tilde{c}$ are non-zero and $|\alpha| \leq 1$, $0 < |\beta| < 1$. Then, after a comparison of the first two coefficients, $(i)$ and $(ii)$ of (3.9) imply $(i)$ and $(ii)$ of the following:

$$(i) \ \begin{cases} 1 = c \\ b = -(2\bar{\alpha})/p \end{cases}, \qquad (ii) \ \begin{cases} 1 = -\beta\tilde{c} \\ b = \left(1 + (2/p - 1)|\beta|^2\right)\tilde{c} \end{cases}. \qquad (3.10)$$

Clearly, the $\alpha$ which satisfies the second equation in $(i)$ of (3.10) is equal to $\tilde{\alpha}(b) = -bp/2$. Further, it can be easily verified that $\beta$ and $\tilde{c}$ which satisfy the equations $(ii)$ of (3.10) are $\tilde{\beta}_j(b)$ and $\tilde{c}_j(b)$, $|b| \geq 2/\sqrt{q}$, $j = 1, 2$. Therefore it follows from Theorem 1 that the only possible candidates for being the best approximations of $z^{-2}\varphi_b$ are $\psi_0(b) = z^{-2}(\varphi_b - \mathcal{P}_0(b))$, $|\tilde{\alpha}(b)| \leq 1 \ (\Leftrightarrow |b| \leq 2/p)$, $\psi_1(b) = z^{-2}(\varphi_b - \mathcal{P}_1(b))$, $|\tilde{\beta}_1(b)| < 1 \ (\Leftrightarrow |b| \geq 2/\sqrt{q})$ and $\psi_2(b) = z^{-2}(\varphi_b - \mathcal{P}_2(b))$, $|\tilde{\beta}_2(b)| < 1 \ (\Leftrightarrow 2/\sqrt{q} \leq |b| < 2/p)$ (see (3.7) for the definitions of $\mathcal{P}_j's$). Here in stating the equivalent inequalities concerning $\tilde{\alpha}_0$ and $\tilde{\beta}_j$, we have used (3.6). It follows that $\psi_0(b)$ (respectively $\psi_1(b)$) is a unique best approximation of $z^{-2}\varphi_b$ if $|b| < 2/\sqrt{q}$ (respectively if $|b| > 2/p$).

The error $e_j(b) (\equiv e_j(|b|))$ in approximating $z^{-2}\varphi_b$ by $\psi_j$ is $\int |\mathcal{P}_j(b)|^p, j = 0, 1, 2$, which is easy to compute. In fact, we have

$$e_0(|b|) = (1 + |\tilde{\alpha}_0(b)|^2), \quad |\tilde{\alpha}_0(b)| \leq 1,$$
$$e_1(|b|) = |\tilde{\beta}_1(b)|^{-p}(1 + |\tilde{\beta}_1(b)|^2), \quad |\tilde{\beta}_1(b)| < 1,$$
$$e_2(|b|) = |\tilde{\beta}_2(b)|^{-p}(1 + |\tilde{\beta}_2(b)|^2), \quad |\tilde{\beta}_2(b)| < 1.$$

To determine which of these functions are best approximations of $z^{-2}\varphi_b$ when $2/\sqrt{q} \le |b| \le 2/p$, we need to compare the errors. To do this, first we observe that the functions $g_1$ and $g_2$ of Lemma 2 satisfy the following relations

$$g_1(v) = q/g_2(v) \quad \text{and} \quad (g_1(v) + g_2(v))(2-p)2^{-1} = v, \quad v \ge \sqrt{\kappa}. \tag{3.11}$$

Now we recall the definitions of the functions $f_j$ from Lemma 1. It follows from (3.11) that $e_j(|b|) = f_j(u(|b|))$, $2/\sqrt{q} \le |b| < 2/p$, $j = 0,1,2$; and $e_j(|b|) = f_j(u(|b|))$, $|b| = 2/p$, $j = 0,1$ (note that $\psi_2$, and, hence, $e_2$ is not defined at $|b| = 2/p$). Hence from Lemma 1 we have $e_0(|b|) < e_j(|b|)$, whenever $2/\sqrt{q} \le |b| < \gamma^*$ and $j = 1,2$. Also $e_1(|b|) < e_j(|b|)$, whenever $\gamma^* < |b| < 2/p$ and $j = 0,2$, and $e_0(|b|) = e_1(|b|)$, when $|b| = \gamma^*$, and finally we have $e_1(|b|) < e_0(|b|)$ when $|b| = 2/p$. Therefore, using this and what we have proved above, we have that $z^{-2}\varphi_b$ has $\psi_0(b)$ (respectively, $\psi_1(b)$) as its unique best approximation if $|b| < \gamma^*$ (respectively, if $\gamma^* < |b|$). Furthermore $z^{-2}\varphi_b$ has two *distinct* best approximations when $|b| = \gamma^*$; namely, $\psi_0(b)$ and $\psi_1(b)$) . This along with the expressions for the errors derived above completes the promised proof.

**Remark 2** Let $0 < p < \infty$ and let $\varphi(z) = \sum_{k=0}^{\infty} a_k z^k \in \mathbf{H}^p, a_0 \ne 0$. Let $N \in \mathbf{N}$. Assume that $P_N(\varphi^{p/2}) \ne 0$ on $\mathbf{D}$, and let $\mathbf{P}_N(\varphi) = z^{-N}\{\varphi - (P_N(\varphi^{p/2}))^{2/p}\}$ where the roots are chosen so that $(P_N(\varphi^{p/2}))^{2/p}(0) = \varphi(0)$. If $p \ge 1$, it is well known that the unique best approximation $\mathbf{b}_N(\varphi)$ of $z^{-N}\varphi$ is equal to $\mathbf{P}_N(\varphi)$ (see [2, 8, 9]). If $0 < p < 1$ , then it is obvious from Theorem 2 that $\mathbf{b}_1(\varphi) = \mathbf{P}_1(\varphi)$ and that $\mathbf{b}_2(\varphi) = \mathbf{P}_2(\varphi)$, provided $a_1 = 0$. However, when $a_1 \ne 0$, even though $\mathbf{P}_2(\varphi)$ is well defined (i.e, $P_N(\varphi^{p/2}) \ne 0$ on $\mathbf{D}$) whenever $|\frac{a_1}{a_0}| \le 2/p$, it is a best approximation of $z^{-2}\varphi$ only when $|\frac{a_1}{a_0}| \le \gamma^* < 2/p$ (and *not* when $\gamma^* < |\frac{a_1}{a_0}| \le 2/p$).

Let $0 < p < 1$. Let $X \equiv \{X(m) = \int_0^{2\pi} e^{im\theta} dM(\theta) : n \in \mathbf{Z}\}$ be a (complex) p-stable harmonizable stochastic process. Then it is shown in [6] that $X$ is regular iff $\mu \ll$ Leb. and $\int \ln \rho > -\infty$, where $\rho = d\mu/d\theta$ (recall that $\mu$ is the (finite) control measure of M). Let us now assume that $X$ is also regular. Let $\varphi(z) = \sum_{k=0}^{\infty} a_k z^k$ be the unique outer function in $\mathbf{H}^p$ satisfying $|\varphi|^p = \rho$ a.e. on $S$ (see [2] or [9], for more details on this); note that $a_0 \ne 0$. Then, (2.3) and two applications of Beurling's Theorem [5, p.110] show that the set of all possible linear predictors of $X(n + N)$ in $\mathcal{L}_n(X)$ is equal to $\{\int z^n\{\frac{\tilde{\varphi}_N}{\varphi}\}dM : \tilde{\varphi}_N \in \mathbf{B}_N(\varphi)\}$ and that $\mathbf{e}_N(X) = \mathbf{e}_N(\varphi)$. (This fact is pointed out in [2] and also in [9] for the case when $p \ge 1$; but, noting that the quoted Beurling's Theorem is also valid for $0 < p < 1$ [5, p.110], the same argument works in the case $0 < p < 1$ ). This fact and Theorem 1 obviously implies that an N-step linear predictor in $\mathcal{L}_n(X)$ always exists and that it is necessarily of the form $\int z^n \{\frac{\bar{\varphi}_N}{\varphi}\} dM$, where $\tilde{\varphi}_N$ is as in (3.1). Further, this fact and Theorem 2 together yield the following result which provides precise formulae for one- and two-step linear predictors. Since, as noted above, $\mathbf{e}_N(X) = \mathbf{e}_N(\varphi)$, the formulae for the corresponding prediction errors are contained in Theorem 2, and, hence, are not repeated in the statement of the following theorem. The notation introduced here (relative to the regular process $X$) and those noted prior to the statement of Theorem 2 are used in the following.

**Theorem 3** *Let $0 < p < 1$ and $n \in \mathbf{Z}$. Let $X \equiv \{X(m) = \int_0^{2\pi} e^{im\theta} dM(\theta) : m \in \mathbf{Z}\}$ be a (complex) p-stable harmonizable regular process. Then the following hold:*

*(a) $X(n+1)$ has a unique linear predictor in $\mathcal{L}_n(X)$ given by*

$$\tilde{X}(n,1) = \int_0^{2\pi} e^{i(n+1)\theta} \left\{ 1 - \overline{\frac{a_0}{\varphi(e^{i\theta})}} \right\} dM(\theta). \tag{3.12}$$

*(b) If either (i) $a_1 = 0$ or (ii) $a_1 \neq 0$ and $|\frac{a_1}{a_0}| \neq \gamma*$, then $X(n+2)$ has a unique linear predictor in $\mathcal{L}_n(X)$ given by*

$$\tilde{X}(n,2) = \int_0^{2\pi} e^{i(n+2)\theta} \left\{ 1 - \overline{\frac{P(e^{i\theta})}{\varphi(e^{i\theta})}} \right\} dM(\theta). \tag{3.13}$$

*In case (i) $\mathcal{P} = a_0$; and in case (ii) $\mathcal{P} = a_0 \mathcal{P}_0(\frac{a_1}{a_0})$, when $|\frac{a_1}{a_0}| < \gamma*$ and $\mathcal{P} = a_0 \mathcal{P}_1(\frac{a_1}{a_0})$, when $|\frac{a_1}{a_0}| > \gamma *$.*

*(c) If $a_1 \neq 0$ and $|\frac{a_1}{a_0}| = \gamma*$, then there are exactly two distinct linear predictors of $X(n+2)$ in $\mathcal{L}_n(X)$. Both are given by the right side of (3.13), one with $\mathcal{P} = a_0 \mathcal{P}_0(\frac{a_1}{a_0})$, and other with $\mathcal{P} = a_0 \mathcal{P}_1(\frac{a_1}{a_0})$.*

**Remark 3** Remark 2 in conjunction with either the comments made above or [2] and [9] imply that the two-step predictor formula (3.13) has a form different from its counterpart for $1 \leq p < 2$ when $a_1 \neq 0$ and $|\frac{a_1}{a_0}| > \gamma*$ and also when $|\frac{a_1}{a_0}| = \gamma*$ provided $\mathcal{P} = a_0 \mathcal{P}_1(\frac{a_1}{a_0})$. In all other cases, the noted two-step predictor formula is identical in form to the corresponding formula for the case $1 \leq p < 2$; the same is true for the one-step predictor formula (3.12).

## References


[1] Cambanis, S. and Miamee, A. G.: On prediction of harmonizable stable processes; *Sankhya (Series A)*. **51** (1989) 269-294


10

[2] Cambanis, S. and Soltani, A. R.: Prediction of stable processes: Spectral and moving average representations; *Z. Wahrs. verw. Gebiete.* **66** (1984) 593-612

[3] Duren, P. L.: *Theory of $H^p$ Spaces.* Academic Press, New York, 1970

[4] Hosoya, Y.: Harmonizable stable processes; *Z.Wahrs. verw. Gebiete.* **60** (1982) 517-533

[5] Koosis, P.: *Introduction to $H^p$ Spaces.* Cambridge University Press, New York, 1980

[6] Makagon, A. and Mandrekar, V.: The spectral representation of stable processes: Harmonizability and regularity; *Probab. Theory Relat. Fields.* **85** (1990) 1-11

[7] Paulsen, V. I.: *Completely Bounded Maps and Dilations.* Pitman Research Notes in Mathematics Series **146**, John Wiley, New York, 1986

[8] Rajput, B. S., Rama-Murthy, K., and Sundberg, C.: An extremal problem in $H^p$ of the upper half plane with application to prediction of stochastic processes; in:*Stable Processes and Related Topics.* S. Cambanis, et al (eds.) Birkhauser, Boston, (1991) 191-252

[9] Rajput, B. S. and Sundberg, C.: On some extremal problems in $H^p$ and the prediction of $L^p$-harmonizable processes; *Probab. Theory Relat. Fields.* (1994) to appear

Alexandru Aleman
Fachbereich Mathematik und Informatik
Fern Universität Hagen
Postfach 940
58084 Hagen, GERMANY
*E-mail*: aleman@mathe.fernuni-hagen.de

Balram S. Rajput
Department of Mathematics
University of Tennessee
Knoxville, TN 37922, U.S.A.
*E-mail*: rajput@novell.math.utk.edu

Stefan Richter
Department of Mathematics
University of Tennessee
Knoxville, TN 37922, U.S.A.
*E-mail*: richter@novell.math.utk.edu

T. BOJDECKI & L. G. GOROSTIZA

# A nuclear space of distributions on Wiener space and application to weak convergence

## 1. Introduction

A nuclear space of distributions on a $\sigma$–finite Wiener space was introduced in Gorostiza and Nualart [10], motivated by the study of weak convergence of fluctuations for a large class of particle systems in $\mathbb{R}^d$ in a way that is hopefully more revealing of the nature of the particle systems than the usual approach is.

By the usual approach, which we refer to as the "temporal" approach, we mean that the particle system is represented by the configuration at each time, given as the sum of the Dirac measures at the locations of the particles in $\mathbb{R}^d$ at that time. It is well–known that appropriate spaces for the analysis of fluctuations of particle systems in this way are $C(T, \mathcal{S}'(\mathbb{R}^d))$ and the Skorokhod space $D(T, \mathcal{S}'(\mathbb{R}^d))$, where $T$ is the underlying time interval and $\mathcal{S}'(\mathbb{R}^d)$ is the space of tempered distributions on $\mathbb{R}^d$. The papers of Martin–Löf [19], Holley and Stroock [13], and Itô [14] were among the first to develop this approach, and many others have followed.

Another approach, which we call the "trajectorial" approach, consists in representing the particle system by the configuration in a path space of $\mathbb{R}^d$–valued functions, given as the sum of the Dirac measures at the trajectories of the particles during the time interval. This representation should be more descriptive because it keeps track of the evolution in time of each particle. Since limit results in the trajectorial approach do not follow automatically from temporal limit results, it is useful to develop a method for the analysis of fluctuations directly in path space. Such a project has been attempted in the papers of Martin–Löf [19], Gorostiza [8], Tanaka and Hitsuda [26], Tanaka [25], and Sznitmann [24]. However, in these papers the trajectorial fluctuation limit is given only in terms of the characteristic functional defined for a class of test functions on path space, and more precise information on the limit is provided only for the corresponding limit process, which can also be obtained by the temporal approach. What is needed in order to complete the trajectorial approach is an identification of the limit as a random element of an appropriate space, which should be a space of distributions, dual to a space of test functions, on path space. Moreover, these spaces should be nuclear so that Lévy's continuity theorem can be used to conclude weak convergence of the random elements from the pointwise convergence of the characteristic functionals (e.g., Boulicaut [5]). This problem was first formulated in Gorostiza [8] in connection with the fluctuations of a branching system. The space of distributions on Wiener space of Watanabe [28] is not suitable for our purpose because it is not nuclear, and, on the other hand, for the applications the standard Wiener measure on the space of $\mathbb{R}^d$–valued continuous functions has to be replaced by the $\sigma$–finite Wiener measure defined below. The space of Hida distributions (e.g. Hida et al. [12]), which are distributions on $\mathcal{S}'(\mathbb{R}^d)$ (with white noise measure) is not necessary for our objective, but it could be used for models where the "particle" paths are $\mathcal{S}'(\mathbb{R}^d)$–valued functions (this might be useful for the study of fluctuations of "superprocesses"). The nuclear Gelfand triple constructed in Gorostiza and Nualart [10] is suitable for the analysis of trajectorial fluctuation limits of many particle systems. Some examples are given in that paper.

Since the trajectorial approach is hoped to be more informative than the temporal one, a basic question is whether, for a given particle system, weak convergence of trajectorial fluctuations implies weak convergence of the corresponding $\mathcal{S}'(\mathbb{R}^d)$–valued processes. This is not an easy problem because there is no natural relationship between the space of distributions on the $\sigma$–finite Wiener space and the space $C(T, \mathcal{S}'(\mathbb{R}^d))$. In this paper we investigate this question and obtain some positive results, and we point out some difficulties which remain.

In Section 2 we recall the nuclear Gelfand triple introduced in Gorostiza and Nualart [10] and the corresponding Lévy continuity theorem. In Section 3 we motivate the weak convergence question relating the trajectorial and the temporal approaches, i.e., given a weakly convergent sequence of random distributions on the $\sigma$–finite Wiener space, are there corresponding $\mathcal{S}'(\mathbb{R}^d)$–processes which also converge weakly? In this respect we identify a class of measures on $C([0,1], \mathbb{R}^d)$ which are distributions on the $\sigma$–finite Wiener space, and which may show up in applications as trajectorial means of some particle systems. There arises the general problem of associating an $\mathcal{S}'(\mathbb{R}^d)$–valued process to a random distribution on the $\sigma$–finite Wiener space, and we also investigate this problem. Finally, we address the weak convergence question and we obtain some partial answers.

This paper is an expanded version of the talk presented by L. G. G. in the U.S.-Japan Bilateral Seminar.

## 2. A Nuclear Gelfand Triple on a $\sigma$–Finite Wiener Space and the Lévy Continuity Theorem

Let $C = C([0,1], \mathbb{R}^d)$, $C_0 = $ the subspace of $C$ of functions which vanish at 0, $\|\omega\|_\infty = \sup_{0 \le t \le 1} |\omega(t)|$, $\omega \in C$, $W_x = $ Wiener measure on $C$ supported on the functions with initial point $x \in \mathbb{R}^d$, and $\lambda = $ Lebesgue measure on $\mathbb{R}^d$. The *$\sigma$–finite Wiener measure* $\mathbb{W}$ on $C$ is defined by

$$\mathbb{W} = \int_{\mathbb{R}^d} W_x \lambda(dx).$$

$\mathbb{W}$ is in a sense a "uniform" measure that can be defined on $C$, as a consequence of the characterization of $W_0$ as the uniform measure on a sphere of dimension $\infty$ with radius $\infty^{1/2}$ (see McKean [20], Cutland and Ng [7]), and it plays on $C$ a rôle analogous to that of $\lambda$ on $\mathbb{R}^d$.

Using the method of Kubo and Yokoi [18] (see also Kondratiev and Samoilenko [16], Meyer and Yan [21, 22], Korezlioglu and Üstünel [17]), in Gorostiza and Nualart [10] a nuclear Gelfand triple

$$\mathcal{H} \subset L^2(C_0, W_0) \subset \mathcal{H}'$$

was constructed, and from it the following result was obtained:

### 2.1 Theorem

(a)
$$\mathcal{S}(\mathbb{R}^d, \mathcal{H}) \subset L^2(C, \mathbb{W}) \subset \mathcal{S}(\mathbb{R}^d, \mathcal{H})'$$
*is a nuclear Gelfand triple.*

(b) $\mathcal{S}(\mathbb{R}^d, \mathcal{H})$ *is Fréchet, reflexive, and its elements are continuous functions on $C$ which are bounded on bounded subsets of $C$.*

(c) *The evaluation maps $\delta_\omega \colon \mathcal{S}(\mathbb{R}^d, \mathcal{H}) \to \mathbb{R}$ defined by $\delta_\omega(F) = F(\omega)$, $\omega \in C$, belong to $\mathcal{S}(\mathbb{R}^d, \mathcal{H})'$.*

Here, $\mathcal{S}(\mathbb{R}^d, \mathcal{H})$ is defined as a space of smooth $\mathcal{H}$–valued functions on $\mathbb{R}^d$ (see Treves [27, p.450]), and since $\mathcal{S}(\mathbb{R}^d, \mathcal{H}) \cong \mathcal{S}(\mathbb{R}^d) \widehat{\otimes} \mathcal{H}$, where $\mathcal{S}(\mathbb{R}^d)$ is the usual space of infinitely differentiable rapidly decreasing functions on $\mathbb{R}^d$, then $\mathcal{S}(\mathbb{R}^d, \mathcal{H})$ can indeed be regarded as a space of functions on $C$.

The standard Gaussian random element $\mathcal{X}$ of $\mathcal{S}(\mathbb{R}^d, \mathcal{H})'$, which is determined by the characteristic functional

$$E(\exp\{i\langle \mathcal{X}, F \rangle\}) = \exp\left\{ -\frac{1}{2} \int_C F^2(\omega) \mathbb{W}(d\omega) \right\}, \qquad F \in \mathcal{S}(\mathbb{R}^d, \mathcal{H}),$$

is called *white noise on $\sigma$–finite Wiener space.*

Due to reflexivity of $\mathcal{S}(\mathbb{R}^d, \mathcal{H})$, Lévy's continuity theorem holds on $\mathcal{S}(\mathbb{R}^d, \mathcal{H})'$ (Boulicaut [5]):

**2.2 Theorem** *Let $\mu_n$, $n = 1, 2, \ldots$ and $\mu$ be probability measures on $\mathcal{S}(\mathbb{R}^d, \mathcal{H})'$. If*

$$\int_{\mathcal{S}(\mathbb{R}^d, \mathcal{H})'} e^{i\langle \xi, F \rangle} \mu_n(d\xi) \to \int_{\mathcal{S}(\mathbb{R}^d, \mathcal{H})'} e^{i\langle \xi, F \rangle} \mu(d\xi) \ as \ n \to \infty$$

*for each $F \in \mathcal{S}(\mathbb{R}^d, \mathcal{H})$, then $\mu_n$ converges weakly to $\mu$ as $n \to \infty$.*

Here and below $\langle \cdot, \cdot \rangle$ stands for the pairing on spaces in duality.

Theorem 2.2 was used in Gorostiza and Nualart [10], and Gorostiza [9] to prove trajectorial fluctuation limits for several particle systems. In particular, the limiting fluctuations of interacting diffusions with respect to the propagation of chaos limit were completely described, thus sharpening the results of Tanaka and Hitsuda [26], and Sznitmann [24].

## 3. Relationship Between the Temporal and the Trajectorial Approaches. The Convergence Problem

Given a particle system on $\mathbb{R}^d$, let $B_i$, $i = 1, 2, \ldots$ denote all the different trajectories. We will assume that these trajectories belong to $C$. Let $N(t) = \sum_i \delta_{B_i(t)}$ denote the configuration of the system at time $t \in [0, 1]$. Then $N = \{N(t), t \in [0, 1]\}$ is a point–measure valued process, and we may assume that its paths are continuous. The process $N$ constitutes the "temporal" description of the system. The "trajectorial" description of the system is defined by $\mathcal{N} = \sum_i \delta_{B_i}$, which is a random point measure on $C$. (Remark: For branching particle systems it is necessary to give suitable weights to the Dirac measures at the particle positions in order to make the two approaches consistent).

For a sequence of particle systems indexed by $n = 1, 2, \ldots$, the *temporal fluctuations* are the signed measure valued processes $X_n = \{X_n(t), t \in [0, 1]\}$, where $X_n(t) = a_n^{-1}(N_n(t) - EN_n(t))$, assuming the mean $EN_n(t)$ exists, and $a_n$ is a suitable normalization. The *trajectorial fluctuations* are the random signed measures on $C$ defined by $\mathcal{X}_n = a_n^{-1}(\mathcal{N}_n - E\mathcal{N}_n)$, assuming the mean $E\mathcal{N}_n$ exists.

For example, if $N_n(0)$ is a Poisson random measure on $\mathbb{R}^d$ with intensity measure $n\lambda$, and from each of its points there arises a Brownian motion and these motions are mutually independent given the initial configuration, then $EN_n(t) = n\lambda$ and

$E\mathcal{N}_n = n\mathbb{W}$. In this case, with $a_n = n^{1/2}$, the process $X_n$ converges weakly in $C([0,1], \mathcal{S}'(\mathbb{R}^d))$ as $n \to \infty$ to an $\mathcal{S}'(\mathbb{R}^d)$–valued continuous Gaussian process $X = \{X(t),\, t \in [0,1]\}$ with covariance functional

$$\mathrm{Cov}(\langle X(s), \varphi \rangle, \langle X(t), \psi \rangle) = \int_{\mathbb{R}^d} \varphi(x) T_{t-s} \psi(x) dx, \quad s \leq t, \quad \varphi, \psi \in \mathcal{S}(\mathbb{R}^d),$$

where $(T_t)_t$ is the Brownian semigroup (Martin–Löf [19], Bojdecki and Gorostiza [3]), and $\mathcal{X}_n$ converges weakly in $\mathcal{S}(\mathbb{R}^d, \mathcal{H})'$ as $n \to \infty$ to $\mathcal{X}$, the white noise on Wiener space (Gorostiza and Nualart [10]). Thus $X$ and $\mathcal{X}$ are the temporal and the trajectorial fluctuation limits for this system, respectively. Note that in order to apply Theorems 2.1 and 2.2, since $\mathcal{N}_n$ is a random element of $\mathcal{S}(\mathbb{R}^d, \mathcal{H})'$, being a convergent sum of evaluations at random paths, to show that $\mathcal{X}_n$ takes values in $\mathcal{S}(\mathbb{R}^d, \mathcal{H})'$ it is necessary to verify that the mean $E\mathcal{N}_n = n\mathbb{W}$ belongs to $\mathcal{S}(\mathbb{R}^d, \mathcal{H})'$, which is obvious in this case.

We have just noted that $\mathbb{W} \in \mathcal{S}(\mathbb{R}^d, \mathcal{H})'$. In the next result we identify a class of $\sigma$–finite measures on $C$ which may also arise as means of particle systems and which also belong to $\mathcal{S}(\mathbb{R}^d, \mathcal{H})'$. (We use the same notation for a continuous process and for its distribution measure on $C$).

**3.1 Proposition**  *Let $Z_0 = \{Z_0(t),\, t \in [0,1]\}$ be a process with paths in $C_0$ such that $E(\|Z_0\|_\infty^n) \leq K^n (n!)^{1/2}$ for all $n = 1, 2, \ldots$, where $K$ is a constant (independent of $n$). Let $\mathbb{Z}$ be the measure on $C$ defined by*

$$\mathbb{Z} = \int_{\mathbb{R}^d} Z_x \mu(dx),$$

*where $Z_x$ denotes the distribution (on $C$) of $Z_0$ translated by $x \in \mathbb{R}^d$ ($Z_x(0) = x$), and $\mu$ is a $\sigma$–finite measure on $\mathbb{R}^d$ such that*

$$\int_{\mathbb{R}^d} (1 + |x|^2)^{-k} \mu(dx) < \infty$$

*for some nonnegative integer $k$. Then $\mathbb{Z} \in \mathcal{S}(\mathbb{R}^d, \mathcal{H})'$.*

**Proof**  We assume $d = 1$ for simplicity of notation. Let $F \in \mathcal{S}(\mathbb{R}, \mathcal{H})$ with Wiener chaos expansion $F(x, \omega) = \sum_{n=0}^\infty I_n(f_n(x))(\omega)$, $x \in \mathbb{R}$, $\omega \in C_0$. Then, using estimates in Gorostiza and Nualart [10] (proof of Theorem A, expression (2.6)), and the seminorms $\| \cdot \|_{H_j^{\odot n}}$ and $\| \cdot \|_{\mathcal{H}_j}$ therein we have

$$|F(x, \omega)| \leq \sum_{n=0}^\infty |I_n(f_n(x))| \leq \sum_{n=0}^\infty \sum_{k=0}^n \frac{n!}{(n-k)!(k!)^{1/2}} \|\omega\|_\infty^{n-k} 2^{-(j-1)n/2} \|f_n(x)\|_{H_j^{\odot n}}$$

16

for all $j = 0, 1, \ldots$ (the right hand side may be infinite). Hence

$$|\langle \mathbb{Z}, F \rangle| \leq \int_C |F(\omega)| \mathbb{Z}(d\omega) = \int_{\mathbb{R}} \int_{C_0} |F(x, \omega)| Z_0(d\omega) \mu(dx)$$

$$\leq \int_{\mathbb{R}} \sum_{n=0}^{\infty} \sum_{k=0}^{n} \frac{n!}{(n-k)!(k!)^{1/2}} E(\|Z_0\|_{\infty}^{n-k}) 2^{-(j-1)n/2} \|f_n(x)\|_{H_j^{\odot n}} \mu(dx)$$

$$\leq \int_{\mathbb{R}} \sum_{n=0}^{\infty} \sum_{k=0}^{n} \frac{n!}{(n-k)!(k!)^{1/2}} K^{n-k}((n-k)!)^{1/2} 2^{-(j-1)n/2} \|f_n(x)\|_{H_j^{\odot n}} \mu(dx)$$

$$\leq \int_{\mathbb{R}} \sum_{n=0}^{\infty} \sum_{k=0}^{n} \binom{n}{k}^{1/2} K^{n-k} 2^{-(j-1)n/2} (n!)^{1/2} \|f_n(x)\|_{H_j^{\odot n}} \mu(dx)$$

$$\leq \int_{\mathbb{R}} \sum_{n=0}^{\infty} (1+K)^n 2^{-(j-1)n/2} (n!)^{1/2} \|f_n(x)\|_{H_j^{\odot n}} \mu(dx)$$

$$\leq \int_{\mathbb{R}} \left[ \sum_{n=0}^{\infty} (1+K)^{2n} 2^{-(j-1)n} \right]^{1/2} \left[ \sum_{n=0}^{\infty} n! \|f_n(x)\|_{H_j^{\odot n}}^2 \right]^{1/2} \mu(dx)$$

$$= \left[ \sum_{n=0}^{\infty} (1+K)^{2n} 2^{-(j-1)n} \right]^{1/2} \int_{\mathbb{R}} \|F(x, \cdot)\|_{\mathcal{H}_j} \mu(dx)$$

$$\leq \left[ \sum_{n=0}^{\infty} (1+K)^{2n} 2^{-(j-1)n} \right]^{1/2} \left[ \sup_{x \in \mathbb{R}} (1+|x|^2)^k \|F(x, \cdot)\|_{\mathcal{H}_j} \right]$$

$$\cdot \int_{\mathbb{R}} (1+|x|^2)^{-k} \mu(dx).$$

The last expression is finite for $j$ sufficiently large, and it converges to 0 if $F \to 0$ in $\mathcal{S}(\mathbb{R}, \mathcal{H})$. $\qquad \square$

**3.2 Corollary** *A sufficient condition for the moment assumption in Proposition 3.1 is $E(\|Z_0\|_{\infty}^n) \leq K_1^n E(\|W_0\|_{\infty}^n)$ for all $n = 1, 2, \ldots$, where $K_1$ is a constant (independent of $n$).*

**Proof** It is known that $E(\|W_0\|_{\infty}^n) \leq J 2^{n/2} (n!)^{1/2} n^{-1/4}$ for every integer $n \geq 2$, where $J$ is a universal constant. $\qquad \square$

**3.3 Examples** 1. Any finite measure $\mu$ meets the condition of the Proposition. It follows in particular that $\delta_\omega \in \mathcal{S}(\mathbb{R}^d, \mathcal{H})'$ for each $\omega \in C$, a fact which we already know from Theorem 2.1, part (c).

2. For $\mathbb{W}$ the hypothesis of Corollary 3.2 is obviously satisfied, and $\lambda$ meets the required condition.

3. Let

$$\mathbb{B} = \int_{\mathbb{R}^d} B_{xx}\lambda(dx),$$

where $B_{xx}$ denotes the distribution of the Brownian bridge from $x$ to $x$. In this case $Z_0 = B_{00}$. Since $B_{00}(t) = W_0(t) - tW_0(1)$, then $\|B_{00}\|_\infty \le 2\|W_0\|_\infty$, and the condition in Corollary 3.2 holds. Hence $\mathbb{B} \in S(\mathbb{R}^d, \mathcal{H})'$. Note that $\mathbb{B}$ is singular with respect to $\mathbb{W}$.

Let us now modify the Poisson system of Brownian motions introduced above by assuming that from each point $x$ of the Poisson random measure $N_n(0)$ there arises a Brownian bridge from $x$ to $x$. Then $EN_n(t) = n\lambda$ and $E\mathcal{N}_n = n\mathbb{B}$, where $\mathbb{B}$ is the $\sigma$-finite Brownian bridge measure on $C$ defined in Example 3 above. In this case, with $a_n = n^{1/2}$, the process $X_n$ converges weakly in $C([0,1], S'(\mathbb{R}^d))$ as $n \to \infty$ to an $S'(\mathbb{R}^d)$-valued continuous Gaussian process $X = \{X(t),\ t \in [0,1]\}$ with covariance functional

$$\mathrm{Cov}(\langle X(s), \varphi\rangle, \langle X(t), \psi\rangle) = \int_{\mathbb{R}^d} \varphi(x)T_{(t-s)[1-(t-s)]}\psi(x)dx, \quad s \le t, \quad \varphi, \psi \in S(\mathbb{R}^d),$$

and $\mathcal{X}_n$ converges weakly in $S(\mathbb{R}^d, \mathcal{H})'$ as $n \to \infty$ to the Gaussian random element $\mathcal{X}$ of $S(\mathbb{R}^d, \mathcal{H})'$ characterized by

$$E(\exp\{i\langle \mathcal{X}, F\rangle\}) = \exp\left\{-\frac{1}{2}\int_C F^2(\omega)\mathbb{B}(d\omega)\right\}, \quad F \in S(\mathbb{R}^d, \mathcal{H}).$$

This example is discussed further in Gorostiza [9]. In particular, it is observed that the process $X$ is not Markovian. (In [9] the proof that $\mathbb{B} \in S(\mathbb{R}^d, \mathcal{H})'$ is not correct).

Note that for each fixed $t \in [0,1]$, the fluctuation limits of the Poisson system of Brownian motions and the Poisson system of Brownian bridges are both equal to the standard white noise measure in $S'(\mathbb{R}^d)$.

In the two previous examples of trajectorial fluctuation limits, as in many applications of this type, one has processes $X_n$, $n = 1, 2, \ldots$ with paths in $C([0,1], S'(\mathbb{R}^d))$, and corresponding random elements $\mathcal{X}_n$, $n = 1, 2, \ldots$ of $S(\mathbb{R}^d, \mathcal{H})'$, both determined by the same particle system. The *convergence problem* is formulated as follows:

**3.4 Question**   *If $\mathcal{X}_n$ converges weakly to $\mathcal{X}$ in $S(\mathbb{R}^d, \mathcal{H})'$ as $n \to \infty$, is there a process $X$ with paths in $C([0,1], S'(\mathbb{R}^d))$ which corresponds to $\mathcal{X}$? If so, does $X_n$ converge weakly to $X$?*

This raises the following more general question:

**3.5 Question** *Given a random element $\mathfrak{X}$ of $\mathcal{S}(\mathbb{R}^d, \mathcal{H})'$, does there correspond to it, in some sense, a process with paths in $C([0,1], \mathcal{S}'(\mathbb{R}^d))$?*

Let us address Question 3.5 first. Given an element $\mathfrak{X}$ of $\mathcal{S}(\mathbb{R}^d, \mathcal{H})'$ (random or not), a natural way to try to associate to it an $\mathcal{S}'(\mathbb{R}^d)$–valued process $X = \{X(t), t \in [0,1]\}$ is to define $X(t)$ by

$$\langle X(t), \varphi \rangle = \langle \mathfrak{X}, F_{\varphi,t} \rangle, \quad \varphi \in \mathcal{S}(\mathbb{R}^d), \text{ where } F_{\varphi,t}(\omega) = \varphi(\omega(t)), \; \omega \in C, \qquad (1)$$

(i.e., testing only on the paths at time $t$). However, expression (1) is not legitimate because $F_{\varphi,t}$ does not belong to $\mathcal{S}(\mathbb{R}^d, \mathcal{H})$. We can see this already restricting $\omega$ to $C_0$ with the Wiener measure $W_0$. Clearly $F_{\varphi,t} \in L^2(C_0, W_0)$, and therefore $F_{\varphi,t}$ has a Wiener chaos expansion $F_{\varphi,t} = \sum_n I_n(f_n)$. The kernels $f_n$ are given by

$$f_n(s_1, \ldots, s_n) = \frac{1}{n!}(E\varphi^{(n)}(W_0(t)))1_{[0,t]^n}(s_1, \ldots, s_n).$$

However, the construction of the space $\mathcal{H} \subset L^2(C_0, W_0)$ in Gorostiza and Nualart [10] implies that any element of $\mathcal{H}$ has Wiener chaos expansion with kernels $f_n$ which are infinitely differentiable and vanish together with all their derivatives on the boundary of $[0,1]^n$. The kernels of $F_{\varphi,t}$ do not satisfy this condition, and therefore $F_{\varphi,t} \notin \mathcal{H}$. Similarly, functionals of the form $F(\omega) = \sum_{j=1}^k \varphi_j(\omega(t_j))$, $0 \le t_1 < \ldots < t_k \le 1$, and $F(\omega) = \int_0^1 \varphi(\omega(t))f(t)dt$, $f \in C([0,1], \mathbb{R})$, $\omega \in C$, do not belong to $\mathcal{S}(\mathbb{R}^d, \mathcal{H})$.

Nevertheless it is possible to justify the use of expression (1) in some cases. Clearly, it holds for elements of the form $\mathfrak{X} = \delta_\omega$ and $X(t) = \delta_{\omega(t)}$, $t \in [0,1]$, for a fixed (or random) $\omega \in C$. It can also be justified for some random elements $\mathfrak{X}$ of $\mathcal{S}(\mathbb{R}^d, \mathcal{H})'$ as follows:

**3.6 Proposition** *Assume the random element $\mathfrak{X}$ of $\mathcal{S}(\mathbb{R}^d, \mathcal{H})'$ is such that $F \mapsto \mathfrak{X}(F)$ is a continuous random linear functional on $L^2(C, \mathbb{W})$. Then for each $t \in [0,1]$ there exists a unique random element $X(t)$ of $\mathcal{S}'(\mathbb{R}^d)$ such that $\langle X(t), \varphi \rangle = \mathfrak{X}(F_{\varphi,t})$ a.s. for all $\varphi \in \mathcal{S}(\mathbb{R}^d)$.*

**Proof** (Continuous random linear functionals are defined in Itô [15]). Fix $\varphi \in \mathcal{S}(\mathbb{R}^d)$ and $t \in [0,1]$. Then $F_{\varphi,t}(\omega) = \varphi(\omega(t))$, $\omega \in C$, belongs to $L^2(C, \mathbb{W})$. Since $\varphi \mapsto F_{\varphi,t}$ is continuous because $\int_C F_{\varphi,t}^2 d\mathbb{W} = \int_{\mathbb{R}^d} \varphi^2(x)dx$, the hypothesis implies that $\varphi \mapsto \mathfrak{X}(F_{\varphi,t})$ is a continuous random linear functional on $\mathcal{S}(\mathbb{R}^d)$. Since $\mathcal{S}(\mathbb{R}^d)$ is nuclear, the conclusion follows by the regularization theorem (Badrikian [1], Itô [15]). $\qquad\square$

**3.7 Corollary** *A sufficient condition for the assumption of Proposition 3.6 (which is equivalent to it in the Gaussian case), is that the mapping $F \mapsto E(\langle \mathfrak{X}, F \rangle^2)$ be continuous at 0 in the topology of $L^2(C, \mathbb{W})$.*

For the Poisson system of Brownian motions introduced above, the trajectorial fluctuation limit $\mathfrak{X}$ (white noise on $\sigma$–finite Wiener space) obviously satisfies the hypothesis of Proposition 3.6. The corresponding $\mathcal{S}'(\mathbb{R}^d)$–valued process $X$ is precisely the temporal fluctuation limit of this system, also described above. Proposition 3.6 also covers others cases.

Once an $\mathcal{S}'(\mathbb{R}^d)$–valued process $X \equiv \{X(t),\ t \in [0,1]\}$ corresponding to $\mathfrak{X}$ is defined, membership of its paths in $C([0,1], \mathcal{S}'(\mathbb{R}^d))$ (or $D([0,1], \mathcal{S}'(\mathbb{R}^d))$) can be studied by means of the criteria of Mitoma [23]. A special case is the following:

**3.8 Corollary** *Let $\mathfrak{X}$ be a centered Gaussian random element of $\mathcal{S}(\mathbb{R}^d, \mathcal{H})'$ such that the mapping $F \mapsto E(\langle \mathfrak{X}, F \rangle^2)$ is continuous at 0 in the topology of $L^2(C, \mathbb{W})$. Then there exists a process $X = \{X(t),\ t \in [0,1]\}$ with paths in $C([0,1], \mathcal{S}'(\mathbb{R}^d))$, such that $\langle X(t), \varphi \rangle = \mathfrak{X}(F_{\varphi,t})$ for all $\varphi \in S(\mathbb{R}^d)$, $t \in [0,1]$, where $\mathfrak{X}(\cdot)$ denotes the continuous extension of $\mathfrak{X}$ on $L^2(C, \mathbb{W})$.*

**Proof** By assumption, $E(\mathfrak{X}(F)^2) \leq K \int_C F^2 d\mathbb{W}$ for all $F \in L^2(C, \mathbb{W})$ and for some constant $K$. Let $X$ be the process given by Corollary 3.7. For any $\varphi \in \mathcal{S}(\mathbb{R}^d)$ and $0 \leq s < t \leq 1$ we have

$$
\begin{aligned}
E(((\langle X(t), \varphi \rangle - \langle X(s), \varphi \rangle)^2) &= E(\mathfrak{X}(F_{\varphi,t} - F_{\varphi,s})^2) \\
&\leq K \int_C (\varphi(\omega(t)) - \varphi(\omega(s)))^2 \mathbb{W}(d\omega) \\
&= 2K \int_{\mathbb{R}^d} \varphi(x)(\varphi(x) - T_{t-s}\varphi(x)) dx \\
&\leq K_1 |t - s|,
\end{aligned}
$$

since $\varphi$ belongs to the domain of the generator of the Brownian semigroup $(T_t)_t$. The real process $\{\langle X(t), \varphi \rangle,\ t \in [0,1]\}$ is clearly centered Gaussian, so $E(((\langle X(t), \varphi \rangle - \langle X(s), \varphi \rangle)^4 \leq K_2(t-s)^2$ for $t, s \in [0,1]$. The existence of a continuous version of $X$ now follows from the Kolmogorov theorem and Mitoma's criterion [23]. $\qquad \square$

The following example brings out two interesting observations on the correspondence of a process in $C([0,1], \mathcal{S}'(\mathbb{R}^d))$ to a random element of $\mathcal{S}(\mathbb{R}^d, \mathcal{H})'$.

Let us apply (1) to obtain elements $f = \{f(t), t \in [0,1]\}$ and $g = \{g(t), t \in [0,1]\}$ of $C([0,1], \mathcal{S}'(\mathbb{R}^d))$ which correspond to the elements of $\mathbb{W}$ and $\mathbb{B}$ of $\mathcal{S}(\mathbb{R}^d, \mathcal{H})'$, respectively. Let $p_t(x, y)$ designate the transition density of the standard Wiener process. Then, for fixed $t \in [0, 1]$ and each $\varphi \in \mathcal{S}(\mathbb{R}^d)$ we have

$$\langle f(t), \varphi \rangle = \langle \mathbb{W}, F_{\varphi,t} \rangle = \int_{\mathbb{R}^d} \int_{C_0} F_{\varphi,t}(x + \omega) W_0(d\omega) dx$$

$$= \int_{\mathbb{R}^d} \int_{\mathbb{R}^d} \varphi(y) p_t(x, y) dy dx = \int_{\mathbb{R}^d} \varphi(y) dx = \langle \lambda, \varphi \rangle,$$

and

$$\langle g(t), \varphi \rangle = \langle \mathbb{B}, F_{\varphi,t} \rangle = \int_{\mathbb{R}^d} \int_{C_0} F_{\varphi,t}(x + \omega) B_{00}(d\omega) dx$$

$$= \int_{\mathbb{R}^d} \int_{\mathbb{R}^d} \varphi(y) \frac{p_t(x, y) p_{1-t}(y, x)}{p_1(x, x)} dy dx$$

$$= \frac{1}{p_1(0, 0)} \int_{\mathbb{R}^d} \varphi(y) \left( \int_{\mathbb{R}^d} p_t(x, y) p_{1-t}(y, x) dx \right) dy$$

$$= \int_{\mathbb{R}^d} \varphi(y) dy = \langle \lambda, \varphi \rangle,$$

where we used the Chapman–Kolmogorov equation. Hence $f(t) = g(t) = \lambda \in \mathcal{S}'(\mathbb{R}^d)$ for all $t \in [0, 1]$. Therefore to the different elements $\mathbb{W}$ and $\mathbb{B}$ of $\mathcal{S}(\mathbb{R}^d, \mathcal{H})'$ there corresponds the same element of $C([0, 1], \mathcal{S}'(\mathbb{R}^d))$. This simple example confirms that the "trajectorial description" is somehow richer than the "temporal description". On the other hand, the definition of $X$ corresponding to $\mathcal{X}$ by means of (1) is not sufficiently precise. For example, we were able to associate to $\mathbb{B}$ an element of $C([0, 1], \mathcal{S}'(\mathbb{R}^d))$ by explicit calculation of the right hand side of (1). But $\mathbb{B}$ is a measure on $C$ which is singular with respect of $\mathbb{W}$, and therefore Proposition 3.6 does not apply to it. Also, there is a question about an intrinsic extension of $\mathbb{B}$ from $\mathcal{S}(\mathbb{R}^d, \mathcal{H})$ to the class of functionals which occur in (1). We do not include here the solution to this question because it is lengthy and it is specific for the particular case of $\mathbb{B}$.

Let us now come back to Question 3.4. Since there is no natural mapping between $\mathcal{S}(\mathbb{R}^d, \mathcal{H})'$ and $C([0, 1], \mathcal{S}'(\mathbb{R}^d))$, we cannot conclude weak convergence in $C([0, 1], \mathcal{S}'(\mathbb{R}^d))$ from weak convergence in $\mathcal{S}(\mathbb{R}^d, \mathcal{H})'$ by means of a "continuous mapping theorem". The following results involve additional assumptions which reduce the generality of the problem but which may hold in special applications.

We will first give a result which in particular allows us to associate an $\mathcal{S}'(\mathbb{R}^d)$-valued process to the weak limit of a sequence of random elements of $\mathcal{S}(\mathbb{R}^d, \mathcal{H})'$.

**3.9 Lemma** *Suppose that $\mathfrak{X}_n$, $n = 1, 2, \ldots$ are random elements of $\mathcal{S}(\mathbb{R}^d, \mathcal{H})'$ such that*

*(i) $\mathfrak{X}_n$ converges weakly to $\mathfrak{X}$ in $\mathcal{S}(\mathbb{R}^d, \mathcal{H})'$ as $n \to \infty$.*

*(ii) $F \mapsto \langle \mathfrak{X}_n, F \rangle$ is a continuous random linear functional in the topology of $L^2(C, \mathbb{W})$ for each $n$, its extension being denoted by $\mathfrak{X}_n(F)$.*

*(iii) $\limsup_n E(\mathfrak{X}_n(F)^2)$ is finite and is continuous at $F = 0$ (in the topology of $L^2(C, \mathbb{W})$).*

*Then $\mathfrak{X}$ extends to a continuous random linear functional $F \mapsto \mathfrak{X}(F)$ on $L^2(C, \mathbb{W})$, and $\mathfrak{X}_n(F)$ converges weakly to $\mathfrak{X}(F)$ as $n \to \infty$ for each $F \in L^2(C, \mathbb{W})$.*

**Proof**  Let $F \in L^2(C, \mathbb{W})$.  There exists a sequence $(F_m)_m$ in $\mathcal{S}(\mathbb{R}^d, \mathcal{H})$ which converges to $F$ in $L^2(C, \mathbb{W})$.  Then

$$E(((\langle \mathfrak{X}, F_m \rangle - \langle \mathfrak{X}, F_{m'} \rangle))^2) \leq \liminf_n E(((\langle \mathfrak{X}_n, F_m \rangle - \langle \mathfrak{X}_n, F_{m'} \rangle))^2)$$
$$\leq \limsup_n E((\langle \mathfrak{X}_n, F_m - F_{m'} \rangle^2)),$$

and therefore $((\langle \mathfrak{X}, F_m \rangle)_m$ is an $L^2$–Cauchy sequence.  Denote the limit by $\mathfrak{X}(F)$. It is clear that $\mathfrak{X}(F)$ does not depend on the choice of $(F_m)_m$, that it is a linear random functional, and that $\langle \mathfrak{X}, F \rangle = \mathfrak{X}(F)$ a.s. if $F \in \mathcal{S}(\mathbb{R}^d, \mathcal{H})$.

Since $\langle \mathfrak{X}_n, F_m \rangle$ converges weakly to $\langle \mathfrak{X}, F_m \rangle$ as $n \to \infty$ for each $m$, in order to prove that $\mathfrak{X}_n(F)$ converges weakly to $\mathfrak{X}(F)$ as $n \to \infty$ it suffices to show that

$$\lim_m \limsup_n E((\mathfrak{X}_n(F_m - F))^2) = 0,$$

due to Chebyshev's inequality and Theorem 4.2 of Billingsley [2], but this holds by assumption.

Now its is easy to verify that $F \mapsto \mathfrak{X}(F)$ is mean–square continuous.  $\square$

By Proposition 3.6 we then have the following consequence:

**3.10 Corollary**  *Under the conditions of Lemma 3.9, for each $t \in [0, 1]$ there is a unique random element $X(t)$ of $\mathcal{S}'(\mathbb{R}^d)$ such that (1) holds a.s. for all $\varphi \in \mathcal{S}(\mathbb{R}^d)$.*

Finally, we have the following temporal–type limits from the limits in $\mathcal{S}(\mathbb{R}^d, \mathcal{H})'$:

**3.11 Proposition**  *Suppose that $\mathfrak{X}_n$, $n = 1, 2, \ldots$, are random elements of $\mathcal{S}(\mathbb{R}^d, \mathcal{H})'$ which satisfy the conditions of Lemma 3.9.  Then*

*(1) There are unique $\mathcal{S}'(\mathbb{R}^d)$–valued processes $X_n = \{X_n(t), t \in [0, 1]\}$ for all $n$ and $X = \{X(t), t \in [0, 1]\}$ which correspond to $\mathfrak{X}_n$ for all $n$ and $\mathfrak{X}$, respectively, by means of (1) for each $t$ and a.s. for all $\varphi \in \mathcal{S}(\mathbb{R}^d)$.*

(2) *The finite–dimensional distributions of $X_n$ converge weakly to those of $X$ as $n \to \infty$, i.e. for every $k = 1, 2, \ldots$, $0 \le t_1 < \ldots < t_k \le 1$ and $\varphi_1, \ldots, \varphi_k \in \mathcal{S}(\mathbb{R}^d)$,*

$$(\langle X_n(t_1), \varphi_1 \rangle, \ldots, \langle X_n(t_k), \varphi_k \rangle) \text{ converges weakly to } (\langle X(t_1), \varphi_1 \rangle, \ldots, \langle X(t_k), \varphi_k \rangle)$$

*as $n \to \infty$.*

(3) *If all $X_n$ and $X$ have paths in $D([0,1], \mathcal{S}'(\mathbb{R}^d))$, then*

$$\int_0^1 \langle X_n(t), \varphi \rangle f(t)dt \text{ converges weakly to } \int_0^1 \langle X(t), \varphi \rangle f(t)dt \text{ as } n \to \infty$$

*for every $\varphi \in \mathcal{S}(\mathbb{R}^d)$ and $f \in C([0,1], \mathbb{R})$.*

**Proof**  Statement 1 follows from Corollary 3.10.

Statement 2 follows from Lemma 3.9 by taking $F$ to be of the form $F(\omega) = \sum_{j=1}^{k} \varphi_j(\omega(t_j))$, $\omega \in C$. Note that $F \in L^2(C, \mathbb{W})$.

For statement 3, it is easy to show that $F$ defined by $F(\omega) = \int_0^1 \varphi(\omega(t))f(t)dt$ belongs to $L^2(C, \mathbb{W})$. Then

$$\mathcal{X}_n(F) = \int_0^1 \langle X_n(t), \varphi \rangle f(t)dt \quad \text{a.s.}$$

for all $n$, and the analogous equality holds for the limits $\mathcal{X}$ and $X$ by Lemma 3.9. The result then follows by Lemma 3.9. $\qquad\square$

**3.12 Remarks**  1. We do not know if $X_n$ converges weakly to $X$ in $C([0,1], \mathcal{S}'(\mathbb{R}^d))$ without additional assumptions (tightness remains an open problem), and if conditions (ii) and (iii) of Lemma 3.9 can be replaced by weaker ones.

2. Each one of the conclusions 2 and 3 of Proposition 3.11 identifies a unique limit process $X$, but they are not equivalent (see Bojdecki, Gorostiza and Ramaswamy [4], Gorostiza and Rebolledo [11]).

3. Weak convergence for a general class of functionals of integral type of $L_p^{E-}$ valued processes is studied by Cremers and Kadelka [6], assuming a "weak tightness" condition and weak convergence of finite–dimensional distributions of the processes. Part 3 of Proposition 3.11 shows that weak convergence in $\mathcal{S}(\mathbb{R}^d, \mathcal{H})'$ implies weak convergence of some functionals of integral type without assuming weak convergence of finite–dimensional distributions (compare with [6], part B).

**Acknowledgements**  The authors thank CIMAT (Guanajuato, México) for hospitality. T.B. thanks CONACyT for support during his visit to CIMAT. The research of L.G.G. is partially supported by CONACyT grant 2059-E9302.

**References**

[1] Badrikian, A.: Séminaire sur les fonctions aléatories linéaires et les mesures Cylindriques; *Lect. Notes in Math.* **139** (1970), Springer–Verlag, Berlin

[2] Billingsley, P.: *Convergence of Probability Measures.* Wiley, New York, 1968

[3] Bojdecki, T. and Gorostiza, L. G.: Langevin equations for $S'$-valued Gaussian processes and fluctuation limits of infinite particle systems; *Probab. Th. Rel. Fileds* **73** (1986) 227–244

[4] Bojdecki, T., Gorostiza, L. G., and Ramaswamy, S.: Convergence of $S'$-valued processes and space–time random fields; *J. Funct. Anal.* **66** (1986) 21–41

[5] Boulicaut, P.: Convergence cylindrique et convergence étroite d'une suite de probabilités de Radon; *Z. Wahrsch. Verw. Gebiete* **28** (1973) 43–52

[6] Cremers, H. and Kadelka, D.: On weak convergence of integral functions of stochastic processes with applications to processes taking paths in $L_p^E$; *Stoch. Proc. Appl.* **21** (1986) 305–317

[7] Cutland, N. and Ng, S.-A.: The Wiener sphere and Wiener measure; *Ann. Probab.* **21** (1993) 1–13

[8] Gorostiza, L. G.: Limites gaussiennes pour les champs aléatoires ramifés super-critiques; in: *Aspects Statistiques et Aspects Physiques des Processus Gaussiens* (1981) 285–398, Éditions du CNRS, Paris

[9] Gorostiza, L. G.: Distributions on Wiener space and fluctuation limits of Poisson systems of Brownian bridges; in: *Chaos expansions, multiple Wiener–Itô integrals and their applications* (1994) 337–347, CRC Press

[10] Gorostiza, L. G. and Nualart, D.: Nuclear Gelfand triples on Wiener space and applications to trajectorial fluctuations of particle systems; *J. Funct. Anal.* (to appear 1994)

[11] Gorostiza, L. G. and Rebolledo, R.: A random field approach to weak convergence of processes; *Statist. Prob. Letters* **18** (1993) 49–55

[12] Hida, T., Kuo, H. -H., Potthoff, J., and Streit, L.: *White Noise: An Infinite Dimensional Calculus.* Kluwer Academic Publishers, 1993

[13] Holley, R. and Stroock, D. W.: Generalized Ornstein–Uhlenbeck processes and infinite particle branching Brownian motions; *Publ. Res. Inst. Math. Sci.* **14** (1981) 741–788

[14] Itô, K.: Distribution valued processes arising from independent Brownian motions; *Math. Z.* **182** (1983) 17–33

[15] Itô, K.: *Foundations of Stochastic Differential Equations in Infinite Dimensional Spaces.* SIAM, Philadelphia, 1984

[16] Kondratiev, Yu. V. and Samoilenko, Yu. S.: The spaces of trial and generalized functions of an infinite number of variables; *Reports Math. Phys.* **14** (1978) 325–350

[17] Korezlioglu, H. and Üstünel, A. S.: a new class of distributions on the Wiener spaces; *Lect. Notes Math.* **1444** (1990) 106–121, Springer–Verlag, Berlin

[18] Kubo, I. and Yokoi, Y.: A remark on the space of testing random variables in the white noise calculus; *Nagoya Math. J.* **115** (1989) 139–149

[19] Martin–Löf, A.: Limit theorems for the motion of a Poisson system of independent Markovian particles with high density; *Z. Wahrsch. Verw. Gebiete* **34** (1976) 205–223

[20] McKean, H. P.: Geometry of differential space; *Ann. Probab.* **1** (1973) 192–206

[21] Meyer, P. A. and Yan, J. A.: Distributions sur l'espace de Wiener (suite). Sém. Proba. XXIII, *Lect. Notes Math.* **1372** (1989) 382–404, Springer–Verlag, Berlin

[22] Meyer, P. A. and Yan, J. A.: Les "fonctions caractéristiques" des distributions sur l'espace de Wiener, Sém. Proba. XXV, *Lect. Notes Math.* **1485** (1991) 61–78, Springer–Verlag, Berlin

[23] Mitoma, I.: Tightness of probabilities on $C([0,1], \mathcal{S}')$ and $D([0,1], \mathcal{S}')$; *Ann. Probab.* **11** (1983) 989–999

[24] Sznitmann, A. S.: A fluctuation result for nonlinear diffusions; in: *Infinite-dimensional analysis and stochastic processes, Research Notes in Math.* **124** (1985) 145–160 Pitman, Boston

[25] Tanaka, H.: Limit theorems for certain diffusions processes with interaction; in: *Proceedings, Taniguchi International Symposium on Stochastic Analysis,* Katata and Kyoto, North Holland, Amsterdam, 1984

[26] Tanaka, H. and Hitsuda, M.: Central limit theorem for a simple diffusion model of interacting particles; *Hiroshima Math. J.* **11** (1981) 415–423

[27] Treves, F.: *Topological Vector Spaces, Distributions and Kernels.* Academic Press, New York, 1967

[28] Watanabe, S.: *Differential Equations and Malliavin Calculus.* Springer–Verlag, Berlin, 1984

Tomasz Bojdecki
Institute of Mathematics
University of Warsaw
ul. Banacha 2
00-097 Warszawa, POLAND
*E-mail*: tobojd@mimuw.edu.pl

Luis G. Gorostiza
Departamento de Matemáticas
Centro de Investigacíon y de Estudios Avanzados
A. P. 14-740
07000 México D. F., MEXICO
*E-mail*: gortega@redvax1.dgsca.unam.mx

A. BUDHIRAJA & G. KALLIANPUR
# Hilbert space valued traces and multiple Stratonovich integrals with statistical applications

## 1. Introduction

Multiple Stratonovich integrals (MSI) and their connections with multiple Wiener integrals and Hilbert space valued traces were introduced for the first time in a paper of Hu and Meyer [3]. Johnson and Kallianpur [4] made the approach in [3] rigorous by defining various Hilbert space valued traces, in particular the limiting trace which was the most relevant for the definition of MSI. Using the ideas of *lifting*, a definition of MSI was also given in [4] and it was shown that these MSI were connected to Multiple Wiener integrals via the limiting traces. While interesting from the point of view of furnishing formulae for certain types of Feynman integrals, these Stratonovich integrals do not meet the requirements of statistical applications since these integrals are invariant over an equivalence class in $L^2[0,1]^p$ and thus do not depend on the value of the integrand on the diagonal.

In the present paper, we take a look at the problem with a view to possible statistical applications. The $k$-traces introduced in Section 2 are defined for a subclass (denoted by $\mathcal{S}_p^1$) of the $L^2$–space of $p$-th order symmetric kernels. $\mathcal{S}_p^1$ is made into a Hilbert space under a new inner product in such a manner that each of the $k$-th order $\tau$-traces ($k = 0, 1, \dots [p/2]$) is a continuous map from $\mathcal{S}_p^1$ to $L^2[0,1]^{p-2k}$.

Results relating the MSI to multiple Wiener integrals similar to those obtained in [4] are derived in Section 2. Statistical applications of our results will be given in Section 3. We study $V$-statistics which are of interest in nonparametric statistics. The proofs of the results will be presented in another paper [1]. In Theorem 3.2 the limiting distribution of a $V$–statistic is derived in terms of MSI. It is natural that Stratonovich integrals are involved since a $V$-statistic (in contrast to a $U$-statistic) allows repeated indices (see Definition 3.1). Hoeffding's pioneering 1948 result [2] is mentioned as a corollary to Theorem 3.2.

## 2. Hilbert Space Valued Traces and Multiple Stochastic Integrals

In this section we will introduce the multiple Stratonovich integral. This integral, in general is different from that considered by Johnson and Kallianpur [4], though the two integrals agree for step functions. The Stratonovich integral is closely tied to certain Hilbert space valued traces. In this section we will introduce these traces and discuss their connection with the 'limiting traces' of Johnson and Kallianpur.

**Definition 2.1** (*The class $S_p$ of step functions*) A real valued symmetric function $f_p$ on $[0,1]^p$ is in $S_p$ if there exists a partition $\{0 = t_1 < \ldots t_m < t_{m+1} = 1\}$, of $[0,1]$ and constants $\{a_{i_1,\ldots,i_p}; i_1, \ldots, i_p = 0, 1, 2, \ldots, m\}$ s.t.

$$f_p(s_1, \ldots, s_p) = a_{i_1,\ldots,i_p} \quad \text{if} \quad (s_1, \ldots, s_p) \in \Delta_{i_1} \times \ldots \times \Delta_{i_p} \tag{2.1}$$

where $\Delta_{i_j} = (t_{i_j}, t_{i_j+1}]$, if $1 \leq i_1 \leq m$ and $\Delta_{i_j} = \{0\}$ if $i_j = 0, j = 1, 2, \ldots p$.

Let $(\Omega, \mathcal{F}, P)$ be a probability space and $\{W_t; 0 \leq t \leq 1\}$ be a Wiener process on this space.

We first define the multiple Stratonovich integral, with respect to the Wiener process for integrands in $S_p$. This integral turns out to be the same as the multiple Stratonovich integral of Johnson and Kallianpur which will be denoted by $\delta_p^s(.)$.

**Definition 2.2** (*Multiple Stratonovich integral for functions in $S_p$*) Let $f_p \in S_p$ be given by (2.1). Define the *multiple Stratonovich integral* (M.S.I.) of $f_p$ as

$$\delta_p(f_p) := \sum_{i_1, i_2, \ldots i_p = 1}^{m} a_{i_1,\ldots,i_p} (W(t_{i_1+1}) - W(t_{i_1})) \ldots (W(t_{i_p+1}) - W(t_{i_p})). \tag{2.2}$$

We will now briefly recall the limiting traces as introduced in [4] and then give a representation for $\delta_p(f_p)$ in terms of those traces.

**Definition 2.3** (*Limiting traces*) Let $f_p$ be a symmetric function in $L^2[0,1]^p$. Fix $k; 1 \leq k \leq [p/2]$. Suppose that for every complete orthonormal system (CONS), $\{\phi_j\}$ for $L^2[0,1]$

$$\sum_{i_1,\ldots i_k=1}^{N} \sum_{i_{2k+1}, i_p=1}^{N} < f_p, \phi_{i_1} \otimes \phi_{i_1} \otimes \ldots \otimes \phi_{i_k} \otimes \phi_{i_k} \otimes \phi_{i_{2k+1}} \ldots \phi_{i_p} > \phi_{i_{2k+1}} \ldots \phi_{i_p} \tag{2.3}$$

converges in $L^2[0,1]^{p-2k}$ to a limit which is independent of the choice of the CONS $\{\phi_i\}$. Then we say that the $k^{th}$-*limiting trace* for $f_p$ exists, which by definition is the limit of the series in (2.3) and is denoted as $\overset{\to k}{Tr} f_p$. $\overset{\to 0}{Tr} f_p$ is defined to be the same as $f_p$.

The following proposition relates the M.S.I, $\delta_p$, with the multiple integral of Johnson and Kallianpur and therefore with multiple Wiener integrals, through the Hu-Meyer formula.

**Proposition 2.4** *Let $f_p \in \mathcal{S}_p$, then $\overrightarrow{Tr}^k f_p$ exists $\forall\ 0 \le k \le [p/2]$ and we have that*

$$\delta_p(f_p) = \delta_p^s(f_p) = \sum_{k=0}^{[p/2]} C_{p,k} I_{p-2k}(\overrightarrow{Tr}^k f_p). \tag{2.4}$$

*Furthermore,*

$$E[\delta_p(f_p)]^2$$

$$\le C \left\{ \int_{[0,1]^p} f_p^2(t_1,\ldots,t_p) dt_1 \ldots dt_p \right.$$

$$\left. + \sum_{k=1}^{[p/2]} \int_{[0,1]^{p-k}} f_p^2(t_1,t_1,\ldots,t_k,t_k,t_{2k+1},\ldots,t_p) dt_1 \ldots dt_k dt_{2k+1} \ldots dt_p \right\} \tag{2.5}$$

*where $C_{p,k} = \frac{p!}{(p-2k)!2^k k!}$, $I_j$ is the $j$–fold multiple integral and $C$ is an appropriate constant.*

The inequality in (2.5) will enable us to extend the domain of definition of the integral to a larger class by a denseness argument.

Let $\mathcal{S}_p^1$ be the class of real valued, measurable, symmetric functions $f_p$, on $[0,1]^p$ s.t.

$$\int_{[0,1]^{p-k}} [^k f_p]^2(t_1,\ldots t_{p-k}) dt_1 \ldots dt_{p-k} < \infty, \forall 0 \le k \le [p/2], \tag{2.6}$$

where $^k f_p : [0,1]^{p-k} \to \mathbf{R}$ for $k = 1,2,\ldots [p/2]$ is defined as:

$$^k f_p(t_1,\ldots t_{p-k}) := f_p(t_1,t_1,\ldots,t_k,t_k,t_{k+1},\ldots t_{p-k}) \tag{2.7}$$

with $^0 f_p \equiv f_p$. We will employ the convention that, $\mathcal{S}_0^1 = \mathbf{R}$ and that $\| \cdot \|_{*,0}$ is the usual Euclidian distance in $\mathbf{R}$.

Consider the following inner product on $\mathcal{S}_p^1$:

$$< f_p, g_p >_{*,p} := \sum_{k=0}^{[p/2]} < {}^k f_p, {}^k g_p >_k, \quad f_p, g_p \in \mathcal{S}_p \tag{2.8}$$

where $< \cdot, \cdot >_k$ is the inner product in $L^2[0,1]^{p-k}$ and $^k f_p$ and $^k g_p$ are as in (2.7).

It can be shown that $(\mathcal{S}_p, <\cdot,\cdot>_{*,p})$ is dense in $(\mathcal{S}_p^1, <\cdot,\cdot>_{*,p})$ and therefore the definition of $\delta_p$ can be extended to $\mathcal{S}_p^1$ by the usual kind of denseness argument.

**Definition 2.5** (*Multiple Stratonovich integral for elements in $\mathcal{S}_p^1$*) Let $f_p \in \mathcal{S}_p^1$ and let $\{f_{p,n}\}$ be a sequence in $\mathcal{S}_p$, s.t., $\| f_{p,n} - f_p \|_{*,p}$ converges to 0 as $n \to \infty$. From (2.5), we have that $\delta_p(f_{p,n})$ converges in $L^2(\Omega)$ as $n \to \infty$. Define the *multiple Stratonovich integral* of $f_p$, denoted by $\delta_p(f_p)$ as:

$$\delta_p(f_p) := L^2(\Omega) - \lim_{n \to \infty} \delta_p(f_{p,n}). \tag{2.9}$$

The multiple Stratonovich integral introduced in this section can be represented as a sum of multiple Wiener integrals through a formula which is very similar to the Hu-Meyer formula (2.4), though the traces appearing in this formula are in general different from the limiting traces that appear in formula (2.4). We discuss below the appropriate notion of a trace for this setup.

**Definition 2.6** Let $f_p \in \mathcal{S}_p^1$. Fix $1 \le k \le [p/2]$. Then the $k$*th-$\tau$ trace* of $f_p$, denoted as $\tau^k f_p$ is an element of $\mathcal{S}_{p-2k}^1$, defined as:

$$\tau^k f_p(\cdot) := \int_{[0,1]^k} f_p(t_1, t_1, \ldots, t_k, t_k, \cdot) dt_1 \ldots dt_k. \tag{2.10}$$

Define $\tau^0 f_p := f_p$.

The following proposition is the analog of the Hu-Meyer formula for $\delta_p$.

**Proposition 2.7** *Let $f_p \in \mathcal{S}_p^1$, then*

$$\delta_p(f_p) = \sum_{k=0}^{[p/2]} C_{p,k} I_{p-2k}(\tau^k(f_p)) \tag{2.11}$$

*and*

$$I_p(f_p) = \sum_{k=0}^{[p/2]} (-1)^k C_{p,k} \delta_{p-2k}(\tau^k(f_p)). \tag{2.12}$$

It is clear from Proposition 2.7 that $\delta_p^s(f_p)$ and $\delta_p(f_p)$ will agree if the limiting traces $\overrightarrow{Tr}^k f_p$ are the same as the traces $\tau^k f_p$. As we remarked earlier that $\tau$–traces are in general not the same as limiting traces, though for functions in $\mathcal{S}_p$, they agree. The following proposition gives another situation in which the two traces are the same.

**Proposition 2.8** *Let $f_p \in L^2[0,1]^p$ be symmetric and continuous.*

(i) *Then the Solé and Utzet trace, $T^k f_p$ ( see [5] for definition of $T^k f_p$) exists and is the same as $\tau^k f_p$, $\forall k = 0, 1, \ldots [p/2]$.*

(ii) *Suppose that for some $k; 1 \le k \le [p/2]$, $\overrightarrow{Tr}^k f_p$ exists, then*

$$\overrightarrow{Tr}^k f_p(\cdot) = \tau^k f_p(\cdot) = \int_{[0,1]^k} f(t_1, t_1, \ldots, t_k, t_k, \cdot) dt_1 \ldots dt_k. \qquad (2.13)$$

**Corollary 2.9** *Let $f_p \in L^2[0,1]^p$ be continuous and symmetric. Suppose that $\exists$ a CONS $\{\phi_i\}$ for $L^2[0,1]$, s.t.:*

$$f_p = \sum_{i_1,\ldots,i_p=1}^{\infty} a_{i_1,\ldots i_p} \phi_{i_1} \otimes \ldots \phi_{i_p} \quad and \quad \sum_{i_1,\ldots i_p=1}^{\infty} \left| a_{i_1,\ldots i_p} \right| < \infty$$

*Then $\overrightarrow{Tr}^k f_p$ exists and we have that $\overrightarrow{Tr}^k f_p = \tau^k f_p$.*

Except for the special situation considered above, the equality of $\overrightarrow{Tr}^k f_p$ and $\tau^k f_p$ is not clear. In general, the following set of sufficient conditions can be given under which the limiting trace equals the $\tau$–trace.

**Proposition 2.10** *Let $f_p \in \mathcal{S}_p^1$ be s.t. $\overrightarrow{Tr}^k f_p$ exists. Suppose that $\exists$ a CONS $\{\phi_i\}$ for $L^2[0,1]$, s.t.*

(a) *The series $\sum_{i_1,\ldots,i_p=1}^{n} < f_p, \phi_{i_1} \otimes \ldots \phi_{i_p} > \phi_{i_1}(t_1) \ldots \phi_{i_p}(t_p)$ converges pointwise on $[0,1]^p$ as $n \to \infty$.*

(b) *$\sum_{i_1,\ldots,i_p=1}^{n} < f_p, \phi_{i_1} \otimes \ldots \phi_{i_p} > \phi_{i_1}(t_1) \phi_{i_2}(t_1) \ldots \phi_{i_{2k-1}}(t_k) \phi_{i_{2k}}(t_k) \phi_{i_{2k+1}}(t_{2k+1})$*

*$\ldots \phi_{i_p}(t_p)$ converges in $L^1[0,1]^{p-k}$ as $n \to \infty$. Then $\tau^k f_p = \overrightarrow{Tr}^k f_p$.*

## 3. Asymptotic Distributions of $V$ Statistics and Stochastic Integrals

In this section we will discuss the asymptotic distribution of a $V$–statistic under certain assumptions of square integrability of the kernel over the diagonal. The limit is a multiple Stratonovich integral w.r.t. the Wiener process.

Let $(\Omega, \mathcal{F}, P)$ be a probability space and $\{W_t; 0 \le t \le 1\}$ be a Wiener process on this space. The integrals $I_p(\cdot), \delta_p(\cdot)$ are as in Section 2. Let $(\Omega_1, \mathcal{F}_1, P_1)$ be another probability space.

**Definition 3.1** Let $X_1, X_2, \ldots$ be i.i.d. random variables defined on $(\Omega_1, \mathcal{F}_1)$, with the common distribution function $F$. Let $f_p : \mathbf{R}^p \to \mathbf{R}$, be a symmetric function.

The $V$-statistic $(V_n(f_p))$, corresponding to the kernel $f$ is defined as:

$$V_n(f_p) \equiv V_n(X_1, \ldots X_n) := n^{-p} \sum_{i_1, \ldots, i_p = 1}^{n} f_p(X_{i_1}, \ldots, X_{i_p}).$$

From now on, we will assume that $\{X_i\}$ is an i.i.d. sequence of $U[0,1]$ variables. We note that all the results in this section, with obvious modifications hold for a general i.i.d. sequence (i.e. not necessarily $U[0,1]$ variables) with a continuous distribution function. Define for $f_p \in L_s^2[0,1]^p$ and $r = 1, 2, \ldots p-1$

$$f_{p,c}^{(r)}(x_1, \ldots x_{p-r}) = f_p^{(r)}(x_1, \ldots x_{p-r}) + (-1)^{p-r} \int_{[0,1]^p} f_p(y_1, \ldots y_p) dy_1 \ldots dy_p$$

$$+ \sum_{\pi_*}' \sum_{s=1}^{p-r-1} \frac{(-1)^s}{s!(p-r-s)!} \int_{[0,1]^s} f_p^{(r)}(x_{\pi_*(1)}, \ldots x_{\pi_*(p-r-s)}, x_s) dz_s,$$

where $\pi_*$ runs over all permutations of the $p - r$ symbols, $\{1, 2, \ldots, p-r\}$. Define $f_{p,c}^{(0)} := f_{p,c}$.

The following theorem gives the asymptotic distribution of a $V$-statistic in terms of MSI.

**Theorem 3.2** *Let $f_p : [0,1]^p \to \mathbf{R}$ be a measurable, symmetric function s.t.,*

$$\int_{[0,1]^r} f_p^2(x_1, x_1 \ldots s_1 - \text{ times}, \ldots x_r, x_r \ldots s_r - \text{ times}) \, dx_1 \ldots dx_r < \infty$$

$$\forall (s_1, \ldots, s_r) \in Z_+^r, \text{ s.t. } s_1 + \ldots + s_r = p, \quad \forall r = 1, 2, \ldots p.$$

*Then,*

$$[V_n(f_p) - \int_{[0,1]^p} f_p(\mathbf{x}) d\mathbf{x}] = \sum_{r=0}^{p-1} \binom{p}{r} V_n(f_{p,c}^{(r)}) + o_p(1)$$

*and*

$$n^{1/2}[V_n(f_p) - \int_{[0,1]^p} f_p(\mathbf{x}) d\mathbf{x}] \xrightarrow{\mathcal{L}} p\delta_1(f_{p,c}^{(p-1)}).$$

*Moreover, if for fixed $0 < k < p$, $f_{p,c}^{(p-j)} = 0$, a.e. $[0,1]^j \forall j = 1, 2, \ldots k$, then*

$$n^{(k+1)/2}[V_n(f_p) - \int_{[0,1]^p} f_p(\mathbf{x}) d\mathbf{x}] \xrightarrow{\mathcal{L}} \binom{p}{k+1}\delta_{k+1}(f_{p,c}^{(p-k-1)}).$$

The next result is due to Hoeffding (see [2], Theorem 7.4 and remarks on page 306). We record it here (in our notation) since it is a natural corollary to Theorem 3.2.

**Corollary 3.3** *Let $f_p$ be as in Theorem 3.2. Suppose that $\| f_{p,c}^{(p-1)} \| \neq 0$. Then both*

$$n^{1/2}[V_n(f_p) - \int_{[0,1]^p} f_p(\mathbf{x})d\mathbf{x}] \quad and \quad n^{1/2}[U_n(f_p) - \int_{[0,1]^p} f_p(\mathbf{x})d\mathbf{x}]$$

*are asymptotically normal with mean 0 and variance $p^2 \| f_{p,c}^{(p-1)} \|^2$.*

**Acknowledgements** Research partially supported by the National Science Foundation and the Air Force Office of Scientific Research Grant No. 91-0030 and the Army Research Office Grant No. DAAL03-92-G-0008.

## References

[1] Budhiraja, A. and Kallianpur, G.: Hilbert space valued traces and multiple Stratonovich integrals with statistical applications; *Probability and Mathematical Statistics*, to appear (1994)

[2] Hoeffding, W.: A class of statistics with asymptotically normal distribution; *Annals of Mathematical Statistics* **19** (1948) 293-325

[3] Hu, Y. Z. and Meyer, P. A.: Sue les intégrales multiples de Stratonovich; *Séminaire de Probabilités XX11, Université de Strasbourg, Lecture Notes in Math.* **1321** (1987) 51-71

[4] Johnson, G. W. and Kallianpur, G.: Homogeneous chaos, $p$–forms, scaling and the Feynman integrals; *Transactions, American Mathematical Society* **340** (1993) 503-548

[5] Solé, J. L. and Utzet, F.: Stratonovich integral and trace; *Stochastics and Stochastic Reports* **29** (1990) 203-220

A. Budhiraja
Department of Statistics
University of North Carolina
Chapel Hill, NC 27599-3260, U.S.A.

G. Kallianpur
Department of Statistics
University of North Carolina
Chapel Hill, NC 27599-3260, U.S.A.
*E-mail*: gk@stat.unc.edu

H. C. CHAE & I. MITOMA
# Invariant nuclear space of $\Gamma$-operator

## 1. Introduction

Two types of fundamental spaces on infinite dimensional topological vector spaces have been studied by [1, 2, 3, 5, 7, 9, 10, 12, 13] in connection with infinite dimensional geometry and analysis. In general, the nuclearity of the fundamental spaces gives us various fruitful results [6]. Until now, it has been known that the fundamental spaces in the Malliavin calculus are not nuclear, while the original Hida space is nuclear.

In the former work [3], to characterize the support of solutions of an infinite dimensional stochastic evolution equation arising from interacting neurons [8], we construct a Hida's type fundamental space which is nuclear and invariant under $d\Gamma$-operator. In this paper, we construct a fundamental space which is nuclear and invariant under $\Gamma$-operator.

First we begin by giving some notations and explanations. Let $E$ be a real locally convex topological vector space and $E'$ the topological dual space of $E$. We denote by $\langle,\rangle$ the pairing of $E$ and $E'$, and by $|\cdot|_E$ the norm of $E$ if $E$ is a Hilbert space. Let $\mathcal{H}$ be a separable real Hilbert space densely and continuously embedded in $E$. Then identifying $\mathcal{H}'$ with $\mathcal{H}$, we have

$$E' \subset \mathcal{H} \subset E.$$

Let $\mu$ be the countably additive Gaussian measure on $E$ whose characteristic functional is given by

$$\int_E \exp[i\langle x, \xi\rangle]d\mu(x) = \exp\left[-\frac{1}{2}|\xi|_{\mathcal{H}}^2\right], \ \xi \in E'.$$

If we replace $E$ by $S'(\mathbf{R})$, $(E, \mu)$ is called the white noise space [9].

Let $\mathbf{A}$ be a self-adjoint operator in Hilbert space $\mathcal{H}$ and $L^2(E, \mu)$ the space of square integrable functions with respect to $\mu$. Further we denote by $\mathcal{H}^{\otimes n}$ the n-fold tensor product space of $\mathcal{H}$, by $S_{\mathbf{A}}$ the Hida space determined by $\mathbf{A}$ and by $\Gamma(\mathbf{A})$ the second quantization operator of $\mathbf{A}$, which will be precisely defined later. From now on we denote the domain of a closed linear operator $\mathbf{T}$ densely defined in $\mathcal{H}$ by $\mathcal{D}(\mathbf{T})$ and define $C^\infty(\mathbf{T}) = \bigcap_{n=1}^\infty \mathcal{D}(\mathbf{T}^n)$. By $\mathcal{D}(\mathbf{T}^n)$, we always mean a Hilbert space $\mathcal{D}(\mathbf{T}^n)$ equipped with the inner product $(\mathbf{T}^n\cdot, \mathbf{T}^n\cdot)_{\mathcal{H}}$. Given $\lambda \in \mathbf{R}$, we mean by $\mathbf{A} \geq \lambda$ that $(\mathbf{A}f, f)_{\mathcal{H}} \geq \lambda(f, f)_{\mathcal{H}}$ for all $f \in \mathcal{D}(\mathbf{A})$.

Now we state our result which is an extension of the commutator theorem [4].

**Theorem 1.1** *Suppose that* $\mathbf{A} \geq 1+\epsilon$, *for some* $\epsilon > 0$ *and there exist a self-adjoint operator* $\mathbf{B} \geq 1 + \epsilon$ *in* $\mathcal{H}$ *and natural numbers* $p$ *and* $q$, *satisfying the following conditions:*

(1) $\mathcal{D}(\mathbf{B}^p) \subset \mathcal{D}(\mathbf{A})$.

(2) *the identity map of* $\mathcal{D}(\mathbf{B}^q)$ *into* $\mathcal{H}$ *is a Hilbert-Schmidt operator.*

(3) $\mathbf{A}C^\infty(\mathbf{B}) \subset C^\infty(\mathbf{B})$.

*Then* $S_{\mathbf{B}}$ *is a nuclear subspace of* $L^2(E, \mu)$ *such that*

$$\Gamma(\mathbf{A})S_{\mathbf{B}} \subset S_{\mathbf{B}}.$$

*Further, suppose that*

(4) *for any nonnegative integers* $n, m$ *and* $k$, *there exist an integer* $\alpha(k) \geq k$ *and a constant* $C_{k,n}^m > 0$ *such that*

$$|(\mathbf{A}^m\mathbf{B}^k \otimes \mathbf{A}^m\mathbf{B}^k \otimes \cdots \otimes \mathbf{A}^m\mathbf{B}^k - \mathbf{B}^k\mathbf{A}^m \otimes \mathbf{B}^k\mathbf{A}^m \otimes \cdots \otimes \mathbf{B}^k\mathbf{A}^m)f|_{\mathcal{H}^{\otimes n}}$$

$$\leq (C_{k,n}^m)^n |\mathbf{B}^{\alpha(k)} \otimes \mathbf{B}^{\alpha(k)} \otimes \cdots \otimes \mathbf{B}^{\alpha(k)} f|_{\mathcal{H}^{\otimes n}},$$

$$\forall f \in \overbrace{C^\infty(\mathbf{B}) \otimes C^\infty(\mathbf{B}) \otimes \cdots \otimes C^\infty(\mathbf{B})}^{n\,\text{times}},$$

*where*

$$\sum_{m=0}^\infty \frac{t^m}{m!}(C_{k,n}^m)^n \leq c(t)^n, \qquad \forall t > 0.$$

*Then*

$$e^{-t\Gamma(\mathbf{A})}S_{\mathbf{B}} \subset S_{\mathbf{B}}.$$

## 2. Space of the Hida Distributions

Before proceeding to construct the Hida space, we introduce the following notations. Let $\mathcal{H}$ be a separable Hilbert space. For $f_i \in \mathcal{H}$, $i = 1, 2, \ldots, n$, we denote the tensor product of them by

$$f_1 \otimes f_2 \otimes \cdots \otimes f_n, \tag{2.1}$$

and define the symmetric tensor product of them by

$$f_1 \hat{\otimes} f_2 \hat{\otimes} \cdots \hat{\otimes} f_n = \frac{1}{n!} \sum_{\sigma \in \pi_n} f_{\sigma(1)} \otimes f_{\sigma(2)} \otimes \cdots \otimes f_{\sigma(n)}, \tag{2.2}$$

where $\pi_n$ is the symmetric group of degree $n$.

Let $\mathcal{H}^{\otimes n}$ and $\mathcal{H}^{\hat{\otimes} n}$ be the $n$-fold tensor product space and the $n$-fold symmetric tensor product space of $\mathcal{H}$, respectively. For $f_i,\ g_i \in \mathcal{H},\ i = 1, 2, \ldots, n$, the inner product $(\cdot, \cdot)_{\mathcal{H}^{\otimes n}}$ is given by

$$(f_1 \otimes f_2 \otimes \cdots \otimes f_n, g_1 \otimes g_2 \otimes \cdots \otimes g_n)_{\mathcal{H}^{\otimes n}} = (f_1, g_1)_{\mathcal{H}}(f_2, g_2)_{\mathcal{H}} \cdots (f_n, g_n)_{\mathcal{H}}, \tag{2.3}$$

and the inner product $(\cdot, \cdot)_{\mathcal{H}^{\hat{\otimes} n}}$ by

$$(f_1 \hat{\otimes} f_2 \hat{\otimes} \cdots \hat{\otimes} f_n, g_1 \hat{\otimes} g_2 \hat{\otimes} \cdots \hat{\otimes} g_n)_{\mathcal{H}^{\hat{\otimes} n}}$$
$$= \left(\frac{1}{n!}\right)^2 \sum_{\sigma, \tau \in \pi_n} (f_{\sigma(1)}, g_{\tau(1)})_{\mathcal{H}}(f_{\sigma(2)}, g_{\tau(2)})_{\mathcal{H}} \cdots (f_{\sigma(n)}, g_{\tau(n)})_{\mathcal{H}}. \tag{2.4}$$

Clearly

$$\mathcal{H}^{\hat{\otimes} n} \subset \mathcal{H}^{\otimes n}, \tag{2.5}$$

and

$$(f_1 \hat{\otimes} f_2 \hat{\otimes} \cdots \hat{\otimes} f_n, g_1 \hat{\otimes} g_2 \hat{\otimes} \cdots \hat{\otimes} g_n)_{\mathcal{H}^{\otimes n}}$$
$$= (f_1 \hat{\otimes} f_2 \hat{\otimes} \cdots \hat{\otimes} f_n, g_1 \hat{\otimes} g_2 \hat{\otimes} \cdots \hat{\otimes} g_n)_{\mathcal{H}^{\hat{\otimes} n}}. \tag{2.6}$$

We review the relation between the Wick ordering $: x^{\otimes n} :$ for $x \in E$ used in [5, 9] and the Wick product $: \langle x, \xi_1 \rangle \langle x, \xi_2 \rangle \cdots \langle x, \xi_n \rangle :$ of random variables $\langle x, \xi_k \rangle,\ x \in E,\ \xi_k \in E^{'},\ k = 1, 2, \ldots, n$, with respect to $\mu$ is defined by the following recursion relation [11];

$$: \langle x, \xi_1 \rangle : = \langle x, \xi_1 \rangle,$$

$$: \langle x, \xi_1 \rangle \langle x, \xi_2 \rangle \cdots \langle x, \xi_n \rangle : = \langle x, \xi_1 \rangle : \langle x, \xi_2 \rangle \cdots \langle x, \xi_n \rangle :$$
$$- \sum_{k=2}^{n} \int_E \langle x, \xi_1 \rangle \langle x, \xi_k \rangle d\mu(x) : \langle x, \xi_2 \rangle \cdots \langle x, \check{\xi}_k \rangle \cdots \langle x, \xi_n \rangle :,\ k \geq 2,$$

where $\langle x, \check{\xi}_k \rangle$ means the term $\langle x, \xi_k \rangle$ is deleted in the product. Then we have

$$\langle : x^{\otimes n} :, \xi_1 \hat{\otimes} \xi_2 \hat{\otimes} \cdots \hat{\otimes} \xi_n \rangle \equiv\ : \langle x, \xi_1 \rangle \langle x, \xi_2 \rangle \cdots \langle x, \xi_n \rangle :. \tag{2.7}$$

The well-known Wiener-Ito theorem states that the space $L^2(E, \mu)$ has the following orthogonal decomposition

$$L^2(E, \mu) = \bigoplus_{n=0}^{\infty} \mathbf{K}_n,$$

where $\mathbf{K}_n$ consists of $n$-homogeneous chaos, i.e. each $\varphi$ in $\mathbf{K}_n$ has the expression

$$\varphi(x) = \langle : x^{\otimes n} :, \hat{f}_n \rangle, \quad \hat{f}_n \in \mathcal{H}^{\hat{\otimes} n}.$$

Thus each $\psi \in L^2(E, \mu)$ can be represented uniquely in the following way;

$$\psi(x) = \sum_{n=0}^{\infty} \langle : x^{\otimes n} :, \hat{f}_n \rangle, \quad \mu - a.e. \ x \in E. \tag{2.8}$$

Moreover, we have

$$|\psi|^2_{L^2(E,\mu)} = \sum_{n=0}^{\infty} n! |\hat{f}_n|^2_{\mathcal{H}^{\hat{\otimes} n}}, \quad [5, 9]. \tag{2.9}$$

Let $\mathbf{A}$ be a positive self-adjoint operator in $\mathcal{H}$. Then there exists a unique positive self-adjoint operator $\Gamma(\mathbf{A})$ in $L^2(E, \mu)$ such that

$$\Gamma(\mathbf{A})1 = 1$$

and for $\xi_i \in \mathcal{D}(\mathbf{A}), i = 1, 2, \ldots, n,$

$$\begin{aligned}
\Gamma(\mathbf{A}) &: \langle x, \xi_1 \rangle \cdots \langle x, \xi_n \rangle : \\
&= : \langle x, \mathbf{A}\xi_1 \rangle \cdots \langle x, \mathbf{A}\xi_i \rangle \cdots \langle x, \mathbf{A}\xi_n \rangle : \\
&= \langle : x^{\otimes n} :, (\mathbf{A} \otimes \cdots \otimes \mathbf{A} \otimes \cdots \otimes \mathbf{A})(\xi_1 \hat{\otimes} \cdots \hat{\otimes} \xi_n) \rangle, \quad [11]
\end{aligned} \tag{2.10}$$

We denote by $\mathcal{P}_{\mathbf{A}}$ the collection of all polynomials of the form

$$\omega(x) = P(\langle x, \xi_1 \rangle, \langle x, \xi_2 \rangle, \ldots, \langle x, \xi_m \rangle), \quad \xi_i \in C^{\infty}(\mathbf{A}),$$

where $P(t_1, \ldots, t_m)$ is a polynomial of $(t_1, \ldots, t_m)$. For each $p \in \mathbf{R}$ we define a semi-norm $_{\mathbf{A}}\| \cdot \|_{2,p}$ by

$$_{\mathbf{A}}\|\omega\|^2_{2,p} = \int_E |\Gamma(\mathbf{A})^p \omega(x)|^2 d\mu(x). \tag{2.11}$$

It is not difficult to see that $\Gamma(\mathbf{A})^p = \Gamma(\mathbf{A}^p)$. By (2.7), each $\omega$ in $\mathcal{P}_{\mathbf{A}}$ has the following expression;

$$\omega(x) = \sum_{n=0}^{\infty} \langle : x^{\otimes n} :, \hat{g}_n \rangle, \qquad \hat{g}_n \in C^{\infty}(\mathbf{A})^{\hat{\otimes} n},$$

where $C^{\infty}(\mathbf{A})^{\hat{\otimes} n} = \overbrace{C^{\infty}(\mathbf{A})\hat{\otimes}C^{\infty}(\mathbf{A})\hat{\otimes}\cdots\hat{\otimes}C^{\infty}(\mathbf{A})}^{n\text{times}}$ is the set of finite linear combinations of the form $\xi_1\hat{\otimes}\xi_2\hat{\otimes}\cdots\hat{\otimes}\xi_n$ with $\xi_i \in C^{\infty}(\mathbf{A})$, $i = 1, 2, \ldots, n$. In fact we note that there exists a natural number $k(\omega)$ such that $\hat{g}_n = 0$ for $n \geq k(\omega)$. Since

$$\hat{g}_n = \sum_{i=1}^{m(n)} a_i(n)\xi_{i_1}\hat{\otimes}\xi_{i_2}\hat{\otimes}\cdots\hat{\otimes}\xi_{i_n}, \quad \xi_{i_k} \in C^{\infty}(\mathbf{A}), \quad k = 1, 2, \ldots, n,$$

by (2.9), ${}_{\mathbf{A}}\|\cdot\|_{2,p}^2$ can be also represented as

$$_{\mathbf{A}}\|\omega\|_{2,p}^2 = \sum_{n=0}^{\infty} n! |(\mathbf{A}^p)^{\otimes n} \hat{g}_n|_{\mathcal{H}^{\hat{\otimes} n}}^2, \tag{2.12}$$

where

$$(\mathbf{A}^p)^{\otimes n} = \mathbf{A}^p \otimes \mathbf{A}^p \otimes \cdots \otimes \mathbf{A}^p.$$

For $p \geq 0$, $(S_{\mathbf{A}})_p$ is the completion of $\mathcal{P}_{\mathbf{A}}$ with respect to the semi-norm ${}_{\mathbf{A}}\|\cdot\|_{2,p}$. We define the fundamental space $S_{\mathbf{A}}$ of the Hida distributions on $E$ by

$$S_{\mathbf{A}} = \bigcap_{p \geq 0} (S_{\mathbf{A}})_p. \tag{2.13}$$

If we take $E = S'(\mathbf{R})$ and $\mathbf{A} = -\left(\frac{d}{dx}\right)^2 + x^2 + 1$, then $S_{\mathbf{A}}$ becomes a nuclear space and originally it is called the fundamental space of the Hida distributions.

## 3. Proof of Theorem 1.1

Consider the fundamental space $(S_{\mathbf{B}})$ of the Hida distributions on $E$ determined by $\mathbf{B}$. Then the condition (2) of Theorem 1.1 yields $S_{\mathbf{B}}$ becomes a nuclear space [2]. Take any $\omega(x) \in \mathcal{P}_{\mathbf{B}}$ then the following expression holds,

$$\omega(x) = \sum_{n=0}^{\infty} \langle : x^{\otimes n} :, \hat{h}_n \rangle, \qquad \hat{h}_n \in C^{\infty}(\mathbf{B})^{\hat{\otimes} n}. \tag{3.1}$$

By the assumption (1) of Theorem 1.1, $\omega(x) \in \mathcal{D}(\Gamma(\mathbf{A}))$, so that

$$\Gamma(\mathbf{A})\omega(x) = \sum_{n=0}^{\infty} \langle : x^{\otimes n} :, \mathbf{A}^{\otimes n}\hat{h}_n \rangle. \tag{3.2}$$

Define the Hilbert space $\mathcal{H}_{\mathbf{B}^l}$ for any natural number $l$ by

$$\mathcal{H}_{\mathbf{B}^l} = \mathcal{D}(\mathbf{B}^l) \equiv \left\{ h \in \mathcal{H} : |\mathbf{B}^l h|_{\mathcal{H}} \langle \infty \right\},$$

where the inner product of the Hilbert space $\mathcal{H}_{\mathbf{B}^l}$ is given by $(\cdot, \cdot)_{\mathcal{H}_{\mathbf{B}^l}} = (\mathbf{B}^l \cdot, \mathbf{B}^l \cdot)_{\mathcal{H}}$.

Noticing that $C^{\infty}(\mathbf{B}) = \bigcap_{n=1}^{\infty} \mathcal{D}(\mathbf{B}^n)$ is a complete metric space equipped with a countable system of semi norms $|\cdot|_n = |\mathbf{B}^n \cdot|_{\mathcal{H}}, \quad n = 1, 2, \ldots$ and the assumption (3) of Theorem 1.1, for any natural number $k$, we have by Baire's category theorem, a natural number $l_1(\geq k)$ and a constant $c_1 \geq 1$ such that

$$|\mathbf{B}^k \mathbf{A} h|_{\mathcal{H}} \leq c_1 |\mathbf{B}^{l_1} h|_{\mathcal{H}}, \quad h \in C^{\infty}(\mathbf{B}). \tag{3.3}$$

Here we prepare a lemma concerning with the operator norm of tensor product of linear operators on Hilbert space, which will be used later. Given Hilbert spaces $\mathcal{H}_1$ and $\mathcal{H}_2$ and a linear operator T from $\mathcal{H}_1$ to $\mathcal{H}_2$, we denote the operator norm of T by

$$\|T\|_{\mathcal{H}_1 \to \mathcal{H}_2} = \sup_{x \in \mathcal{H}_1} \frac{|Tx|_{\mathcal{H}_2}}{|x|_{\mathcal{H}_1}}.$$

We note that (3.3) implies

$$\|\mathbf{A}\|_{\mathcal{H}_{\mathbf{B}^{l_1}} \to \mathcal{H}_{\mathbf{B}^k}} \leq c_1. \tag{3.4}$$

For simplicity we use the notation $\mathcal{H}_{\mathbf{B}^l}^{\hat{\otimes} n}$ instead of $(\mathcal{H}_{\mathbf{B}^l})^{\hat{\otimes} n}$.

Since $\mathbf{A}$ is bounded linear operators from $\mathcal{H}_{\mathbf{B}^l}$ to $\mathcal{H}_{\mathbf{B}^k}$, we have the following lemma, which is an extended version of Proposition [11, p.299].

**Lemma 3.1** $\quad \|\mathbf{A}^{\otimes n}\|_{\mathcal{H}_{\mathbf{B}^l}^{\hat{\otimes} n} \to \mathcal{H}_{\mathbf{B}^k}^{\hat{\otimes} n}} \leq \|\mathbf{A}\|_{\mathcal{H}_{\mathbf{B}^l} \to \mathcal{H}_{\mathbf{B}^k}}^n.$

In the proof of the lemma, the following inequality is essential.

$$\begin{aligned}
\|\mathbf{A}^{\otimes n}\|_{\mathcal{H}_{\mathbf{B}^l}^{\hat{\otimes} n} \to \mathcal{H}_{\mathbf{B}^k}^{\hat{\otimes} n}}^2 &= \sup_{\hat{f} \in \mathcal{H}_{\mathbf{B}^l}^{\hat{\otimes} n}} \frac{(\mathbf{A}^{\otimes n}\hat{f}, \mathbf{A}^{\otimes n}\hat{f})_{\mathcal{H}_{\mathbf{B}^k}^{\hat{\otimes} n}}}{(\hat{f}, \hat{f})_{\mathcal{H}_{\mathbf{B}^l}^{\hat{\otimes} n}}} \\
&= \sup_{\hat{f} \in \mathcal{H}_{\mathbf{B}^l}^{\hat{\otimes} n}} \frac{(\mathbf{A}^{\otimes n}\hat{f}, \mathbf{A}^{\otimes n}\hat{f})_{\mathcal{H}_{\mathbf{B}^k}^{\otimes n}}}{(\hat{f}, \hat{f})_{\mathcal{H}_{\mathbf{B}^l}^{\otimes n}}} \\
&\leq \sup_{f \in \mathcal{H}_{\mathbf{B}^l}^{\otimes n}} \frac{(\mathbf{A}^{\otimes n} f, \mathbf{A}^{\otimes n} f)_{\mathcal{H}_{\mathbf{B}^k}^{\otimes n}}}{(f, f)_{\mathcal{H}_{\mathbf{B}^l}^{\otimes n}}} \\
&= \|\mathbf{A}^{\otimes n}\|_{\mathcal{H}_{\mathbf{B}^l}^{\otimes n} \to \mathcal{H}_{\mathbf{B}^k}^{\otimes n}}^2. \tag{3.5}
\end{aligned}$$

Now we return to the proof of Theorem 1.1. Observe that

$$|\Gamma(\mathbf{B})^k\Gamma(\mathbf{A})\langle : x^{\otimes n} :, \hat{h}_n\rangle|^2_{L^2(E,d\mu)}$$

$$= n!|(\mathbf{B}^k\mathbf{A} \otimes \cdots \otimes \mathbf{B}^k\mathbf{A} \otimes \cdots \otimes \mathbf{B}^k\mathbf{A})\hat{h}_n|^2_{\mathcal{H}^{\hat{\otimes}n}}$$

$$= n!|(\mathbf{A} \otimes \cdots \otimes \cdots \otimes \mathbf{A})\hat{h}_n|^2_{\mathcal{H}^{\hat{\otimes}n}_{\mathbf{B}^k}}$$

$$\leq n!\|\mathbf{A} \otimes \cdots \otimes \cdots \otimes \mathbf{A}\|^2_{\mathcal{H}^{\hat{\otimes}n}_{\mathbf{B}^{l_1}} \to \mathcal{H}^{\hat{\otimes}n}_{\mathbf{B}^k}}|\hat{h}_n|^2_{\mathcal{H}^{\hat{\otimes}n}_{\mathbf{B}^{l_1}}}$$

$$\leq n!\|\mathbf{A}\|^{2n}_{\mathcal{H}_{\mathbf{B}^{l_1}} \to \mathcal{H}_{\mathbf{B}^k}}|(\mathbf{B}^{l_1} \otimes \cdots \otimes \mathbf{B}^{l_1})\hat{h}_n|^2_{\mathcal{H}^{\hat{\otimes}n}} \quad \text{(by Lemma 3.1)}$$

$$\leq n!c_1^{2n}|(\mathbf{B}^{l_1} \otimes \cdots \otimes \mathbf{B}^{l_1})\hat{h}_n|^2_{\mathcal{H}^{\hat{\otimes}n}} \quad \text{(by (3.4))}$$

and hence

$$_\mathbf{B}\|\Gamma(\mathbf{A})\omega\|^2_{2,k} = |\Gamma(\mathbf{B})^k\Gamma(\mathbf{A})\omega|^2_{L^2(E,d\mu)}$$

$$= \sum_{n=0}^{\infty}|\Gamma(\mathbf{B})^k\Gamma(\mathbf{A})\langle : x^{\otimes n} :, \hat{h}_n\rangle|^2_{L^2(E,d\mu)}$$

$$\leq \sum_{n=0}^{\infty}n!c_1^{2n}|(\mathbf{B}^{l_1} \otimes \cdots \otimes \mathbf{B}^{l_1})\hat{h}_n|^2_{\mathcal{H}^{\hat{\otimes}n}}. \tag{3.6}$$

Since for the natural number $q$ given in the assumption (2) of Theorem 1.1, we choose a natural number $m_1$ such that $(\mathbf{B}^q)^{-m_1} \leq (c_1)^{-1}$, the right hand side of (3.6) is dominated by

$$\sum_{n=0}^{\infty}n!c_1^{2n}(\frac{1}{c_1})^{2n}|(\mathbf{B}^{l_1+qm_1} \otimes \cdots \otimes \mathbf{B}^{l_1+qm_1})\hat{h}_n|^2_{\mathcal{H}^{\hat{\otimes}n}}$$

$$\leq \sum_{n=0}^{\infty}n!|(\mathbf{B}^{l_1+qm_1} \otimes \cdots \otimes \mathbf{B}^{l_1+qm_1})\hat{h}_n|^2_{\mathcal{H}^{\hat{\otimes}n}}$$

$$= |\Gamma(\mathbf{B})^{l_1+qm_1}\omega|^2_{L^2(E,d\mu)} = {}_\mathbf{B}\|\omega\|^2_{2,l_1+qm_1}.$$

Therefore, we have $\Gamma(\mathbf{A})S_\mathbf{B} \subset S_\mathbf{B}$. Thus the proof of the first half of Theorem 1.1 is completed.

Now we will prove the second half of the theorem. By (2.10) and (3.1), we have the following expression;

$$e^{-t\Gamma(\mathbf{A})}\langle : x^{\otimes n} :, \hat{h}_n\rangle = \sum_{m=0}^{\infty}\frac{(-t)^m}{m!}\langle : x^{\otimes n} :, (\mathbf{A}^m \otimes \cdots \otimes \mathbf{A}^m)\hat{h}_n\rangle.$$

By the assumption (4) of Theorem 1.1 and the manner similar to that in (3.5) we get

$$\|(\mathbf{B}^k\mathbf{A}^m)^{\otimes n} - (\mathbf{A}^k\mathbf{B}^m)^{\otimes n}\|^2_{\mathcal{H}^{\hat{\otimes} n}_{\mathbf{B}^{\alpha(k)}}\to\mathcal{H}^{\hat{\otimes} n}} \le (C^m_{k,n})^n. \tag{3.7}$$

Then we have

$$\left|\Gamma(\mathbf{B})^k\sum_{m=0}^{N}\frac{(-t)^m}{m!}\langle: x^{\otimes n}:,(\mathbf{A}^m\otimes\cdots\otimes\mathbf{A}^m)\hat{h_n}\rangle\right|^2_{L^2(E,d\mu)}$$

$$=\left|\sum_{m=0}^{N}\frac{(-t)^m}{m!}\langle: x^{\otimes n}:,(\mathbf{B}^k\mathbf{A}^m\otimes\cdots\otimes\mathbf{B}^k\mathbf{A}^m)\hat{h_n}\rangle\right|^2_{L^2(E,d\mu)}$$

$$=\left|\sum_{m=0}^{N}\frac{(-t)^m}{m!}\langle: x^{\otimes n}:,(\mathbf{A}^m\mathbf{B}^k\otimes\mathbf{A}^m\mathbf{B}^k\otimes\cdots\otimes\mathbf{A}^m\mathbf{B}^k\right.$$

$$\left.+\mathbf{B}^k\mathbf{A}^m\otimes\mathbf{B}^k\mathbf{A}^m\otimes\cdots\otimes\mathbf{B}^k\mathbf{A}^m-\mathbf{A}^m\mathbf{B}^k\otimes\mathbf{A}^m\mathbf{B}^k\otimes\cdots\otimes\mathbf{A}^m\mathbf{B}^k)\hat{h_n}\rangle\right|^2_{L^2(E,d\mu)}$$

$$\le n!\left(|(\mathbf{B}^k\otimes\cdots\otimes\mathbf{B}^k)\hat{h_n}|_{\mathcal{H}^{\hat{\otimes} n}}+\sum_{m=0}^{N}\frac{t^m}{m!}\left|((\mathbf{B}^k\mathbf{A}^m)^{\otimes n}-(\mathbf{A}^m\mathbf{B}^k)^{\otimes n})\hat{h_n}\right|_{\mathcal{H}^{\hat{\otimes} n}}\right)^2$$

$$\le n!\left(|\hat{h_n}|_{\mathcal{H}^{\hat{\otimes} n}_{\mathbf{B}^k}}+\sum_{m=0}^{N}\frac{t^m}{m!}\left|(\mathbf{B}^k\mathbf{A}^m)^{\otimes n}-(\mathbf{A}^m\mathbf{B}^k)^{\otimes n}\right|_{\mathcal{H}^{\hat{\otimes} n}_{\mathbf{B}^{\alpha(k)}}\to\mathcal{H}^{\hat{\otimes} n}}|\hat{h_n}|_{\mathcal{H}^{\hat{\otimes} n}_{\mathbf{B}^{\alpha(k)}}}\right)^2$$

$$\le n!\left(|\hat{h_n}|_{\mathcal{H}^{\hat{\otimes} n}_{\mathbf{B}^k}}+\sum_{m=0}^{N}\frac{t^m}{m!}(C^m_{k,n})^n|\hat{h_n}|_{\mathcal{H}^{\hat{\otimes} n}_{\mathbf{B}^{\alpha(k)}}}\right)^2\quad (\text{by}(3.7))$$

and hence by the condition (4) of Theorem 1.1,

$$|\Gamma(\mathbf{B})^k e^{-t\Gamma(\mathbf{A})}\langle: x^{\otimes n}:,\hat{h_n}\rangle|^2_{L^2(E,d\mu)}$$

$$\le n!(1+c(t))^{2n}|(\mathbf{B}^{\alpha(k)}\otimes\cdots\otimes\mathbf{B}^{\alpha(k)})\hat{h_n}|^2_{\mathcal{H}^{\hat{\otimes} n}},$$

so that the rest of the proof is similar to that in the first half, which completes the proof of Theorem 1.1.

## References


[1] Arai, A. and Mitoma, I.: De Rham-Hodge-Kodaira decomposition in $\infty$-dimensions; *Math. Ann.* **291** (1991) 51–73

[2] Arai, A. and Mitoma, I.: Comparison and Nuclearity of spaces of differential forms on topological vector spaces; *J. Funct. Anal.* **111** (1993) 278–294

[3] Chae, H. C., Handa, K., Mitoma, I., and Okazaki, Y.: Invariant nuclear space of a second quantization operator; *Preprint* (1994)

[4] Fröhlich, J.: Application of commutator theorems to the integration of representations of Lie algebra and commutation relations; *Commun. Math. Phys.* **54** (1977) 135–150

[5] Hida, T., Potthoff, J., and Streit, L.: White noise analysis and applications; in: *Mathematics + Physics* 3, World Scientific 1989)

[6] Ito, K.: Infinite dimensional Ornstein-Uhlenbeck processes; in: *Taniguchi Symp. Stochastic Analysis*, Katata, Kinokuniya, Tokyo, (1984) 197–224

[7] Jaffe, A., Lesniewski, A., and Weitsman, J.: Index of a family of Dirac operators on loop space; *Commun. Math. Phys.* **112** (1987) 75–88

[8] Kallianpur, G. and Mitoma, I: A Segal-Langevin type stochastic differential equation on a space of generalized functionals; *Can. J. Math.* **44** (1992) 524–552

[9] Kuo, H. -H.: Lecture on white noise analysis; *Soochow J. Math.* **18** (1992) 229–300

[10] Mitoma, I.: De Rham-Kodaira decomposition and fundamental spaces of Wiener functionals; in: *Gaussian Random Fields*, World Scientific, (1991) 285–297

[11] Reed, M. and Simon, B.: *Methods of Modern Mathematical Physis*, Vol.1 and 2. Academic Press, 1980

[12] Shigekawa, I.: Sobolev spaces over the Wiener space based on an Ornstein-Uhlenbeck operator; *J. Math. Kyoto Univ.* **32** (1992) 731–748

[13] Watanabe, S.: *Lectures on Stochastic Differential Equations and Malliavin's Calculus*. Springer-Verlag, Berlin, 1984

Hong Chul Chae
Department of Mathematics
Saga university
Saga 840, JAPAN

Itaru Mitoma
Department of Mathematics
Saga university
Saga 840, JAPAN
*E-mail*: mitoma@math.ms.saga-u.ac.jp

P. -L. CHOW

# Stationary solutions of some parabolic Itô equations

## 1. Introduction

Consider the formal solution of a random heat equation in $D = (0, \pi)$ given by

$$
\begin{cases}
\dfrac{\partial u(t, x)}{\partial t} & = \tfrac{1}{2}\Delta u(t, x) + \dot{B}_t(x), \quad t > 0, x \in D, \\
u|_{\partial D} & = 0, \\
u(0, x) & = \xi(x),
\end{cases}
\tag{1.1}
$$

where $\Delta = \frac{\partial^2}{\partial x^2}, \dot{B}_t(x)$ is a space-time white noise [5], $\partial D = \{0, \pi\}$ and $\xi$ is a deterministic or random function independent of the white noise. Let $\{e_k\}$ be a complete orthonormal system (CONS) in $H = L^2(D)$, and let $\{b_t^k\}$ be a sequence of iid Brownian motions in $\mathbf{R}^1$. Then $\dot{B}_t$ can be represented as

$$
\dot{B}_t(x) = \sum_{k=1}^{\infty} \dot{b}_t^k e_k
\tag{1.2}
$$

in an $L_2$-sense, where $\dot{b}_t^k$ is a white noise in $\mathbf{R}^1$. In particular we take $e_k(x) = \sqrt{\frac{1}{2}} \sin kx, k = 1, 2, \ldots, n, \ldots$, and seek a series solution of Eq. (1.1) in the form:

$$
u(t, x) = \sum_{k=1}^{\infty} u_t^k e_k(x).
\tag{1.3}
$$

Upon substituting (1.2) and (1.3) into Eq. (1.1), one gets a sequence of independent Langevin equations:

$$
\begin{cases}
du_t^k & = -\tfrac{1}{2}k^2 u_t^k dt + db_t^k, \\
u_0^k & = \xi_k, \quad k = 1, 2, \ldots,
\end{cases}
\tag{1.4}
$$

where $\xi_k = (\xi, e_k) = \int_0^\pi \xi(x)e_k(x)dx$. It is well known that, for each $k$, the Langevin equation (1.4) has a unique invariant measure $\mu_k = N(0, 1/k^2)$, the centered Gaussian distribution with variance $1/k^2$. Furthermore, if the initial distribution $\mathcal{D}(\xi_k) = \mu_k$ is independent of $b_t^k$, then the solution $u_t^k$ is a stationary Ornstein-Uhlenbeck process [1]. Hence, formally, since $\xi = \sum_{k=1}^{\infty} \xi^k e_k$, we see that $\xi$ is

Gaussian with mean zero and covariance operator $Q$ defined by

$$E(\xi, g)(\xi, h) = E(\sum_{k=1}^{\infty} g_k \xi_k)(\sum_{\ell=1}^{\infty} h_\ell \xi_\ell)$$

$$= \sum_{k=1}^{\infty} (1/k^2) g_k h_k$$

$$= (Qg, h), \quad g, h \in H,$$

where $g_k = (g, e_k)$, $h_\ell = (h, e_k)$, and $Q = -\Delta^{-1}$. Therefore it seems clear that the solution $u(t, x)$ of Eq. (1.1) is stationary if the initial random variable with distribution $\mathcal{D}(\xi) = N(0, -\Delta^{-1})$ is independent of $\dot{B}_t$.

Next let us consider the nonlinear problem:

$$\begin{cases} \dfrac{\partial u(t, x)}{\partial t} &= \frac{1}{2}\Delta u + f(u) + \dot{B}_t(x), t > 0, x \in D, \\ \dfrac{\partial u}{\partial x}\big|_{\partial D} &= 0, \\ u(0, x) &= \xi(x), \end{cases} \tag{1.5}$$

where, in contrast with Eq. (1.1), a Neumann boundary condition is imposed and a nonlinear $f(u)$ is added. The stationary solution of a nonlinear parabolic equation of the form (1.5) was first considered by Markus [7]. Under the conditions that $f$ is Lipschitz continuous and $f(u) = F'(u)$ such that $F(u)$ is of linear growth as $|u| \to +\infty$, it was shown that, as $t \to \infty$, the solution $u(t, x)$ approaches a stationary solution in an $L_2$-sense, and the associated invariant measure $\mu_F$ can be obtained explicitly as the Radon-Nikodym derivative:

$$\frac{d\mu_F}{d\mu}(u) = e^{F(u)} \bigg/ \int e^F d\mu,$$

where $\mu$ is the invariant measure for the linear equation when $f = 0$. The result was extended to the case where $f$ is monotone [8].

We note that the above result is limited to one space dimension. For the domain $D \subset \mathbf{R}^d$ with $d \geq 2$, a solution to Eq. (1.5) exists only in the sense of distributions [10]. To obtain a regular solution, we need to replace the space-time white noise by $\dot{W}_t(x)$, the formal derivative of a Wiener process $W_t(x)$ in $L_2(D)$. By regarding Eq. (1.5) as s stochastic equation in a Hilbert or Banach space, the problem of stationary solutions and the associated invariant measures has been studied by several authors, including DaPrato and Zabczyk [4]. Their results pertain to the so-called mild solutions of stochastic equations. Recently, in a joint work with Khasminskii [3], we proved a general theorem concerning the existence of invariant measures for the strong solutions. In the context of parabolic Itô equations, our theorem seems to have a wider applicability. This paper gives a detailed account for the application of our theorem to a class of parabolic Itô equations with some illustrative examples. Since the stationary solution of a parabolic Itô equation is closely related to the question of invariant measures, we shall be mainly concerned with the existence of the latter. For other related papers on this subject, one can consult the bibliography in [4].

## 2. Invariant Measures

Let $D \subset \mathbf{R}^d$ be a bounded domain with smooth boundary $\partial D$, Denote $L_2(D)$ by $H$ and $H_0^1(D)$ by $V$, where $H_0^1(D)$ is the Sobolev space of functions on $D$ which have first-order partial derivatives in $H$ and vanish on $\partial D$. The norms in $H$ and $V$ are denoted by $|\cdot|$ and $||\cdot||$, respectively, and the inner product in $H$, by $(\cdot, \cdot)$. Let $W_t(\cdot)$ be a Wiener process in $H$ defined on a probability space $(\Omega, \mathcal{F}, P)$ with covariance operator $Q$.

Now we consider the parabolic Itô equation:

$$\begin{cases} \dfrac{\partial u(t, x)}{\partial t} &= \Delta u + f(u, \nabla u) + \sigma(u, \nabla u)\dot{W}_t(x), \\ u|_{\partial D} &= 0, \\ u(0, x) &= \xi(x), \end{cases} \tag{2.1}$$

where $\Delta$ denotes the Laplacian operator, $\nabla u$ is the gradient of $u$, and $\xi$ is a random function independent of $W_t$. The properties of the functions $f$ and $\sigma$ will be specified later. Suppose that $G(t, x, y)$ denotes the Green's function for the heat equation, Eq. (2.1) with $f = \sigma \equiv 0$. In terms of $G$, Eq. (2.1) can be rewritten as an integral equation:

$$\begin{aligned} u_t = T_t \xi &+ \int_0^t T_{t-s} f(u_s, \nabla u_s) ds \\ &+ \int_0^t T_{t-s} \sigma(u_s, \nabla u_s) dW_t, \end{aligned} \tag{2.2}$$

where we set $u_t = u(t, \cdot)$ and

$$(T_t \varphi)(x) = \int_D G(t, x, y)\varphi(y) dy.$$

A solution $u_t$ of Eq. (2.2) with $u_t \in C([0,T], H)$ a.s. is called a *mild solution*. Let us impose the following conditions:

(A.1) $E|\xi|^2 < \infty$ and the kernel $q(x,y)$ of the covariance operator $Q$ of $W_t$ is essentially bounded so that

$$\text{ess.sup}_{x,y \in D} \ |q(x,y)| = q_\infty < \infty.$$

(A.2) The coefficients $f(u,p) = f(u)$ and $\sigma(u,p) = \sigma(u)$ are independent of $p \in \mathbf{R}^d$.

(A.3) The functions $f(r)$ and $\sigma(r)$ are Lipschitz continuous so that

$$\exists K > 0 : |f(r) - f(s)| + |\sigma(r) - \sigma(s)| \le K|r - s|, \quad \forall r, s \in \mathbf{R}^1.$$

(A.4) There exists $\beta < \lambda_1$ such that

$$([f(u) - f(v)], u - v) + \frac{1}{2} q_\infty |\sigma(u) - \sigma(v)|^2 \le \beta |u - v|^2, \quad \forall u, v \in V,$$

where

$$\lambda_1 = \inf_{v \in V} \frac{|\nabla v|^2}{|v|} > 0.$$

Under conditions (A.1) - (A.3), Eq. (1.2) has a unique mild solution $u_t$ which is a time-homogeneous diffusion process in $H$ [4]. Let $P(t, \varphi, \cdot)$ denote the transition probability function given by

$$P(t, \varphi, B) = P\{u_t \in B | u_0 = \varphi\}$$

for any $\varphi \in H$ and Borel set $B \in \mathcal{B}(H)$. For any bounded continuous function $\Phi \in C_b(H)$, the transition operator $P_t$ is defined by

$$(P_t \Phi)(\varphi) = \int_H \Phi(v) P(t, \varphi, dv). \tag{2.3}$$

Then a probability measure $\mu$ on $(H, \mathcal{B}(H))$ is said to be an *invariant measure* of $P_t$ if $\mu$ satisfies the equation:

$$\mu(B) = \int_H P(t, v, B) \mu(dv), \quad \forall B \in \mathcal{B}(H). \tag{2.4}$$

As a realization of a stochastic evolution equation, we can deduce from Theorem 11.22 in [4] that

**Theorem 2.1** *Suppose that conditions (A.1) - (A.4) hold. Then there exists a unique invariant measure $\mu$ for Eq. (2.2).*

The theorem can be easily verified by showing that (A.1) - (A.4) imply the conditions for the above-mentioned Thm. 11.22 are satisfied. However the conditions (A.2) and (A.4) are rather stringent and, because of them, Thm. 2.1 does not apply to the parabolic Itô equation (1.1) in general. In what follows, we shall present a more refined result based on our recent paper [3].

Returning to Eq. (2.1), it can be rewritten as an Itô equation in $V^\star = H^{-1}(D)$:

$$u_t = \xi + \int_0^t [\Delta u_s + f(u_s, \nabla u_s)]ds + \int_0^t \sigma(u_s, \nabla u_s)dW_s. \tag{2.5}$$

A $V$-valued process $u_t$ which satisfies the above equation for a.e. $(t, \omega) \in [0, T] \times \Omega$ is known as a *strong solution*. Here let us assume that

(B.1) Condition (A.1) holds.

(B.2) The functions $f, \sigma : \mathbf{R}^1 \times \mathbf{R}^d \to \mathbf{R}^1$ are Lipschitz continuous so that $\exists K > 0$:

$$|f(r,p) - f(s,q)| + |\sigma(r,p) - \sigma(s,q)| \leq K(|r-s| + |p-q|) \ \forall \ r, s \in \mathbf{R}^1 \ \text{and} \ p, q \in \mathbf{R}^d.$$

Then the following fact is well known [9]:

**Lemma 2.1** *If conditions (B.1) and (B.2) hold, then Eq. (2.5) has a unique strong solution $u_t \in L_2(\Omega, C([0, T], H)) \cap L_2(\Omega \times [0, T], V)$. Moreover it satisfies:*

$$E|u_t|^2 + 2E \int_0^t |\nabla u_s|^2 ds = E|\xi|^2 + 2E \int_0^t (u_s, f_s)ds$$

$$+ E \int_0^t |Q^{1/2}\sigma_s u_s|^2 ds, \ 0 < t < T, \tag{2.6}$$

*where $f_s = f(u_s, \nabla u_s)$ and $\sigma_s = \sigma(u_s, \nabla u_s)$.*

For the existence of an invariant measure, in place of condition (A.4), another condition is needed. To this end, for a fixed $\varphi \in H$, we define

$$\mu_T(B) = \frac{1}{T} \int_0^T P(t, \varphi, B)dt \ \text{for} \ B \in \mathcal{B}(H), \ T > 0, \tag{2.7}$$

which is a probability measure on $(H, \mathcal{B}(H))$. For $R > 0$, let $B_R = \{v \in H : \|v\| < R\}$. Then our main result can be stated as a theorem, which is the generalization of a finite-dimensional case by Khasminskii [6].

46

**Theorem 2.2** *Let the conditions* (B.1) *and* (B.2) *hold. If, for a fixed* $\varphi \in H$, *there exists* $T_0 > 0$ *such that*

$$\forall \; T > T_0 : \mu_T(H \setminus B_R) \leq q_R \to 0 \;\; \text{as} \;\; R \to \infty, \tag{2.8}$$

*then an invariant measure* $\mu$ *for Eq.* (2.1) *exists and has its support in* $V$.

The proof can be found in our paper [3]. In fact condition (2.8) implies that $\{\mu_T\}$ is weakly compact so that there exists a sequence $\mu_n = \mu_{T_n}$ converges weakly to $\mu$, which can be shown to be an invariant measure. Also we have

$$
\begin{aligned}
\mu_T(H \setminus B_R) &= \mu_T(V \setminus B_R) \\
&\leq \frac{1}{T} \int_0^T E\|u_t\|^2 dt / R^2,
\end{aligned}
\tag{2.9}
$$

which follows from the Chebyshev inequality. On the other hand, by conditions (B.1) and (B.2), Eq. (2.6) yields the following inequality:

$$\int_0^T E\|u_t\|^2 dt \leq c_1(1+T) + c_2 \int_0^T E|u_t|^2 dt, \tag{2.10}$$

for some $c_1, c_2 > 0$. Now, in view of (2.9) and (2.10), if

$$\sup_{T > T_0} \frac{1}{T} \int_0^T E|u_t|^2 dt = M < \infty, \tag{2.11}$$

then condition (2.8) holds. Thus we have

**Corollary 2.1** *The conclusion of Theorem 2.2 holds if condition* (2.8) *is replaced by condition* (2.11).

## 3. Lyapunov Method

The condition (2.8) or (2.9), unlike condition (A.4), is not explicit and difficult to verify. Here we introduce a method based on Lyapunov functions, which seems easier to apply. Recall that for strong solutions, we have the Itô formula. Let $\Phi : H \to \mathbf{R}^1$ be a twice Fréchet differentiable function. Then, under certain regularity conditions on the first two derivatives $\Phi'$ and $\Phi''$, the following Itô formula holds [9]:

$$\Phi(u_t) = \Phi(\xi) + \int_0^t \mathcal{L}\Phi(u_s)ds + \int_0^t (\Phi'(u_s), \sigma_s dW_s), \tag{3.1}$$

where $\sigma_s = \sigma(u_s, \nabla u_s)$,

$$\mathcal{L}\Phi(v) = \frac{1}{2}Tr[\Phi''(v)\Gamma(v)] + <B(v), \Phi'(v)> \quad \text{for} \quad v \in V,$$
$$\Gamma(v) = (\sigma Q \sigma)(v),$$
$$B(v) = \Delta v + f(v, \nabla v),$$

$$(3.2)$$

$Tr$ denotes the trace and $< \cdot, \cdot >$ is the duality pairing between $V^\star$ and $V$. A $C_2(H)$-functional $\Phi$ for which (3.1) holds is said to be an *Itô functional*. In terms of such a functional, the following existence theorem for invariant measures is useful.

**Theorem 3.1** *Let conditions* (B.1) *and* (B.2) *hold. Suppose that Eq.* (2.1) *satisfies condition* $C : \exists$ *Itô functional* $\Phi : H \to \mathbf{R}^+$ *such that*

(i)
$$\exists M \in \mathbf{R}^1 : \mathcal{L}\Phi(v) \leq M \;\; \forall v \in V.$$

(ii)
$$\sup_{v \in V, \|v\| \geq R} \mathcal{L}\Phi(v) = -\varphi_R \to -\infty \;\; as \;\; R \to \infty.$$

*Then there exists an invariant measure $\mu$ given as in Thm. 2.2.*

With the aid of the Itô formula (3.1), the proof follows from the fact that condition $C$ for Thm. 3.1 implies the condition (2.8) for Thm. 2.2. The functional $\Phi$ satisfying condition $C$ in Thm. 3.1 will be called a *Lyapunov functional*, which is important in stability analysis [2]. Also, in view of (3.2), condition C is more explicit in the sense that, given $\Phi$, it depends only on the coefficients of Eq. (2.1).

Before giving some examples, we will show that the existence part of Thm. 2.1 is a special case of Thm. 3.1 by choosing $\Phi(v) = |v|^2$. Then we have

$$\mathcal{L}\Phi(v) = Tr[\Gamma(v)] + 2 < B(v), v >$$
$$\leq 2\{-|\nabla v|^2 + (f(v), v) + \frac{1}{2}q_\infty|\sigma(v)|^2\}.$$

$$(3.3)$$

By condition (A.4), it is easy to show that the above yields

$$\mathcal{L}\Phi(v) \leq -\varepsilon\|v\|^2 + C_\varepsilon,$$

for some constants $\varepsilon > 0$ and $C_\varepsilon$. Therefore, it follows from Lemma 2.1 and Thm. 3.1 that

**Corollary 3.1** *Under conditions* (A.1) - (A.4), *there exists an invariant measure $\mu$ for Eq.* (2.1) *which has its support in $V$.*

# 4. Examples

Now we give three concrete examples to show the application of the previous theorems to Eq. (2.1). Again we suppose

$$\lambda_1 = \inf_{v \in V} \frac{\|v\|^2}{|v|^2} > 0. \tag{4.1}$$

Then, since $|\nabla v|^2 \geq \alpha \|v\|^2$ for some $\alpha > 0$, we see that $|\nabla v|$ is an equivalent norm on $V$. All the following equations are subject to the same initial-boundary conditions as in Eq. (2.1).

**Example 1** Consider the bilinear equation:

$$\frac{\partial u}{\partial t} = \Delta u + (\beta u + \gamma)\dot{W}_t, \quad t > 0, \quad x \in D, \tag{4.2}$$

where $\beta, \gamma > 0$ are constants. Then we have $f \equiv 0$ and $\sigma(v) = \beta v + \gamma$ in Eq. (2.1). Take $\Phi(v) = |v|^2$. It is easy to calculate

$$\mathcal{L}\Phi(v) = -2|\nabla v|^2 + \beta^2(\hat{q}v, v) + \gamma^2 Tr.Q,$$

where $\hat{q}(x) = q(x, x)$. Define

$$\lambda_q = \inf_{v \in H_0^1} \frac{|\nabla v|^2}{(\hat{q}v, v)}.$$

Then $\lambda_q \geq \lambda_1/q_\infty > 0$ by (4.1). Thus, if $\lambda_q > \beta^2/2$,

$$\mathcal{L}\Phi(v) \leq -\delta\|v\|^2 + \gamma^2 Tr.Q, \quad \text{for some} \ \delta > 0,$$

so that condition C for Thm. 3.1 is satisfied. In particular, if the smallest eigenvalue $\lambda_1$ of $(-\Delta)$ is greater than $q_\infty \beta^2/2$, Eq. (4.2) has an invariant measure $\mu$ on $H_0^1$ which gives rise to a stationary solution.

**Example 2** For $\beta, k, \gamma > 0$, consider the nonlinear equation:

$$\frac{\partial u}{\partial t} = \Delta u + \beta \sin ku + \gamma \dot{W}_t(x), t > 0, x \in D. \tag{4.3}$$

Again we take $\Phi(v) = |v|^2$. Then

$$\mathcal{L}\Phi(v) = -2|\nabla v|^2 + 2\beta(v, \sin kv) + \gamma^2 Tr.Q$$
$$\leq -2|\nabla v|^2 + \beta|v|^2 + C,$$

for some constant $C > 0$. Similar to (Ex. 1), if the smallest eigenvalue of $(-\Delta)$ is greater than $\beta^2/2$, the equation (4.3) has a stationary solution associated with an initial invariant distribution $\mu$.

**Example 3** Consider the nonlinear reaction-diffusion problem:

$$\frac{\partial u}{\partial t} = \Delta u + f(u) + \beta|\nabla u(t,x)|\dot{W}_t(x) + \dot{\widetilde{W}}_t, \quad t > 0, \quad x \in D, \tag{4.4}$$

where $\widetilde{W}_t$ is a $\widetilde{Q}$-Wiener process independent of $W_t$. Suppose there exists a smooth Lyapunov function $\varphi(\zeta)$ for the associated ordinary differential equation:

$$\frac{d\zeta}{dt} = f(\zeta), t > 0, \zeta \in \mathbf{R}^d,$$

such that $f(\zeta)\varphi'(\zeta) \leq 0$, $\varphi''(\zeta) \geq \delta$ for some $\delta > 0$, and $\varphi(\zeta) \to \infty$ as $|\zeta| \to \infty$. Let us define

$$\Phi(v) = \int_D \varphi[v(x)]dx, \quad v \in H. \tag{4.5}$$

Then it is easy to check that

$$\mathcal{L}\Phi(v) = -\int_D [1 - \frac{1}{2}\beta^2 q(x,x)]\varphi''[v(x)]|\nabla v(x)|^2 dx$$
$$+ \int_D \varphi'[v(x)]f[v(x)]dx + Tr.\widetilde{Q}$$
$$\leq -\delta(1 - \frac{1}{2}\beta^2 q_\infty)|\nabla v|^2 + Tr.\widetilde{Q}.$$

Therefore, if $\beta^2 q_\infty < 2$, $\Phi$ as defined by (4.5), is a Lyapunov functional so that Thm. 4.1 ensures the existence of an invariant measure $\mu$ for Eq. (4.4).

We remark in passing that a sufficient condition for the uniqueness of an invariant measure can be given [3]. However the condition is somewhat stringent and will not be presented here.

**Acknowledgements** The author wishes to express his deep gratitude to Professors H. Kunita and H. -H. Kuo for their hard work in organizing this highly stimulating Seminar. The work was supported by NSF Grant DMS-91-01360.

## References

[1] Breiman, L. : *Probability Theory*. Reading, Mass.: Addison-Wesley, 1968
[2] Chow, P. -L.: Stability of Nonlinear Stochastic Evolution Equations; *J. Math. Analy. Appl.* **89** (1982) 400–419

[3] Chow, P. -L. and Khasminskii, R. Z.: Stationary solutions of nonlinear stochastic evolution equation; *Preprint* (1994)

[4] DaPrato, G. and Zabczyk, J.: *Stochastic Equations in Infinite Dimensions.* Cambridge: Cambridge Univ. Press, 1992

[5] Hida, T., Kuo, H. -H., Potthoff, J., and Streit, L.: *White Noise: An Infinite Dimensional Calculus.* Kluwer Academic Publishers, 1993

[6] Khasminskii, R. Z.: *Stochastic Stability of Differential Equations.* Alphen aan den Rijn, The Netherlands: Sijthoff & Noordhoff, 1980

[7] Marcus, R.: Parabolic Itô Equations; *Trans. Amer. Math. Soc.* **198** (1974) 177–190

[8] Marcus, R. : Parabolic Itô Equations with Monotone Nonlinearities; *J. Funct. Analy.* **29** (1978) 257–287

[9] Pardoux, E.: Stochastic Partial Differential Equations and Filtering of Diffusion Processes; *Stochastics* **3** (1979) 127–167

[10] Walsh, J. B.: An Introduction to Stochastic Partial Differential Equations; *Lecture Notes in Math.* **1180**, Springer-Verlag (1984) 265–439

Pao-Liu Chow
Department of Mathematics
Wayne State University
Detroit, MI 48202, U.S.A.
*E-mail*: plchow@waynest1.bitnet

M. DE FARIA & L. STREIT
# Some recent advances in white noise analysis

We generalize the construction of the Hida test function and distribution spaces from White Noise Analysis to general Gaussian spaces. As an application we study the chaos expansion of intersection local times for higher dimensional Brownian motion.

## 1. Introduction

White Noise Analysis has seen considerable development in the past two decades. We cite [3, 9, 10, 11, 12, 14] as examples, and refer to [4] for an overview. It provides a framework for infinite dimensional analysis, particularly on the Gel'fand triple of Hida test functions and Hida distributions

$$(S) \subset L^2(S^*, d\mu) \subset (S)^*.$$

Various generalizations suggest themselves. On the one hand there are of course alternate choices of test and generalized functions surrounding the same Hilbert space $L^2(S^*, d\mu)$ [7, 9], with interesting applications [14]. On the other hand one may want to generalize from White Noise to other Gaussian processes - such as e.g. vector valued White Noise or the multi-parameter Gaussian fields relevant for the construction of time ordered expectation functionals for gauge field theories [13] - maintaining however the essential features of the Hida spaces. Here we shall follow [6] in presenting the latter generalization, and shall then apply it to a discussion of higher dimensional intersection local times [1].

## 2. Gaussian Spaces and Construction of the Triple

Our starting point will be a real nuclear $\sigma$-Hilbert space $N$

$$N \subset \cdots \subset H_{p+1} \subset H_p \subset \cdots \subset H_0 = H. \tag{1}$$

We denote the corresponding, increasing Hilbert norms by $|\cdot|_p$, resp. $|\cdot| = |\cdot|_0$, further down we shall use the same notation for the norm induced on complexified symmetric tensor products $H_{c,p}^{\widehat{\otimes} n}$.

On the dual space $N^*$ one defines a Gaussian measure $\mu$ by its characteristic function

$$C(f) = \int_{N^*} e^{i<\omega,f>} d\mu(\omega) = e^{-\frac{1}{2}|f|^2}.$$

By the Ito-Segal-Wiener isomorphism the corresponding $L^2-$space is isomorphic to the Fock space $\Gamma$ over H

$$L^2(N^*, d\mu) \simeq \Gamma(H), \tag{2}$$

where

$$\Gamma(H) = \bigoplus_{n=0}^{\infty} H_c^{\widehat{\otimes}n}$$

consists of direct sums of symmetric tensor products $F_n \in H_c^{\widehat{\otimes}n}$ and has its norm defined by

$$\|F\|^2 = \sum_{n=0}^{\infty} n! \, |F_n|^2 . \tag{3}$$

To construct a Gel'fand triple around $L^2(N^*, d\mu)$ we introduce a family of norms

$$\|F\|_{p,q}^2 = \sum (n!)\, 2^{nq} \, |F_n|_p^2 \tag{4}$$

on $\Gamma(H)$. For the corresponding norms on (dense subspaces of) $L^2(d\mu)$ induced by the Ito-Segal-Wiener isomorphism we shall use the same notation, i.e., for $\varphi \simeq F$ we write

$$\|\varphi\|_{p,q} \equiv \|F\|_{p,q} .$$

Now we define

$$(H_p)_q = \left\{ \varphi \in L^2(d\mu) : \|\varphi\|_{p,q} < \infty \right\}$$

and finally the test function space

$$(N) = \bigcap_{p,q}(H_p)_q .$$

We denote its dual by $(N)^*$ and have

$$(N)^* = \bigcup (H_{-p})_{-q} . \tag{5}$$

It is worth noting that the spaces $(N)$ and $(N)^*$ are indeed only functions of the underlying nuclear space $N$, and not of the particular choice of $\sigma-$Hilbert spaces

$H_p$ used in the construction. Different Hilbert norms $|\cdot|_p$ will give rise to the same triple if they give rise to the same $N$. In particular one obtains the usual spaces $(S)$ and $(S)^*$ of White Noise Analysis if one chooses $N = S(R)$ above.

As usual the functionals $\varphi_f$ given for $\omega \in N^*$ and $f \in N$ by

$$\varphi_f(\omega) = e^{i<\omega,f>} \tag{6}$$

are elements of $(N)$ and total in $L^2(N^*, d\mu)$. This allows definition of the T-transform of $\Phi \in (N)^*$ as the dual product between generalized and test function

$$(T\Phi)(f) = \langle\langle \Phi, \varphi_f \rangle\rangle \tag{7}$$

and likewise for the S-transform with

$$\varphi_f(\omega) = e^{<\omega,f>}C(f) \equiv: e^{<\omega,f>} : .$$

The latter is the generating functional for the so-called Wick powers of the field $\omega$

$$: \omega^{\otimes n} :\, \in (N_c^*)^{\widehat{\otimes}n} .$$

They are obtained from

$$: e^{<\omega,f>} :\, = \sum_{n=0}^{\infty} \frac{1}{n!} \left\langle : \omega^{\otimes n} :, f^{\otimes n} \right\rangle$$

by polarization.

With these and the functions $F_n$ from (3) we have the chaos expansion

$$\Phi(\omega) = \sum \left\langle : \omega^{\otimes n} :, F_n \right\rangle \tag{8}$$

while the S-transform is

$$S\Phi(f) = \sum \left\langle F_n, f^{\otimes n} \right\rangle . \tag{9}$$

We close this section with an example that is only a very slight generalization of the usual White Noise Analysis framework, but of some importance in applications as we shall have occasion to see below.

**Example 1** For any $d \geq 1$, let

$$N = R^d \otimes S(R) \subset H = R^d \otimes L^2(R, dx) \subset N^* = R^d \otimes S^*(R).$$

The characteristic function is

$$C(f) = \int_{N^*} e^{i<\omega,f>} d\mu(\omega) = e^{-\frac{1}{2}|f|^2} = e^{-\frac{1}{2}\sum_{i=1}^{d}\int f_i^2(t)dt}$$

and $\omega = (\omega_1, \ldots, \omega_d)$ is d-dimensional white noise. In terms of these components $\omega_i$ we state for later use the chaos expansion (8)

$$\Phi(\omega) = \sum_{(n)} \left\langle : \omega^{\otimes(n)} :, F_{(n)} \right\rangle$$

and S-transform (9)

$$S\Phi(f) = \sum_{(n)} \left\langle F_{(n)}, f^{\otimes(n)} \right\rangle.$$

Here we have used the multi-index

$$(n) = (n_1, \ldots, n_d), \quad (n)! = \prod_{i=1}^{d} n_i!, \quad n = \sum_{i=1}^{d} n_i$$

and

$$f^{\otimes(n)} = \bigotimes_{i=1}^{d} f_i^{\otimes n_i}, \quad F_{(n)} \in \bigotimes_{i=1}^{d} \widehat{S^*}(R^d).$$

## 3. The Characterization of Generalized Functionals

In many applications one is confronted with heuristic expressions for which one wants to decide whether they are generalized functionals from a suitable space such as $(N)^*$. This turns out to be quite straightforward if one controls their S- or T-transform. S- and T-transform of any $\Phi \in (N)^*$ are U-functionals in the sense of the following definition.

**Definition 2**   A complex valued function $U$ on the nuclear space $N$ is called a *U-functional* if for all $f, g \in N$, the following conditions are satisfied:
(1)  The mapping $\lambda \to U(\lambda f + g)$ has an analytic extension to all complex $\lambda$.
(2)  There are constants $C_1$ and $C_2$ such that for all complex $z$ and some continuous quadratic form $f \to B(f)$ on $N$

$$|U(zf)| < C_1 e^{C_2|z|^2|B(f)|}. \tag{10}$$

The remarkable fact is that conversely any U-functional is the S- and T-transform of generalized functionals.

**Theorem 3** ([6]) *The following are equivalent* :

1. U *is a U-functional,*
2. U=$S\Phi$ *for* $\Phi \in (N)^*$,
3. U=$T\widehat{\Phi}$ *for* $\widehat{\Phi} \in (N)^*$.

The following corollaries illustrate the usefulness of describing $(N)^*$ in terms of S-transforms. The first one deals with convergence in the space $(N)^*$ of generalized functionals.

**Corollary 4** *Let* $\{U_k\}_{k\in N}$ *denote a sequence of U-functionals with the following properties* :

(1) *For all* $f \in N$, $\{U_k(f)\}_{k\in N}$ *is a Cauchy sequence.*
(2) *There exist* $C_i > 0$ *such that* $|U_k(zf)| \leq C_1 \exp(C_2 |z^2 B(f)|)$ *uniformly in* $k$. *Then there is a unique* $\Phi \in (N)^*$ *such that* $S^{-1}U_k$ *converges strongly to* $\Phi$.


In the second one we control the (Bochner-)integration of families of generalized functionals.

**Corollary 5** *Let* $(\Omega, B, m)$ *be a measure space, and* $\Phi_\lambda$ *a mapping defined on* $\Omega$ *with values in* $(N)^*$. *We assume that the S-transform of* $\Phi_\lambda$
(1) *is an m-measurable function of* $\lambda$ *for any test function vector* $f \in N$,
(2) *obeys a U-functional estimate*

$$| (S\Phi_\lambda)(z\mathbf{f}) |\leq C_1(\lambda) \exp(C_2(\lambda)|z^2 B(f)|)$$

*for some fixed B and for* $C_1 \in L^1(m), C_2 \in L^\infty(m)$.
*Then* $\Phi_\lambda$ *is Bochner-integrable in the Hilbert space* $(H_{-p})_{-q}$ *for* $p, q$ *large enough and*

$$\int_\Omega dm(\lambda)\Phi_\lambda \tag{11}$$

*is an element in* $(N)^*$ *satisfying*

$$\int_\Omega dm(\lambda)\Phi_\lambda)(f) = \int_\Omega dm(\lambda)(S(\Phi_\lambda)(f)). \tag{12}$$


We close this section with an example of such integrals which will be of use further down.

**Example 6** Let $g \in H, a \in C$. Then the following is an element in $(N)^*$

$$\delta(<\omega, g> -a) = \frac{1}{2\pi} \int_{-\infty}^{\infty} d\lambda e^{i\lambda(<\omega,g>-a)}$$

Evidently, these functionals generalize (and include) Donsker's $\delta$-function.

## 4. Intersection Local Times

In this section we shall study intersection local times of d-dimensional Brownian motion and their chaos expansion in the White Noise Analysis framework, following [1]. Related investigations can be found in [2, 5, 16, 17].

For functionals of d-dimensional Brownian motion Example 1 provides the proper starting point. We obtain Brownian motion **B** through

$$\mathbf{B}(t) \equiv <\omega, 1 \otimes 1_{[0,t]}> = \left( \int_0^t \omega_1(s)ds, \ldots, \int_0^t \omega_d(s)ds \right)$$

in terms of d-dimensional white noise. Accordingly the Bochner integral

$$\delta(\mathbf{B}(t) - \mathbf{B}(s)) \equiv (\frac{1}{2\pi})^d \int_{R^d} d\lambda e^{i\lambda(\mathbf{B}(t)-\mathbf{B}(s))}$$

is a generalized white noise functional if $t \neq s$, with S-transform given by

$$S(\delta(\mathbf{B}(t) - \mathbf{B}(s)))(f) = \left( \frac{1}{\sqrt{2\pi|t-s|}} \right)^d e^{-\frac{1}{2|t-s|} \sum_{i=1}^d (\int_s^t f_i(u)du)^2}$$

$$= \left( \frac{1}{\sqrt{2\pi|t-s|}} \right)^d \sum_{(n)} \frac{\left(-\frac{1}{2|t-s|}\right)^n}{(n)!} \prod_{i=1}^d (\int_s^t f_i(u)du)^{2n_i}. \quad (13)$$

Comparison with (9) and (8) yields

$$F_{2(n)}(u_1, \ldots, u_{2n}) = \frac{1}{(n)!}(2\pi)^{-d/2}(\frac{1}{|t_2 - t_1|})^{\frac{d}{2}+n}(-\frac{1}{2})^n \prod_{k=1}^{2n} 1_{[t_1,t_2]}(u_k) \quad (14)$$

while all the F with one or more odd indices vanish.

We denote truncated chaos expansions which include only $(n)$ with $n \geq m$, by

$$\Phi_m(\omega) \equiv \sum_{(n):n \geq m} \left\langle : \omega^{\otimes(n)} :, F_{(n)} \right\rangle.$$

**Theorem 7** *Let $d = 1, 2, 3, \ldots$, and $2m > d - 2$. Then the Bochner integral*

$$L^{(2m)} \equiv \int_{0 < t_1 < t_2 < 1} d^2 t\, \delta^{(2m)}(\mathbf{B}(t_2) - \mathbf{B}(t_1))$$

*is an element in $\in (S)^*$ with kernel functions given by*

$$G_{2(n)}(u_1, \ldots, u_{2n}) = (-)^n \left( (n + \tfrac{d}{2} - 1)(n + \tfrac{d}{2} - 2)(2\pi)^{d/2}\, 2^n\, (n)! \right)^{-1} \cdot$$

$$\cdot \Theta(u)\Theta(1 - v) \cdot (1 - v^{-n - \frac{d}{2} + 2} - (1 - u)^{-n - \frac{d}{2} + 2} + (v - u)^{-n - \frac{d}{2} + 2}), \quad (15)$$

*where $v \equiv \max_{1 \le k \le 2n} u_k$ and $u \equiv \min_{1 \le k \le 2n} u_k$, if $n \ge m$. All other $G'_{(n)}s$ vanish.*

The proof is essentially an application of Theorem 5, plus an explicit calculation of the kernel functions $G_{(n)}$. For more details and further results we refer to [1].

**Acknowledgements** Support through the program STRIDE is gratefully acknowledged.

# References

[1] de Faria, M., Hida, T., Streit, L., and Watanabe, H.: Intersection local times as generalized white noise functionals; (in preparation)

[2] He, S. W., Wang, J. -G., Yang, W. -Q., and Yao, R. -Q.: Local times of self-intersection for multidimensional Brownian motion; *Shanghai preprint*

[3] Hida, T,: *Analysis of Brownian Functionals.* Carleton Math. Lecture Notes **13**, 1975

[4] Hida, T., Kuo, H. -H., Potthoff, J., and Streit, L.: *White Noise Analysis: An infinite dimensional calculus.* Kluwer, Dordrecht, 1993

[5] Imkeller, P., Perez-Abreu, V., and Vives, J.: Chaos expansions of double intersection local times of Brownian motion in $R^d$ and renormalization; *Preprint* (1993)

[6] Kondratiev, Yu. G., Leukert, P., Potthoff, J., Streit, L., and Westerkamp, W.: Generalized functionals in Gaussian spaces - the characterization theorem revisited; *Mannheim University preprint, no. 175* (1994)

[7] Kondratiev, Yu. G., Leukert, P., and Streit, L.: Wick calculus in Gaussian spaces; *BiBoS preprint* (1994)

[8] Kondratiev, Yu. G. and Samoilenko, Yu. S.: Spaces of trial and generalized functions of an infinite number of variables; *Rep. Math. Phys.* **14** (1978) 325–350

[9] Kondratiev, Yu. G. and Streit, L.: Spaces of white noise distributions: constructions, descriptions, applications I; *Rep. Math. Phys.* **33** (1993) 341–366

[10] Kubo, I. and Takenaka, S.: Calculus on Gaussian white noise I-IV; *Proc. Japan Acad.* **56** (1980) 376–380, **56** (1980) 411–416, **57** (1981) 433–437, **58** (1982) 186–189

[11] Kuo, H. -H.: Lectures on White Noise Analysis; *Soochow J. Math.* **18** (1992) 229–300

[12] Lee, Y. -J.: Analytic version of test functionals, Fourier transform and a characterization of measures in white noise calculus; *Funct. Anal.* **100** (1991) 359–380

[13] Leukert, P.: Diplomarbeit, Bielefeld, 1994, in preparation

[14] Øksendal, B.: Stochastic partial differential equations an applications in hydrodynamics; in: "Stochastic Analysis and Applications in Physics", A. I. Cardoso et al. (eds.), Kluwer, Dordrecht, 1994

[15] Potthoff, J. and Streit, L.: A characterization of Hida distributions; *J. Funct. Anal.* **101** (1991) 212–229

[16] Streit, L. and Westerkamp, W.: A generalization of the characterization theorem for generalized functionals of white noise; in: "Dynamics of Complex and Irregular Systems", Ph. Blanchard et al. (eds.), World Scientific, 1993

[17] Watanabe, H.: The local time of self-intersections of Brownian motions as generalized Brownian functionals; *Lett. Math. Phys.* **23** (1991) 1–9

Margarida de Faria
CCM, Universidade da Madeira
P-9000 Funchal, PORTUGAL
*E-mail*: streit%ptifm2.ifm.pt@ptearn.fc.ul.pt
and
BiBoS, Universität Bielefeld
D-33501 Bielefeld, GERMANY
*E-mail*: streit@physf.uni-bielefeld.de

Ludwig Streit
CCM, Universidade da Madeira
P-9000 Funchal, PORTUGAL
*E-mail*: streit%ptifm2.ifm.pt@ptearn.fc.ul.pt
and
BiBoS, Universität Bielefeld
D-33501 Bielefeld, GERMANY
*E-mail*: streit@physf.uni-bielefeld.de

I. DÔKU, H. -H. KUO, and Y. -J. LEE

# Fourier transform and heat equation in white noise analysis

## 1. Introduction

Let $(H, B)$ be an abstract Wiener space [1, 9]. Suppose $\varphi$ is a real-valued twice $H$-differentiable function on $B$ such that the second $H$-derivative $D^2\varphi(x)$ at $x$ is a trace class operator of $H$. Then the Gross Laplacian $\Delta_G\varphi$ of $\varphi$ [2] is defined by

$$\Delta_G\,\varphi(x) = \mathrm{Trace}_H\,D^2\varphi(x).$$

Let $\mu$ be the standard Gaussian measure on $B$. Suppose $f$ is a bounded Lipschitz function on $B$. It has been shown in [2] (see also [9]) that the function

$$u(t, x) = \int_B f\left(x + \sqrt{t}\,y\right) d\mu(y) \tag{1.1}$$

satisfies the following heat equation associated with the Gross Laplacian

$$\frac{\partial u(t, x)}{\partial t} = \frac{1}{2}\Delta_G\,u(t, x), \quad u(0, \cdot) = f. \tag{1.2}$$

On the other hand, consider the heat equation on the finite dimensional space $\mathbb{R}^n$. It is well-known that $u(t, x)$ can be derived by using the Fourier transform, i.e., let $v(t, y) = \widehat{u}(t, y)$. Then $v(t, y)$ satisfies the equation:

$$\frac{\partial v(t, y)}{\partial t} = -\frac{1}{2}|y|^2 v(t, y), \quad v(0, \cdot) = \widehat{f}.$$

Hence $v(t, y)$ is given by

$$v(t, y) = \widehat{f}(y) \exp\left[-\frac{t}{2}|y|^2\right]. \tag{1.3}$$

By taking the inverse Fourier transform of $v(t, y)$ we get the solution $u(t, x)$.

The finite dimensional Fourier transform has an infinite dimensional counterpart within the framework of white noise analysis [4, 10, 11, 12, 14]. Thus it is natural to ask whether $u(t, x)$ in Eq. (1.1) can be derived as above by using the white noise Fourier transform. It turns out that there are several difficulties. For instance, the white noise analogue of Eq. (1.3) has no meaning (see the next section). This is due mainly to the fact that the Gross Laplacian is a singular operator given by the trace kernel function [4, p. 205]. We will approximate this kernel function by smooth functions to overcome these difficulties. The parabolic equations arising from the approximation have been studied previously, e.g., in [16]. But we will use the white noise Fourier transform to solve these equations. This paper can be regarded as a step toward the application of Fourier transform to partial differential equations associated with white noise kernel operators [4, 5].

## 2. White Noise Calculus

Let $\mathcal{S}(\mathbb{R})$ be the Schwartz space of real-valued rapidly decreasing functions on $\mathbb{R}$ and $\mathcal{S}'(\mathbb{R})$ its dual space. Then $\mathcal{S}(\mathbb{R}) \subset L^2(\mathbb{R}) \subset \mathcal{S}'(\mathbb{R})$ is a Gel'fand triple. In white noise calculus the space $\mathcal{S}(\mathbb{R})$ is often regarded as a nuclear space with the family $\{|\cdot|_p;\, p > 0\}$ of norms given by

$$|\xi|_p = |A^p \xi|, \quad \xi \in \mathcal{S}(\mathbb{R}),$$

where the operator $A = -d^2/du^2 + u^2 + 1$ and $|\cdot|$ is the $L^2(\mathbb{R})$-norm. Let $\mathcal{S}_p(\mathbb{R})$ be the completion of $\mathcal{S}(\mathbb{R})$ with respect to the norm $|\cdot|_p$, $p > 0$. Its dual space $\mathcal{S}_p'(\mathbb{R})$ is the completion of $L^2(\mathbb{R})$ with respect to the norm $|f|_{-p} = |A^{-p} f|$ and we have $\mathcal{S}_p(\mathbb{R}) \subset L^2(\mathbb{R}) \subset \mathcal{S}_p'(\mathbb{R})$. Moreover, $(L^2(\mathbb{R}), \mathcal{S}_p'(\mathbb{R}))$ is an abstract Wiener space if and only if $p > \frac{1}{2}$.

The standard Gaussian measure $\mu$ on $\mathcal{S}'(\mathbb{R})$ is the probability measure with the characteristic function

$$\int_{\mathcal{S}'(\mathbb{R})} e^{i\langle x, \xi \rangle}\, d\mu(x) = e^{-\frac{1}{2}|\xi|^2}, \quad \xi \in \mathcal{S}(\mathbb{R}).$$

Let $(L^2)$ be the Hilbert space of complex-valued $\mu$-square integrable functions with norm denoted by $\|\cdot\|$. By the Wiener-Itô theorem every $\varphi$ in $(L^2)$ can be represented uniquely by

$$\varphi = \sum_{n=0}^{\infty} I_n(f_n), \quad f_n \in \widehat{L_c^2}(\mathbb{R}^n),$$

where $I_n$ denotes the multiple Wiener integral of order $n$ and $\widehat{L_c^2}(\mathbb{R}^n)$ the space of symmetric complex-valued $L^2$-functions on $\mathbb{R}^n$. Moreover, for the $(L^2)$-norm $\|\varphi\|$ of $\varphi$, we have

$$\|\varphi\|^2 = \sum_{n=0}^{\infty} n!\, |f_n|^2_{L_c^2(\mathbb{R}^n)}.$$

The *second quantization operator* $\Gamma(A)$ is densely defined on $(L^2)$ as follows. For $\varphi = \sum_{n=0}^{\infty} I_n(f_n)$,

$$\Gamma(A)\varphi = \sum_{n=0}^{\infty} I_n(A^{\otimes n} f_n). \tag{2.1}$$

The domain of $\Gamma(A)$ consists of all $\varphi$ in $(L^2)$ such that $\sum_{n=0}^{\infty} n!\, |A^{\otimes n} f_n|^2_{L^2_c(\mathbb{R}^n)} < \infty$. For $p \in \mathbb{N}$, define

$$\|\varphi\|_p = \|\Gamma(A)^p \varphi\|$$

and let $(S)_p \equiv \{\varphi \in (L^2);\ \|\varphi\|_p < \infty\}$. Note that $(S)_p$ is a complex Hilbert space with norm $\|\cdot\|_p$. Let $(S)$ be the projective limit of $\{(S)_p;\ p \in \mathbb{N}\}$. It serves as a space of *test functions* in white noise calculus. The elements in the dual space $(S)^*$ of $(S)$ are called *generalized functions* (or *Hida distributions*). It is a fact that $(S) \subset (L^2) \subset (S)^*$ is a Gel'fand triple. We will use the following convention for the relationship between the natural bilinear pairing $\langle\!\langle \cdot, \cdot \rangle\!\rangle$ of $(S)^*$ and $(S)$ and the inner product $(\cdot, \cdot)$ of $(L^2)$:

$$\langle\!\langle f, \varphi \rangle\!\rangle = (f, \overline{\varphi}), \quad f \in (L^2),\ \varphi \in (S).$$

The dual space $(S)^*_p$ of $(S)_p$ is the completion of $(L^2)$ with respect to the norm $\|\varphi\|_{-p} = \|\Gamma(A)^{-p}\varphi\|$ and we have $(S)_p \subset (L^2) \subset (S)^*_p$.

For some calculations in this paper, it is more convenient to use another family $\{[\![ \cdot ]\!]_p;\ p \in \mathbb{N}\}$ of norms on $(S)$. It is well-known that each $\varphi$ in $(S)$ has a continuous version [7]. In fact, it has an analytic version $\widetilde{\varphi}$ [14]. Moreover, for any $p \in \mathbb{N}$, the following norm $[\![ \varphi ]\!]_p$ of $\varphi$ is finite

$$[\![ \varphi ]\!]_p = \sup \left\{ |\widetilde{\varphi}(x)| \exp\left[-2^{-1}|x|^2_{-p}\right];\ x \in S'_{p,c}(\mathbb{R}) \right\}, \tag{2.2}$$

where $S'_{p,c}(\mathbb{R})$ is the complexification of $S'_p(\mathbb{R})$. This family $\{[\![ \cdot ]\!]_p;\ p \in \mathbb{N}\}$ of norms is equivalent to the above family $\{\|\cdot\|_p;\ p \in \mathbb{N}\}$ of norms arising from the second quantization operator $\Gamma(A)$ [4]. For simplicity, we will use $\varphi$ to denote also its analytic version and elements in $(S)$ are assumed to be the analytic versions.

A fundamental tool in white noise calculus is the $S$-transform defined on the space $(S)^*$ of Hida distributions. The $S$-transform $S\Phi$ of $\Phi \in (S)^*$ is the function on $S(\mathbb{R})$ given by

$$(S\Phi)(\xi) = e^{-\frac{1}{2}|\xi|^2} \langle\!\langle \Phi, e^{\langle \cdot, \xi \rangle} \rangle\!\rangle, \quad \xi \in S(\mathbb{R}).$$

The white noise analogue of finite dimensional Fourier transform can be defined in terms of the $S$-transform [4, 11, 12]. The *Fourier transform* $\mathcal{F}\Phi$ of $\Phi \in (\mathcal{S})^*$ is defined to be the Hida distribution with the $S$-transform given by

$$S(\mathcal{F}\Phi)(\xi) = \langle\!\langle \Phi, e^{-i\langle \cdot, \xi \rangle} \rangle\!\rangle = (S\Phi)(-i\xi)e^{-\frac{1}{2}|\xi|^2}, \quad \xi \in \mathcal{S}(\mathbb{R}).$$

An equivalent definition is given in [14]. This Fourier transform is continuous from $(\mathcal{S})^*$ into itself [3].

For each $x \in \mathcal{S}'(\mathbb{R})$ the Gâteaux differentiation $D_x$ in the direction $x$ is a continuous linear operator from $(\mathcal{S})$ into itself. Its adjoint $D_x^*$ is continuous from $(\mathcal{S})^*$ into itself. In particular, the *white noise differentiation* $\partial_t \equiv D_{\delta_t}$ is continuous from $(\mathcal{S})$ into itself. Its adjoint operator $\partial_t^*$ is continuous from $(\mathcal{S})^*$ into itself. If $\xi \in \mathcal{S}(\mathbb{R})$, then $D_\xi$ extends by continuity to a continuous linear operator $\widetilde{D}_\xi$ from $(\mathcal{S})^*$ into itself. For each $\xi \in \mathcal{S}(\mathbb{R})$, let $q_\xi$ denote the multiplication by $\langle \cdot, \xi \rangle$. The operator $q_\xi$ is continuous from $(\mathcal{S})$ into itself. It extends by continuity to a continuous linear operator $\widetilde{q}_\xi$ from $(\mathcal{S})^*$ into itself. Moreover, we have $\widetilde{q}_\xi = \widetilde{D}_\xi + D_\xi^*$.

The white noise Fourier transform has the same properties with respect to the differentiation and multiplication as the finite dimensional Fourier transform, i.e., for any $\xi \in \mathcal{S}(\mathbb{R})$,

$$\mathcal{F}\widetilde{D}_\xi = i\widetilde{q}_\xi \mathcal{F}, \quad \mathcal{F}\widetilde{q}_\xi = i\widetilde{D}_\xi \mathcal{F}.$$

By letting $\xi = \delta_t$ informally (since $\delta_t \notin \mathcal{S}(\mathbb{R})$), we get the symbolic expression:

$$\mathcal{F}(\partial_t \Phi)(y) = iy(t)(\mathcal{F}\Phi)(y), \quad y \in \mathcal{S}'(\mathbb{R}). \tag{2.3}$$

Let $\theta(\mathbf{s}, \mathbf{t}) \in \mathcal{S}'(\mathbb{R}^{j+k})$, $\mathbf{s} = (s_1, \ldots, s_j)$, $\mathbf{t} = (t_1, \ldots, t_k)$. The following white noise kernel operator was introduced in [5] (see also [4, 15])

$$\Xi_{j,k}(\theta) = \int_{\mathbb{R}^{j+k}} \theta(\mathbf{s}, \mathbf{t}) \partial_{s_1}^* \cdots \partial_{s_j}^* \partial_{t_1} \cdots \partial_{t_k} \, d\mathbf{s} d\mathbf{t}.$$

The operator $\Xi_{j,k}(\theta)$ is continuous from $(\mathcal{S})$ into $(\mathcal{S})^*$. It is continuous from $(\mathcal{S})$ into itself if and only if $\theta \in \mathcal{S}(\mathbb{R}^j) \otimes \mathcal{S}'(\mathbb{R}^k)$. For example, let $\tau$ be the element in $\mathcal{S}'(\mathbb{R}^2)$ defined by

$$\langle \tau, f \rangle = \int_{\mathbb{R}} f(t, t) \, dt, \quad f \in \mathcal{S}(\mathbb{R}^2).$$

Then the associated operator $\Xi_{0,2}(\tau)$ is continuous from $(\mathcal{S})$ into itself. This operator can be identified as the Gross Laplacian, i.e.,

$$\Delta_G = \Xi_{0,2}(\tau) = \int_{\mathbb{R}} \partial_s^2 \, ds.$$

The Gross Laplacian is a singular operator in the sense that the kernel function is a delta function. On the other hand, if $\theta \in \mathcal{S}'(\mathbb{R}^j) \otimes \mathcal{S}(\mathbb{R}^k)$, then $\Xi_{j,k}(\theta)$ extends to a continuous linear operator from $(\mathcal{S})^*$ into itself. This fact will be used in the next section.

Now, let us consider the heat equation (1.2) in the white noise setup:

$$\frac{\partial u(t,x)}{\partial t} = \frac{1}{2} \int_{\mathbb{R}} \partial_s^2 \, u(t,x) \, ds, \quad u(0,\cdot) = f.$$

Let $v(t,y) = (\mathcal{F}u(t,\cdot))(y)$. Then, by Eq. (2.3), $v(t,y)$ satisfies the equation

$$\frac{\partial v(t,y)}{\partial t} = -\frac{1}{2}|y|^2 v(t,y), \quad v(0,\cdot) = \mathcal{F}f$$

and it appears that $v(t,y)$ is given by

$$v(t,y) = (\mathcal{F}f)(y) \exp\left[-\frac{t}{2}|y|^2\right].$$

But there are two difficulties: (1) $\exp[-\frac{t}{2}|y|^2]$ is not in $(\mathcal{S})^*$ without renormalization and (2) multiplication on $(\mathcal{S})^*$ is not defined. Thus it is natural to take a kernel function $\kappa(s_1, s_2)$ and consider the following parabolic equation:

$$\frac{\partial u(t,x)}{\partial t} = \frac{1}{2} \int_{\mathbb{R}^2} \kappa(s_1, s_2) \partial_{s_1} \partial_{s_2} u(t,x) \, ds_1 ds_2, \quad u(0,\cdot) = f. \tag{2.4}$$

Then the Fourier transform $v(t,\cdot)$ of $u(t,\cdot)$ is given by

$$v(t,y) = \widehat{f}(y) \exp\left[-\frac{t}{2} \int_{\mathbb{R}^2} \kappa(s_1, s_2) y(s_1) y(s_2) \, ds_1 ds_2\right]. \tag{2.5}$$

In the next section we will find conditions on $\kappa$ and $f$ such that the multiplication in Eq. (2.5) is meaningful and the inverse Fourier transform $u(t,\cdot)$ of $v(t,\cdot)$ is given by a convolution. In §4 we will give some results related to the convergence of $u(t,\cdot)$ to the function in Eq. (1.1) as $\kappa \to \tau$ in $\mathcal{S}'(\mathbb{R}^2)$.

## 3. Main Results

The Fourier transform $\mathcal{F}$ is the adjoint of a continuous operator $\mathcal{G}$ from $(\mathcal{S})$ into itself [3, 4]. For $\varphi \in (\mathcal{S})$, $\mathcal{G}\varphi$ is defined by

$$\mathcal{G}\varphi(y) = \langle\!\langle : e^{-i\langle \cdot, y \rangle} :, \varphi \rangle\!\rangle, \quad y \in \mathcal{S}'(\mathbb{R})$$

where $: e^{-i\langle\cdot,y\rangle}: \in (S)^*$ has $S$-transform $(S : e^{-i\langle\cdot,y\rangle}:)(\xi) = e^{-i\langle y,\xi\rangle}$, $\xi \in \mathcal{S}(\mathbb{R})$.

In this paper we will use the Fourier-Gauss transform $\mathcal{G}_{a,b}$, $a, b \in \mathbb{C}$, introduced in [13]. For $\varphi \in (S)$, $\mathcal{G}_{a,b}\varphi$ is defined by

$$\mathcal{G}_{a,b}\varphi(y) = \int_{\mathcal{S}'(\mathbb{R})} \varphi(ax + by)\, d\mu(x).$$

The operator $\mathcal{G}_{a,b}$ is continuous from $(S)$ into itself with respect to the family $\{[\![\,\cdot\,]\!]_p;\ p \in \mathbb{N}\}$ of norms in Eq. (2.2) [14]. But this family is equivalent to the family $\{\|\cdot\|_p;\ p \in \mathbb{N}\}$ of norms in $(S)$. Therefore, $\mathcal{G}_{a,b}$ is continuous from $(S)$ into itself.

**Proposition 1** $\mathcal{G} = \mathcal{G}_{1,-i}$ and $\mathcal{G}^{-1} = \mathcal{G}_{1,i}$.

**Proof** For each $\xi$ in $\mathcal{S}(\mathbb{R})$, let $\theta_\xi$ be the function $\theta_\xi = \exp[\langle\cdot,\xi\rangle - \frac{1}{2}|\xi|^2]$. Then

$$\mathcal{G}\theta_\xi(y) = \langle\!\langle : e^{-i\langle\cdot,y\rangle}:, \theta_\xi \rangle\!\rangle = (S : e^{-i\langle\cdot,y\rangle}:)(\xi) = e^{-i\langle y,\xi\rangle}. \qquad (3.1)$$

On the other hand, note that $\int_{\mathcal{S}'(\mathbb{R})} e^{\langle x,\xi\rangle}\, d\mu(x) = \exp[\frac{1}{2}|\xi|^2]$. Hence

$$\mathcal{G}_{1,-i}\theta_\xi(y) = e^{-\frac{1}{2}|\xi|^2} \int_{\mathcal{S}'(\mathbb{R})} e^{\langle x-iy,\xi\rangle}\, d\mu(x) = e^{-i\langle y,\xi\rangle}.$$

Therefore, $\mathcal{G}\theta_\xi = \mathcal{G}_{1,-i}\theta_\xi$ for all $\xi \in \mathcal{S}(\mathbb{R})$. But the linear span of $\{\theta_\xi;\ \xi \in \mathcal{S}(\mathbb{R})\}$ is dense in $(S)$. Thus $\mathcal{G} = \mathcal{G}_{1,-i}$ since both of them are continuous from $(S)$ into itself. The second assertion is obvious in view of the fact that $\mathcal{G}^{-1}\theta_\xi = e^{i\langle\cdot,\xi\rangle}$, which can be easily derived from Eq. (3.1). $\qquad\square$

In some lemmas below, $L$ is a continuous linear operator from $\mathcal{S}'(\mathbb{R})$ into $\mathcal{S}(\mathbb{R})$. Such an operator is continuous from $\mathcal{S}'_p(\mathbb{R})$ into $\mathcal{S}_p(\mathbb{R})$ for any $p > 0$. But $(L^2(\mathbb{R}), \mathcal{S}'_p(\mathbb{R}))$ is an abstract Wiener space for any $p > \frac{1}{2}$. Therefore, $L$ is a trace class operator of $L^2(\mathbb{R})$ [9, p. 83].

**Lemma 2** *Let $L$ be a continuous linear operator from $\mathcal{S}'(\mathbb{R})$ into $\mathcal{S}(\mathbb{R})$ such that it is positive and self-adjoint on $L^2(\mathbb{R})$. Let $h(x) = e^{-\frac{1}{2}\langle x, Lx\rangle}$. Suppose $f$ is a measurable function on $\mathcal{S}'(\mathbb{R})$ such that $\int_{\mathcal{S}'(\mathbb{R})} |f((I+L)^{\frac{1}{2}}x)|^2\, d\mu(x) < \infty$. Then for any $\psi \in (S)$, $\mathcal{G}_{1,-i}(h\psi)$ is defined and the following holds for any $p > \frac{1}{2}$*

$$\left| \int_{\mathcal{S}'(\mathbb{R})} f(y)\mathcal{G}_{1,-i}(h\psi)(y)\, d\mu(y) \right| \le C_{p,L}[\![\psi]\!]_p \left( \int_{\mathcal{S}'(\mathbb{R})} \left| f((I+L)^{\frac{1}{2}}x) \right|^2 d\mu(x) \right)^{\frac{1}{2}},$$

*where* $C_{p,L} = \int_{\mathcal{S}'(\mathbb{R})} e^{\frac{1}{2}|x|^2_{-p}}\, d\mu(x) \big[\det(I+L) \int_{\mathcal{S}'(\mathbb{R})} e^{|Lx|^2 + |\sqrt{I+L}\,x|^2_{-p}}\, d\mu(x)\big]^{\frac{1}{2}}$.

**Proof**  By using the definition of $\mathcal{G}_{1,-i}$ and the positivity of $L$, we can check by direct calculation

$$|\mathcal{G}_{1,-i}(h\psi)(y)| \le \alpha_p [\![\psi]\!]_p \, e^{\frac{1}{2}(\langle y,Ly\rangle + |y|^2_{-p})},$$

where $\alpha_p = \int_{\mathcal{S}'(\mathbb{R})} e^{\frac{1}{2}|x|^2_{-p}} d\mu(x)$. This shows that $\mathcal{G}_{1,-i}(h\psi)$ is defined and

$$\left| \int_{\mathcal{S}'(\mathbb{R})} f(y)\mathcal{G}_{1,-i}(h\psi)(y) \, d\mu(y) \right| \le \alpha_p [\![\psi]\!]_p \int_{\mathcal{S}'(\mathbb{R})} |f(y)| e^{\frac{1}{2}(\langle y,Ly\rangle + |y|^2_{-p})} \, d\mu(y).$$

By the Girsanov theorem for the transformation $y = \sqrt{I+L}\,x$ [9, p. 141], we have

$$\int_{\mathcal{S}'(\mathbb{R})} |f(y)| e^{\frac{1}{2}(\langle y,Ly\rangle + |y|^2_{-p})} \, d\mu(y) =$$

$$\sqrt{\det(I+L)} \int_{\mathcal{S}'(\mathbb{R})} \left| f\big((I+L)^{\frac{1}{2}}x\big) \right| e^{\frac{1}{2}(|Lx|^2 + |\sqrt{I+L}\,x|^2_{-p})} \, d\mu(x).$$

Then apply the Schwarz inequality to finish the proof.  $\square$

**Lemma 3**  *Let $L$ be a continuous linear operator from $\mathcal{S}'(\mathbb{R})$ into $\mathcal{S}(\mathbb{R})$ such that it is positive and self-adjoint on $L^2(\mathbb{R})$. Then for any $\psi \in (\mathcal{S})$, we have*

$$\int_{\mathcal{S}'(\mathbb{R})} e^{-\frac{1}{2}\langle y-ix, L(y-ix)\rangle} \psi(y-ix) \, d\mu(y) =$$

$$\frac{1}{\sqrt{\det(I+L)}} \, e^{\frac{1}{2}\langle x,(I+L)^{-1}Lx\rangle} \int_{\mathcal{S}'(\mathbb{R})} \psi\big((I+L)^{-\frac{1}{2}}y - i(I+L)^{-1}x\big) \, d\mu(y).$$

**Proof**  Let $\mu_1$ be the standard Gaussian measure on $\mathbb{R}$. Let $f$ be an analytic function such that $|f(z)| \le Ce^{\frac{1}{4}|z|^2}$, $z \in \mathbb{C}$, for some constant $C$. It is easy to check that for any nonnegative number $\alpha$, we have

$$\int_{\mathbb{R}} e^{-\frac{1}{2}\alpha(y-ix)^2} f(y-ix) \, d\mu_1(y) =$$

$$\frac{1}{\sqrt{1+\alpha}} \, e^{\frac{1}{2}\alpha(1+\alpha)^{-1}x^2} \int_{\mathbb{R}} f\big((1+\alpha)^{-\frac{1}{2}}y - i(1+\alpha)^{-1}x\big) \, d\mu_1(y).$$

The conclusion follows from this identity and the representation of $L$ in terms of its eigenvalues and eigenvectors.  $\square$

**Lemma 4**  *Let $L$ be a continuous linear operator from $\mathcal{S}'(\mathbb{R})$ into $\mathcal{S}(\mathbb{R})$ such that it is positive and self-adjoint on $L^2(\mathbb{R})$. Then for any measurable function $g$ on $\mathcal{S}'(\mathbb{R})$ such that $\int_{\mathcal{S}'(\mathbb{R})} |g((I+L)^{\frac{1}{2}}x)| \, d\mu(x) < \infty$, we have*

$$\int_{\mathcal{S}'(\mathbb{R})} e^{\frac{1}{2}\langle x,(I+L)^{-1}Lx\rangle} g(x) \, d\mu(x) = \sqrt{\det(I+L)} \int_{\mathcal{S}'(\mathbb{R})} g\big((I+L)^{\frac{1}{2}}x\big) \, d\mu(x).$$

**Proof** Just apply the Girsanov theorem for the transformation $Tx = \sqrt{I+L}\,x$ [9, p. 141]. $\qquad\square$

**Remark** Note that a function $g$ satisfying the condition in this lemma must be in $L^1(\mu)$. In fact, since $L$ is positive, the conclusion in the lemma implies immediately that

$$\int_{\mathcal{S}'(\mathbb{R})} |g(x)|\, d\mu(x) \leq \sqrt{\det(I+L)} \int_{\mathcal{S}'(\mathbb{R})} \left| g\big((I+L)^{\frac{1}{2}}x\big)\right| d\mu(x).$$

**Lemma 5** *Let $A, B,$ and $C$ be continuous linear operators from the complexification $\mathcal{S}_c(\mathbb{R})$ of $\mathcal{S}(\mathbb{R})$ into itself such that $A^*A + B^*B = C^*C$. Then for any $\psi$ in $(\mathcal{S})$, the following equality holds:*

$$\int_{\mathcal{S}'(\mathbb{R})} \int_{\mathcal{S}'(\mathbb{R})} \psi(A^*x + B^*y)\, d\mu(x)d\mu(y) = \int_{\mathcal{S}'(\mathbb{R})} \psi(C^*z)\, d\mu(z). \tag{3.2}$$

**Proof** Note that both sides of Eq. (3.2) define continuous linear functionals on $(\mathcal{S})$. Hence it suffices to show that Eq. (3.2) holds for all $\psi_\xi = e^{\langle \cdot, \xi\rangle}$, $\xi \in \mathcal{S}(\mathbb{R})$, since the linear span of $\{\psi_\xi;\ \xi \in \mathcal{S}(\mathbb{R})\}$ is dense in $(\mathcal{S})$. But for $\psi_\xi$, we have

$$\int_{\mathcal{S}'(\mathbb{R})} \int_{\mathcal{S}'(\mathbb{R})} \psi_\xi(A^*x + B^*y)\, d\mu(x)d\mu(y) = e^{\frac{1}{2}(|A\xi|^2 + |B\xi|^2)},$$

$$\int_{\mathcal{S}'(\mathbb{R})} \psi_\xi(C^*z)\, d\mu(z) = e^{\frac{1}{2}|C\xi|^2}.$$

Hence we get the assertion since $A^*A + B^*B = C^*C$. $\qquad\square$

Let $Q : \mathcal{S}(\mathbb{R}) \to \mathcal{S}(\mathbb{R})$ be a continuous linear operator. The second quantization operator $\Gamma(Q)$ of $Q$ as defined in Eq. (2.1) is a continuous linear operator from $(\mathcal{S})$ into itself [6]. If $Q$ extends to a bounded linear operator from $L^2(\mathbb{R})$ into itself such that $\|Q\|_{L^2(\mathbb{R}), L^2(\mathbb{R})} \leq 1$, then $\Gamma(Q)$ extends to a bounded linear operator from $(L^2)$ into itself and $\|\Gamma(Q)\|_{(L^2),(L^2)} \leq 1$. If, in addition, $Q$ satisfies $\langle Qf, \xi\rangle = \langle f, Q\xi\rangle$ for all $f \in L^2(\mathbb{R})$ and $\xi \in \mathcal{S}(\mathbb{R})$, then we have

$$\langle\!\langle \Gamma(Q)F, \varphi\rangle\!\rangle = \langle\!\langle F, \Gamma(Q)\varphi\rangle\!\rangle, \quad F \in (L^2),\ \varphi \in (\mathcal{S}). \tag{3.3}$$

**Lemma 6** *Let $L$ be a continuous linear operator from $\mathcal{S}'(\mathbb{R})$ into $\mathcal{S}(\mathbb{R})$ such that it is positive and self-adjoint on $L^2(\mathbb{R})$. Then for any $\psi \in (\mathcal{S})$, we have*

$$\Gamma\big((I+L)^{-\frac{1}{2}}\big)\psi(x) = \int_{\mathcal{S}'(\mathbb{R})} \psi\big((I+L)^{-\frac{1}{2}}L^{\frac{1}{2}}y + (I+L)^{-\frac{1}{2}}x\big)\, d\mu(y). \tag{3.4}$$

**Proof** Let $T\psi$ denote the function in the right-hand side of Eq. (3.4). It is easy to check that $T$ is continuous from $(S)$ into itself. Therefore, it suffices to show that Eq. (3.4) holds for $\theta_\xi = \exp[\langle \cdot, \xi \rangle - \frac{1}{2}|\xi|^2]$, $\xi \in S(\mathbb{R})$. By direct calculation we get

$$T\theta_\xi = \exp\left[\langle \cdot, (I+L)^{-\frac{1}{2}}\xi \rangle - \frac{1}{2}\langle \xi, (I+L)^{-1}\xi \rangle\right].$$

On the other hand, note that $\theta_\xi = \sum_{n=0}^\infty (n!)^{-1} I_n(\xi^{\otimes n})$. Therefore,

$$\Gamma\left((I+L)^{-\frac{1}{2}}\right)\theta_\xi = \sum_{n=0}^\infty \frac{1}{n!} I_n\left(\left((I+L)^{-\frac{1}{2}}\right)^{\otimes n}\xi^{\otimes n}\right)$$

$$= \exp\left[\langle \cdot, (I+L)^{-\frac{1}{2}}\xi \rangle - \frac{1}{2}\langle \xi, (I+L)^{-1}\xi \rangle\right].$$

Thus $T\theta_\xi = \Gamma\left((I+L)^{-\frac{1}{2}}\right)\theta_\xi$ for any $\xi \in S(\mathbb{R})$ and hence $T = \Gamma\left((I+L)^{-\frac{1}{2}}\right)$. $\qquad\square$

Now, we can study the parabolic equation in Eq. (2.4). First we state the conditions on $\kappa$ and $f$:

- The kernel function $\kappa$ belongs to $S(\mathbb{R}^2)$ and is symmetric such that $\langle x\otimes x, \kappa \rangle \geq 0$ for all $x \in S'(\mathbb{R})$;
- The function $f$ is measurable on $S'(\mathbb{R})$ with $\int_{S'(\mathbb{R})} \left|f\left((I+tK)^{\frac{1}{2}}x\right)\right|^2 d\mu(x) < \infty$ for all $t > 0$, where $K$ is the integral operator with kernel function $\kappa$.

Then the white noise kernel operator $\Xi_{0,2}(\kappa) = \int_{\mathbb{R}^2} \kappa(s_1, s_2)\partial_{s_1}\partial_{s_2}\, ds_1 ds_2$ is continuous from $(S)$ into itself. Moreover, $\Xi_{0,2}(\kappa)$ extends to a continuous linear operator $\widetilde{\Xi}_{0,2}(\kappa)$ from $(S)^*$ into itself [4, 5]. We can rewrite Eq. (2.4) as follows:

$$\frac{\partial u(t,x)}{\partial t} = \frac{1}{2}\widetilde{\Xi}_{0,2}(\kappa)u(t,x), \quad u(0,\cdot) = f. \tag{3.5}$$

Here we use $\widetilde{\Xi}_{0,2}(\kappa)$ instead of $\Xi_{0,2}(\kappa)$ since we will use the Fourier transform to find a weak solution in the space $(S)^*$. The Fourier transform $v(t,\cdot)$ of $u(t,\cdot)$ is given as in Eq. (2.5) by

$$v(t,y) = \widehat{f}(y)\exp\left[-\frac{t}{2}\langle y, Ky \rangle\right]. \tag{3.6}$$

An essential difficulty is how to interpret $v(t,\cdot)$ as a Hida distribution since the multiplication of two Hida distributions is not defined in general. Observe that informally, for any $\psi \in (S)$

$$\langle\!\langle v(t,\cdot), \psi \rangle\!\rangle = \langle\!\langle \widehat{f}\, h, \psi \rangle\!\rangle = \langle\!\langle \widehat{f}, h\psi \rangle\!\rangle = \langle\!\langle f, \mathcal{G}_{1,-i}(h\psi) \rangle\!\rangle$$

$$= \int_{S'(\mathbb{R})} f(y)\mathcal{G}_{1,-i}(h\psi)(y)\, d\mu(y).$$

Here and below $h$ denotes the function $h(y) \equiv e^{-t\langle y, Ky \rangle/2}$. By Lemma 2, $v(t, \cdot)$ is a Hida distribution and

$$\langle\!\langle v(t, \cdot), \psi \rangle\!\rangle = \int_{\mathcal{S}'(\mathbb{R})} f(y)\mathcal{G}_{1,-i}(h\psi)(y)\, d\mu(y). \tag{3.7}$$

Thus we *define* the function $v(t, \cdot)$ given in Eq. (3.6) as the Hida distribution defined by Eq. (3.7).

Assuming $u(t, x)$ is a weak solution of Eq. (3.5), we can use the Fourier transform to find $u(t, x)$ as follows. The Fourier transform $v(t, \cdot)$ of $u(t, \cdot)$ is given by Eq. (3.6). For any $\varphi \in (\mathcal{S})$, by Proposition 1 and Eq. (3.7), we have

$$\langle\!\langle u(t, \cdot), \varphi \rangle\!\rangle = \langle\!\langle \mathcal{F}^{-1}v(t, \cdot), \varphi \rangle\!\rangle = \langle\!\langle v(t, \cdot), \mathcal{G}_{1,i}\varphi \rangle\!\rangle$$

$$= \int f(y)\mathcal{G}_{1,-i}(h\mathcal{G}_{1,i}\varphi)(y)\, d\mu(y). \tag{3.8}$$

Apply Lemma 3 with $L = tK$ and $\psi = \mathcal{G}_{1,i}\varphi$ to get

$$\mathcal{G}_{1,-i}(h\mathcal{G}_{1,i}\varphi)(x) =$$
$$\frac{1}{\sqrt{\det(I + tK)}} e^{\frac{1}{2}\langle x, (I+tK)^{-1}tKx \rangle} \int \mathcal{G}_{1,i}\varphi\big((I + tK)^{-\frac{1}{2}}y - i(I + tK)^{-1}x\big)\, d\mu(y).$$

By using this equality and then applying Lemma 4 with $L = tK$ and

$$g(x) = f(x) \int \mathcal{G}_{1,i}\varphi\big((I + tK)^{-\frac{1}{2}}y - i(I + tK)^{-1}x\big)\, d\mu(y),$$

we can derive that

$$\int f(x)\mathcal{G}_{1,-i}(h\mathcal{G}_{1,i}\varphi)(x)\, d\mu(x) =$$
$$\int f\big((I + tK)^{\frac{1}{2}}x\big) \int \mathcal{G}_{1,i}\varphi\big((I + tK)^{-\frac{1}{2}}y - i(I + tK)^{-\frac{1}{2}}x\big)\, d\mu(y)d\mu(x). \tag{3.9}$$

Use Lemma 5 with $A = I$, $B = i(I + tK)^{-\frac{1}{2}}$, and $C = (I + tK)^{-\frac{1}{2}}(tK)^{\frac{1}{2}}$ to simplify the last integral in the $y$-variable:

$$\int \mathcal{G}_{1,i}\varphi\big((I + tK)^{-\frac{1}{2}}y - i(I + tK)^{-\frac{1}{2}}x\big)\, d\mu(y)$$

$$= \int\!\!\int \varphi\big(z + i(I + tK)^{-\frac{1}{2}}y + (I + tK)^{-\frac{1}{2}}x\big)\, d\mu(z)d\mu(y)$$

$$= \int \varphi\big((I + tK)^{-\frac{1}{2}}(tK)^{\frac{1}{2}}y + (I + tK)^{-\frac{1}{2}}x\big)\, d\mu(y). \tag{3.10}$$

On the other hand, by Lemma 6 with $L = tK$ and $\psi = \varphi$, we have

$$\Gamma\big((I+tK)^{-\frac{1}{2}}\big)\varphi(x) = \int \varphi\big((I+tK)^{-\frac{1}{2}}(tK)^{\frac{1}{2}}y + (I+tK)^{-\frac{1}{2}}x\big)d\mu(y). \qquad (3.11)$$

It follows from Eqs. (3.8)–(3.11) that

$$\langle\!\langle u(t,\cdot),\varphi\rangle\!\rangle = \int f\big((I+tK)^{\frac{1}{2}}x\big)\Gamma\big((I+tK)^{-\frac{1}{2}}\big)\varphi(x)\, d\mu(x).$$

Therefore, by Eq. (3.3) with $Q = (I+tK)^{-\frac{1}{2}}$, we have

$$\langle\!\langle u(t,\cdot),\varphi\rangle\!\rangle = \int \varphi(x)\Gamma\big((I+tK)^{-\frac{1}{2}}\big)f\big((I+tK)^{\frac{1}{2}}\cdot\big)(x)\, d\mu(x).$$

But it is easy to derive from Lemma 6 that

$$\Gamma\big((I+tK)^{-\frac{1}{2}}\big)f\big((I+tK)^{\frac{1}{2}}\cdot\big)(x) = \int f\big(x+(tK)^{\frac{1}{2}}y\big)\, d\mu(y).$$

Hence we have shown that the following holds for all $\varphi \in (\mathcal{S})$:

$$\langle\!\langle u(t,\cdot),\varphi\rangle\!\rangle = \int \varphi(x)\int f\big(x+(tK)^{\frac{1}{2}}y\big)\, d\mu(y)d\mu(x).$$

Thus a weak solution $u(t,x)$ of Eq. (3.5), if it exists, must be given by

$$u(t,x) = \int f\big(x+(tK)^{\frac{1}{2}}y\big)\, d\mu(y). \qquad (3.12)$$

Next we prove that this function $u(t,x)$ is indeed a weak solution. But we need to show first that $u(t,\cdot)$ is a Hida distribution for any $t$. In fact, the next lemma with $L = tK$ implies that $u(t,\cdot) \in (L^2)$.

**Lemma 7** *Let $L$ be a continuous linear operator from $\mathcal{S}'(\mathbb{R})$ into $\mathcal{S}(\mathbb{R})$ such that it is positive and self-adjoint on $L^2(\mathbb{R})$. Suppose $f$ is a measurable function on $\mathcal{S}'(\mathbb{R})$ such that $\int_{\mathcal{S}'(\mathbb{R})} |f((I+L)^{\frac{1}{2}}x)|^2\, d\mu(x) < \infty$. Then*

$$\int_{\mathcal{S}'(\mathbb{R})} \int_{\mathcal{S}'(\mathbb{R})} \left|f\big(x+L^{\frac{1}{2}}y\big)\right|^2 d\mu(y)d\mu(x) = \int_{\mathcal{S}'(\mathbb{R})} \left|f\big((I+L)^{\frac{1}{2}}x\big)\right|^2 d\mu(x).$$

**Proof** By integrating $x$-variable first and make a translation $x = z - \sqrt{L}\,y$, we get

$$\iint \left|f\big(x+L^{\frac{1}{2}}y\big)\right|^2 d\mu(y)d\mu(x) = \iint |f(z)|^2 e^{\langle z,\sqrt{L}\,y\rangle - \frac{1}{2}\langle y,Ly\rangle}\, d\mu(z)d\mu(y).$$

70

Then apply the Girsanov theorem for $y = (I + L)^{-\frac{1}{2}} x$ to get

$$\int e^{\langle z, \sqrt{L}\, y \rangle - \frac{1}{2}\langle y, Ly \rangle}\, d\mu(y) = \frac{1}{\sqrt{\det(I + L)}}\, e^{\frac{1}{2}\langle z, (I+L)^{-1} Lz \rangle}.$$

The assertion follows immediately from the above two equalities and Lemma 4 with $g = f^2$. $\qquad\qquad\qquad\qquad\qquad\qquad\qquad\qquad\qquad\qquad\qquad\qquad\qquad\square$

**Theorem 8** *Let $K$ be a continuous linear operator from $S'(\mathbb{R})$ into $S(\mathbb{R})$ such that it is positive and self-adjoint on $L^2(\mathbb{R})$. Suppose $f$ is a function on $S'(\mathbb{R})$ with*

$$\int_{S'(\mathbb{R})} \left| f\big((I + tK)^{\frac{1}{2}}\, x\big)\right|^2 d\mu(x) < \infty.$$

*Then Eq. (3.5) has a unique weak solution. Moreover, the solution is given by $u(t,x)$ in Eq. (3.12).*

**Proof** The uniqueness of a weak solution is obvious since the Fourier transform $\mathcal{F}$ is one-to-one from $(S)^*$ onto itself. Thus we need only to show that the function $u(t,x)$ in Eq. (3.12) is a weak solution of Eq. (3.5).

For simplicity, let $\vartheta_\kappa(x) \equiv \langle x, Kx \rangle - \mathrm{Trace}_{L^2(\mathbb{R})} K$ and let $\nabla_\kappa$ be the operator $\nabla_\kappa \varphi(x) = \langle Kx, \varphi'(x) \rangle$ on $(S)$. Define

$$\Lambda_\kappa = \Xi_{0,2}(\kappa) - 2\nabla_\kappa + \vartheta_\kappa. \tag{3.13}$$

It is easy to check that $\Lambda_\kappa$ is continuous from $(S)$ into itself and $\widetilde{\Xi}_{0,2}(\kappa) = \Lambda_\kappa^*$. Therefore, for any $\varphi \in (S)$

$$\langle\!\langle \widetilde{\Xi}_{0,2}(\kappa) u(t, \cdot), \varphi \rangle\!\rangle = \langle\!\langle u(t, \cdot), \Lambda_\kappa \varphi \rangle\!\rangle. \tag{3.14}$$

Now, for $u(t,x)$ given in Eq. (3.12), the following holds for any $\varphi \in (S)$

$$\langle\!\langle u(t, \cdot), \varphi \rangle\!\rangle = \iint f(x)\varphi\big(x - \sqrt{tK}\, y\big) w(t, x, y)\, d\mu(y) d\mu(x), \tag{3.15}$$

where $w(t, x, y) = \exp\left[ \sqrt{t}\langle x, \sqrt{K}\, y \rangle - \frac{1}{2} t \langle y, Ky \rangle \right]$. Use Eq. (3.15) and the integration by parts formula [8] to derive

$$
\begin{aligned}
&\langle\!\langle u(t, \cdot), \Xi_{0,2}(\kappa)\varphi \rangle\!\rangle \\
&= -\frac{1}{\sqrt{t}} \iint f(x) \Big\langle \varphi'\big(x - \sqrt{tK}y\big), \sqrt{K}\, y \Big\rangle w(t, x, y)\, d\mu(y) d\mu(x) \\
&\quad - \frac{1}{\sqrt{t}} \iint f(x)\varphi\big(x - \sqrt{tK}y\big)\big(\langle x, \sqrt{K}\, y \rangle - \sqrt{t}\langle y, Ky \rangle\big) w(t, x, y)\, d\mu(y) d\mu(x) \\
&\quad + \iint f(x)\varphi\big(x - \sqrt{tK}y\big)\vartheta_\kappa\big(x - \sqrt{tK}y\big) w(t, x, y)\, d\mu(y) d\mu(x)
\end{aligned}
\tag{3.16}
$$

and

$$\langle\!\langle u(t,\cdot),\nabla_\kappa\varphi\rangle\!\rangle$$

$$= -\frac{1}{\sqrt{t}}\iint f(x)\varphi(x - \sqrt{tK}y)\big(\langle x,\sqrt{K}y\rangle - \sqrt{t}\langle y,Ky\rangle\big)w(t,x,y)\,d\mu(y)d\mu(x)$$

$$+ \iint f(x)\varphi(x - \sqrt{tK}y)\vartheta_\kappa(x - \sqrt{tK}y)w(t,x,y)\,d\mu(y)d\mu(x). \tag{3.17}$$

It follows from Eqs. (3.13), (3.14), (3.16), and (3.17) that

$$\langle\!\langle \widetilde{\Xi}_{0,2}(\kappa)u(t,\cdot),\varphi\rangle\!\rangle$$

$$= -\frac{1}{\sqrt{t}}\iint f(x)\big\langle\varphi'(x - \sqrt{tK}y),\sqrt{K}y\big\rangle w(t,x,y)\,d\mu(y)d\mu(x)$$

$$+ \frac{1}{\sqrt{t}}\iint f(x)\varphi(x - \sqrt{tK}y)\big(\langle x,\sqrt{K}y\rangle - \sqrt{t}\langle y,Ky\rangle\big)w(t,x,y)\,d\mu(y)d\mu(x). \tag{3.18}$$

By differentiating Eq. (3.15) in $t$, we see easily that $2\frac{\partial}{\partial t}\langle\!\langle u(t,\cdot),\varphi\rangle\!\rangle$ equals to the right hand side of Eq. (3.18). Therefore, we have proved that for all $\varphi \in (\mathcal{S})$

$$\Big\langle\!\!\Big\langle \frac{\partial u(t,x)}{\partial t},\varphi\Big\rangle\!\!\Big\rangle = \Big\langle\!\!\Big\langle \frac{1}{2}\widetilde{\Xi}_{0,2}(\kappa)u(t,x),\varphi\Big\rangle\!\!\Big\rangle,$$

i.e., $u(t,x)$ given in Eq. (3.12) is a weak solution of Eq. (3.5).

Finally, we show that $u(t,\cdot)$ converges weakly to $f$ as $t \to 0$, i.e.,

$$\lim_{t\to 0}\langle\!\langle u(t,\cdot),\varphi\rangle\!\rangle = \langle\!\langle f,\varphi\rangle\!\rangle, \quad \forall\varphi \in (\mathcal{S}).$$

But this follows easily from Eq. (3.15), the Lebesgue dominated convergence theorem, and the following inequality for all $t \in [0,1]$:

$$\big|f(x)\varphi(x - \sqrt{tK}\,y)w(t,x,y)\big| \le [\![\varphi]\!]_p |f(x)|\,e^{|x|^2_{-p}+|Ky|^2_{-p}+|\langle x,\sqrt{K}y\rangle|}. \qquad \square$$

## 4. Convergence of the Solutions

For the convergence of the solutions in Eq. (3.12) as $K \to I$ strongly, the integrability condition on $f$ does not seem to be appropriate. Instead, we take $f$ from the space $\mathcal{A}_p$ $(p > 0)$ consisting of analytic functions $\varphi$ on $\mathcal{S}'_{p,c}(\mathbb{R})$ such that $[\![\varphi]\!]_p < \infty$ (see Eq. (2.2)).

Let $K$ be the same operator as in the previous section. For a function $f$, let

$$u_{f,K}(t,x) = \int f(x + \sqrt{tK}\,y)\,d\mu(y), \quad u_f(t,x) = \int f(x + \sqrt{t}\,y)\,d\mu(y).$$

We have the following results. The proofs will appear elsewhere.

1. If $f \in \mathcal{A}_p$ $(p > 0)$, then $u_f(t, \cdot) \in \mathcal{A}_{q(t)}$ for any $t < 2^{2p} - 1$, where $q(t) = \frac{1}{2}\log_2\left(2^{2p} - t\right)$;

2. If $f \in \mathcal{A}_p$ $(p > \frac{1}{2})$, then $u_{f,K}(t, \cdot) \in \mathcal{A}_{p-\frac{1}{2}}$ for all $t < (2\sigma_{p,K})^{-1}$, where $\sigma_{p,K}$ is the largest eigenvalue of the operator $\sqrt{K}A^{-2p}\sqrt{K}$;

3. Suppose $f$ is an analytic function on $\mathcal{S}'_{p,c}(\mathbb{R})$ for some $p > 1$ and there exists a constant $C$ such that $|f(x) - f(y)| \leq C|x - y|_{-p}\exp\left[\frac{1}{2}\left(|x|^2_{-p} + |y|^2_{-p}\right)\right]$. Then $u_{f,K}(t, \cdot)$ converges to $u_f(t, \cdot)$ in $\mathcal{A}_{p-1}$ as $K \to I$ strongly on $\mathcal{S}'_p(\mathbb{R})$ for any $t \leq \min\{2^{2p-1}, (2\sigma_p)^{-1}\}$, where $\sigma_p$ is the maximum of $\sigma_{p,K}$ over $K$.

**Acknowledgements** Research of I. Dôku is supported by Ministry of Education of Japan grant SR05640089. He is grateful for the hospitality of the Mathematics Department of Louisiana State University during his visit. Part of the results was obtained by H. -H. Kuo and Y. -J. Lee during Kuo's visit to Cheng Kung University. Kuo wants to thank the warm hospitality of the CKU Mathematics Department and the financial support of the National Science Council of Taiwan.


# References

[1] Gross, L.: Abstract Wiener spaces; in: *Proc. 5th Berkeley Symp. Math. Stat. Probab.* **2**, 31–42. Berkeley: Univ. Berkeley (1965)

[2] Gross, L.: Potential theory on Hilbert space; *J. Functional Analysis* **1** (1967) 123–181

[3] Hida, T., Kuo, H. -H. and Obata, N.: Transformations for white noise functionals; *J. Functional Analysis* **111** (1993) 259–277

[4] Hida, T., Kuo, H. -H., Potthoff, J. and Streit, L.: *White Noise: An Infinite Dimensional Calculus.* Kluwer Academic Publishers, 1993

[5] Hida, T., Obata, N. and Saitô: Infinite dimensional rotations and Laplacians in terms of white noise calculus; *Nagoya Math. J.* **128** (1992) 65–93

[6] Kubo, I. and Kuo, H. -H.: Finite dimensional Hida distributions; *Preprint* (1993)

[7] Kubo, I. and Yokoi, Y.: A remark on the space of testing random variables in the white noise calculus; *Nagoya Math. J.* **115** (1989) 139–149

[8] Kuo, H. -H.: Integration by parts for abstract Wiener measures; *Duke Math. J.* **41** (1974) 373–379

[9] Kuo, H. -H.: *Gaussian Measures in Banach Spaces.* Lecture Notes in Math., Vol. 463. Berlin, Heidelberg, New York: Springer-Verlag, 1975

[10] Kuo, H. -H.: On Fourier transform of generalized Brownian functionals; *J. Multivariate Analysis* **12** (1982) 415–431

[11] Kuo, H. -H.: The Fourier transform in white noise calculus; *J. Multivariate Analysis* **31** (1989) 311-327

[12] Kuo, H. -H.: Lectures on white noise analysis; *Soochow J. Math.* **18** (1992) 229–300

[13] Lee, Y. -J.: Unitary operators on the space of $L^2$-functions over abstract Wiener spaces; *Soochow J. Math.* **13** (1987) 165–174

[14] Lee, Y. -J.: Analytic version of test functionals, Fourier transform and a characterization of measures in white noise calculus; *J. Functional Analysis* **100** (1991) 359–380

[15] Obata, N.: An analytic characterization of symbols of operators on white noise functionals; *J. Math. Soc. Japan* **45** (1993) 421–445

[16] Piech, M. A.: A fundamental solution of the parabolic equation on Hilbert space; *J. Functional Analysis* **3** (1969) 85–114

Isamu Dôku
Department of Mathematics
Saitama University
Urawa 338, JAPAN
*E-mail*: h00060@sinet.ad.jp
and
Department of Mathematics
Louisiana State University
Baton Rouge, LA 70803, U.S.A.
*E-mail*: doku@marais.math.lsu.edu

Hui-Hsiung Kuo
Department of Mathematics
Louisiana State University
Baton Rouge, LA 70803, U.S.A.
*E-mail*: mmkuo@lsuvax.sncc.lsu.edu

Yuh-Jia Lee
Department of Mathematics
Cheng Kung University
Tainan 700, TAIWAN
*E-mail*: yjlee@mail.ncku.edu.tw

B. K. DRIVER

# The non-equivalence of Dirichlet forms on path spaces

## 1. Introduction

To each finite dimensional Riemannian manifold $M$ with covariant derivative $\nabla$ is an associated Dirichlet form $(\mathcal{E}^\nabla)$ on the path space $W(M)$ of $M$. If $M = \mathbb{R}^d$ and $\nabla$ is the Levi-Civita covariant derivative on $\mathbb{R}^d$, then $Q \equiv \mathcal{E}^\nabla$ is the "usual" Dirichlet form on the classical Wiener space. These lecture notes will discuss the relationship (or lack of) between $\mathcal{E}^\nabla$ and $Q$ as $(M, \nabla)$ varies.

## 2. Smooth Preliminaries

Let $(M^d, \langle \cdot, \cdot \rangle, \nabla, o)$ be given, where $M$ is a compact connected manifold (without boundary) of dimension $d$, $\langle \cdot, \cdot \rangle$ is a Riemannian metric on $M$, $\nabla$ is a $\langle \cdot, \cdot \rangle$-compatible covariant derivative, and $o$ is a fixed base point in $M$. Let $T = T^\nabla$ and $R = R^\nabla$, denote the torsion and curvature of $\nabla$ respectively. We denote parallel translation up to time "$s$" along a smooth path $\sigma : [0, 1] \to M$ by $P_s(\sigma) = P_s^\nabla(\sigma)$.

**Standing Assumption:** The covariant derivative $(\nabla)$ is assumed to be *Torsion Skew Symmetric* or *TSS* for short. That is to say, if $T = T^\nabla$ is the torsion tensor of $\nabla$, then $\langle T(X, Y), Y \rangle \equiv 0$ for all vector fields $X$ and $Y$ on $M$. (With the TSS condition, the Laplacian on functions ($\Delta f = \mathrm{tr}(\nabla \mathrm{grad} f)$) associated to $\nabla$ is the same as the usual Levi-Civita Laplacian.)

## 2.1. Examples of $(M, \nabla)$

**Example 2.1** Let $M = SO(n)$ – the $n \times n$ real orthogonal matrices $g$ with $\det(g) = 1$. (In this case $d = n(n-1)/2$.) Take $o = I$, $\langle A, B \rangle \equiv \mathrm{tr}(A^t B)$ for $A, B \in T_g G$ — the set of $n \times n$ real matrices $(A)$ such that $g^{-1}A$ is skew symmetric. There are three natural covariant derivatives on $G$ : namely the *left* ($\nabla^L$), *right* ($\nabla^R$), and *Levi-Civita* ($\nabla^{Levi}$) covariant derivative. The left and right covariant derivatives may be described by describing how they act on a vector field $A(t) \in T_{g(t)}G$ along a curve $g(t) \in G$. The formulas are:

$$\frac{\nabla^L A(t)}{dt} = g(t) \frac{d(g(t)^{-1}A(t))}{dt}$$

and

$$\frac{\nabla^R A(t)}{dt} = \frac{d(A(t)g(t)^{-1})}{dt} g(t).$$

The Levi-Civita covariant derivative is the average ($\nabla^{Levi} = (\nabla^L + \nabla^R)/2$) of the left and the right covariant derivatives.

With these definitions we have the following table

| $\nabla$ | $P_s^\nabla(\sigma)A$ | $R^\nabla < A, B > C$ | $T^\nabla < A, B >$ |
|---|---|---|---|
| $\nabla^L$ | $\sigma(s)A$ | $0$ | $-g[g^{-1}A, g^{-1}B]$ |
| $\nabla^R$ | $A\sigma(s)$ | $0$ | $[Ag^{-1}, Bg^{-1}]g$ |
| $\nabla^{Levi}$ | $* * *$ | $g[[g^{-1}A, g^{-1}B], g^{-1}C]/4$ | $0$ |

where $A, B, C \in T_g G$ and $[A, B] \equiv AB - BA$ when $A$ and $B$ are square matrices. The entry for parallel translation for the Levi-Civita covariant derivative is left blank, since no explicit formula for $P_s^\nabla(g)$ can in general be given when $\nabla$ has nonzero curvature.

**Example 2.2** Let $M$ be an oriented hyper-surface in $\mathbb{R}^d$ and $\langle \cdot, \cdot \rangle$ be the usual inner product on $\mathbb{R}^d$. Let $N : M \to S^{d-1} \subset \mathbb{R}^d$ be a smoothly varying unit normal vector on $M$. If $X(t)$ is a smoothly varying tangent vector to $M$ along a smooth curve $\sigma(t)$ in $M$, define

$$\nabla X(t)/dt = Q(\sigma(t))dX(t)/dt = dX(t)/dt + \{X(t) \cdot dN(\sigma(t))/dt\}N(\sigma(t)),$$

where $Q(\sigma(t))$ denotes orthogonal projection onto the tangent plane to $M$ at $\sigma(t)$. This covariant derivative is the Levi-Civita covariant derivative on $M$. The torsion tensor ($T^\nabla$) is zero and the curvature tensor ($R \equiv R^\nabla$) is given by

$$R\langle v, w \rangle z = \langle dN\langle w \rangle, z \rangle dN\langle v \rangle - \langle dN\langle v \rangle, z \rangle dN\langle w \rangle,$$

where $v, w, z$ are tangent vectors to $M$ at (say) $x \in M$ and

$$dN\langle v \rangle \equiv \frac{d}{dt}\bigg|_0 N(\sigma(t)),$$

where $\sigma : (-1, 1) \to M$ is any smooth curve in $M$ such that $\sigma(0) = x$ and $\dot{\sigma}(0) = v$.

# 3. Stochastic Preliminaries

Let

$$W(M) \equiv \{\sigma \in C([0,1], M) | \upsilon(0) = o\},$$

$$W \equiv \{\omega \in C([0,1], T_oM) | \omega(0) = 0 \in T_oM\},$$

(i.e. $W \equiv W(T_oM)$), $\mu$ be Wiener measure on $W$, and $\nu$ be Wiener measure on $W(M)$. As usual $H \subset W$ will denote the Cameron-Martin subspace of $W$ consisting of those paths $h \in W$ such that $(h,h)_H \equiv \int_0^1 |h'(s)|^2 ds < \infty$, where $|h'(s)|^2 \equiv \langle h'(s), h'(s)\rangle$. Let $\{P_s^\nabla\}_{s\in[0,1]}$ denote a fixed version of the stochastic parallel transport process on $W(M)$. So

$$P_s^\nabla(\sigma) : T_oM \to T_{\sigma(s)}M$$

is an isometry for all $s \in [0,1]$.

**Example 3.1** Take $M = G = SO(n)$ and $\nabla$ to be either $\nabla^L$ or $\nabla^R$, then $P_s^L(\sigma)A \equiv P_s^{\nabla^L}(\sigma)A = \sigma(s)A$ and $P_s^R(\sigma)A \equiv P_s^{\nabla^R}(\sigma)A = A\sigma(s)$ respectively.

**Example 3.2** If $M$ is an embedded hypersurface of $\mathbb{R}^d$, then $P_s^\nabla(\sigma)$ may be thought of as a version of the $d \times d$–matrix valued solution to the stochastic differential equation

$$dP_s^\nabla(\sigma) + N(\sigma(s))\{\delta(N(\sigma(s))\}^t P_s^\nabla(\sigma) = 0 \text{ with } P_0^\nabla(\sigma) = I, \qquad (3.1)$$

where $\delta$ denotes the Stratonovich differential.

## 3.1. Stochastic development

There is a well known measure theoretic isomorphism $(\Psi^\nabla)$ between $(W, \mu)$ and $(W(M), \nu)$. The map $\Psi \equiv \Psi^\nabla$ is defined uniquely up to $\mu$–equivalence as the solution to the stochastic (functional) differential equation:

$$\delta\Psi_s(\omega) = P_s^\nabla(\Psi_\cdot(\omega))\delta\omega \text{ with } \Psi_0(\omega) = o, \qquad (3.2)$$

where $\delta$ is used to denote the Stratonovich differential and $\omega$ is a Wiener path in $W$, see Eells and Elworthy [8] and Malliavin [16]. The measure theoretic inverse to $\Psi^\nabla$ will be denoted by $b^\nabla$. Notice that $b^\nabla$ is an $\mathbb{R}^d$–valued Brownian motion defined on the probability space $(W(M), \nu)$.

It is well known that $\Psi$ carries the measure $\mu$ to $\nu$ and that $\Psi$ is invertible up to equivalence. However, as pointed out by Malliavin [17, 18], the map $\Psi$ does **not** preserve the natural "Riemannian metrics" on $W$ and $W(M)$ except in the case that $(M, \nabla)$ itself has trivial geometry.

**Example 3.3** Let $M = G = SO(n)$, $\nabla = \nabla^L$ or $\nabla^R$, and $\Psi^L \equiv \Psi^{\nabla^L}$ or $\Psi^R \equiv \Psi^{\nabla^R}$ respectively. Then the functions $\Psi^L$ and $\Psi^R$ are solutions to the stochastic differential equations

$$\delta\Psi_s^L(\omega) = \Psi_s^L(\omega)\delta\omega(s) \text{ with } \Psi_0^L(\omega) = I$$

and

$$\delta\Psi_s^R(\omega) = \delta\omega(s)\Psi_s^R(\omega) \text{ with } \Psi_0^R(\omega) = I$$

respectively.

**Remark 3.4** For embedded surfaces, bounding an open convex subset of $\mathbb{R}^3$, equipped with the induced Riemannian structure, the map $\Psi$ has the interpretation of transferring the path $\omega$ in $\mathbb{R}^2$ to a path on the surface by "rolling" the surface along $\omega$ without slipping.

## 3.2. Flows, quasi-invariance, and integration by parts

The space $W(M)$ is to be thought of as an infinite dimensional manifold. To understand the notion of a tangent vector to $W(M)$ at $\sigma \in W(M)$, consider a differentiable curve $(t \in (-1, 1) \to f(t, \cdot) \in W(M))$ such that $f(0, \cdot) = \sigma(\cdot)$. The derivative $X(s) \equiv \frac{d}{dt}|_0 f(t, s)$ of such a curve in $W(M)$ is a vector-field along $\sigma$, i.e. $X \in C([0, 1], TM)$ such that $X(s) \in T_{\sigma(s)}M$ for all $s \in [0, 1]$. So it is reasonable to say that a vector-field $X$ along $\sigma \in W(M)$ is in the tangent space $(T_\sigma W(M))$ to $W(M)$ at $\sigma$.

**Example 3.5** For each $h \in H$ let $X^h$ be the vector-field on $W(M)$ defined by

$$X^h(\sigma)(s) = P_s^\nabla(\sigma)h(s) \ \forall s \in [0, 1]. \tag{3.3}$$

Notice that $X^h(\sigma) \in T_\sigma W(M)$.

**Theorem 3.6** **(Quasi-invariant Flow)** *Let $h \in H$ and $X^h$ be the vector-field in (3.3). Then $X^h$ admits a flow $(e^{tX^h})_{t \in \mathbb{R}}$ on $W(M)$ which leaves Wiener measure $(\nu)$ quasi-invariant. That is for each $t \in \mathbb{R}$, $\nu \circ e^{tX^h}$ and $\nu$ are mutually absolutely continuous relative to each other.*

This theorem was first proved in Driver [4] (see Theorem 8.5.) under the hypothesis that $h \in H \cap C^1$. The extension to the $h \in H$ was proved by Elton Hsu in [13, 14]. In the case that the manifold $M = G$ is a Lie group or a homogeneous space, the quasi-invariance of left and right multiplication by finite energy paths has been extensively studied, see [1, 10, 19, 20, 21].

**Definition 3.7** A *smooth cylinder function* on $W(M)$ is a function $f$ of the form

$$f(\sigma) = F(\sigma(s_1), \sigma(s_2), \ldots, \sigma(s_n)),$$

where $F : M^n \to \mathbb{R}$ is smooth and $\{s_1, s_2, \ldots, s_n\} \subset [0,1]$. Let $\mathcal{F}$ denote the class of smooth cylinder functions on $W(M)$.

**Theorem 3.8  (Integration by Parts)** *Let $f \in \mathcal{F}$ and $X \equiv X^h$ be as in (3.3). Define the action of $X$ on $f$ by the formula:*

$$Xf = L^2 - \lim_{t \to 0}(f \circ e^{tX} - f)/t.$$

*Then the $L^2$–adjoint $X^*$ of $X$ is densely defined. Furthermore, $X^*$ acting on cylinder functions is given by:*

$$X^* = -X + z^h,$$

*where $z^h$ is a certain function on $W(M)$ constructed from the curvature tensor, the torsion tensor, and the parallel translation operator $P_s^\nabla$. (See Theorem 9.1 of Driver [4] for the explicit formula for $z^h$.)*

This integration by parts theorem is a Corollary of the previous theorem, see [4]. It is also possible to prove and extend Theorem 3.8 by using ideas of Bismut [2]; see Leandre [15], and Fang and Malliavin [9].

**Definition 3.9** The *gradient operator* $D^\nabla : \mathcal{F} \to L^2(\nu, H)$ is defined by

$$D^\nabla f \equiv \sum_{h \in S} X^h f \otimes h, \quad f \in \mathcal{F},$$

where $S \subset H$ is an orthonormal basis for $H$.

Using Theorem 3.8, it is easy to see that $D^\nabla$ is a closable operator. We will continue to denote the closure by $D^\nabla$. Associated to $D^\nabla$ is the closed quadratic form $\mathcal{E}^\nabla(\cdot, \cdot)$ on $L^2(\nu)$ defined by

$$\mathcal{E}^\nabla(f, g) \equiv \int_{W(M)} (D^\nabla f, D^\nabla g) d\nu \ \forall f, g \in \mathcal{D}(D^\nabla), \tag{3.4}$$

where $\mathcal{D}(D^\nabla)$ is the domain of $D^\nabla$. The following theorem is from Driver and Röckner [7].

**Theorem 3.10** *The form $\mathcal{E}^\nabla$ is a local quasi-regular Dirichlet form and hence there exists an associated diffusion process on $W(M)$. (This process is the Ornstein Uhlenbeck process when $M = \mathbb{R}^d$.)*

## 4. Non-Comparability of the Pull-Back Form

Using the stochastic development map $\Psi^\nabla$ and its inverse $b^\nabla$ it is possible to "pull-back" Dirichlet forms on $W(M)$ to Dirichlet forms on $W$. It is natural to compare these pulled back forms with the "usual" Dirichlet form $Q$ on $W$.

**Definition 4.1** The *usual Dirichlet* form on $W$ is the closed symmetric quadratic form $Q$ on $L^2(W, \mu)$ determined by

$$Q(F, F) = \int_W (DF, DF)_H d\mu, \tag{4.1}$$

where $D : L^2(W, \mu) \to L^2(W, H)$ is the closed operator determined on cylinder functions $(F : W \to \mathbb{R})$ by

$$DF \equiv \sum_{h \in S} \partial_h F \otimes h.$$

Here $\partial_h F(\omega) = L^2 - \lim_{t \to 0} (F(\omega + th) - F(\omega)/t)$ and $S \subset H$ is an orthonormal basis of $H$.

**Definition 4.2** Let $\mathcal{E}^\nabla$ be as in (3.4), the *pull-back* of $\mathcal{E}^\nabla$ by the development map $\Psi^\nabla$ is the symmetric quadratic form $Q^\nabla$ on $L^2(W, \mu)$ determined by:

$$Q^\nabla(F, F) \equiv \mathcal{E}^\nabla(F \circ b^\nabla, F \circ b^\nabla) \ \forall F \in \mathcal{D}(Q^\nabla),$$

where

$$\mathcal{D}(Q^\nabla) \equiv \{F \in L^2(W, \mu) : F \circ b^\nabla \in \mathcal{D}(\mathcal{E}^\nabla) = \mathcal{D}(D^\nabla)\}.$$

(Recall that $b^\nabla \equiv (\Psi^\nabla)^{-1}$.)

**Definition 4.3** Let $Q$ and $Q'$ be two closed symmetric non-negative quadratic forms on $L^2(W, \mu)$. $Q$ and $Q'$ are said to be *comparable* if $\mathcal{D}(Q) = \mathcal{D}(Q')$ and there is a constant $0 < C < \infty$ such that

$$C^{-1}Q(f, f) \le Q'(f, f) \le CQ(f, f) \ \forall f \in \mathcal{D}(Q) = \mathcal{D}(Q').$$

**Question 1** Are $Q$ and $Q^\nabla$ comparable?

**Question 2** Given two covariant derivative $\nabla^{(1)}$ and $\nabla^{(2)}$ on $M$, are the Dirichlet forms $\mathcal{E}^1 \equiv \mathcal{E}^{\nabla^{(1)}}$ and $\mathcal{E}^2 \equiv \mathcal{E}^{\nabla^{(2)}}$ comparable?

We have the following negative answer in the case that $M = G = SO(n)$, $\nabla^{(1)} = \nabla^L$ and $\nabla^{(2)} = \nabla^R$.

**Theorem 4.4** *Let $M = SO(n)$ (more generally a compact Lie group), $Q^L \equiv Q^{\nabla^L}$, $Q^R \equiv Q^{\nabla^R}$, $\mathcal{E}^L \equiv \mathcal{E}^{\nabla^L}$, and $\mathcal{E}^R \equiv \mathcal{E}^{\nabla^R}$. Then*

*1. $\sup_F Q(F)/Q^L(F) = \infty$ and $\sup_F Q^L(F)/Q(F) = \infty$.*
*2. $\sup_F Q(F)/Q^R(F) = \infty$ and $\sup_F Q^R(F)/Q(F) = \infty$.*
*3. $\sup_f \mathcal{E}^R(f)/\mathcal{E}^L(f) = \infty$ and $\sup_f \mathcal{E}^L(f)/\mathcal{E}^R(f) = \infty$.*

*In the above expressions, the supremum is taken over the intersection of the domain of the quadratic form in the numerator with that in the denominator.*

For a proof and a more detailed statement of the above theorem, see [6]. I conjecture that analogues of the above result hold for arbitrary Riemannian manifolds ($M$) with non-trivial geometry.

## 5. The (Non) Equivalence of the Forms

Because of the above theorem, it is natural to look for a weaker notion for the equivalence of quadratic forms.

**Definition 5.1** Let $(W, \mu)$ and $(W, \mu')$ be two probability spaces and let $Q$ and $Q'$ be closed symmetric non-negative forms on $L^2(\mu)$ and $L^2(\mu')$, respectively. Then $Q$ and $Q'$ are said to be *equivalent* if there exists measurable maps $\phi : W \to W'$ and $\Psi : W' \to W$ such that:

1. $\Psi \circ \phi = id_W$ $\mu$–a.s. and $\phi \circ \Psi = id_{W'}$ $\mu'$–a.s.
2. $\mathcal{D}(Q) = \{f \in L^2(\mu) | f \circ \Psi \in \mathcal{D}(Q')\}$ and $\mathcal{D}(Q') = \{f \in L^2(\mu') | f \circ \phi \in \mathcal{D}(Q)\}$.
3. There is a constant $0 < C < \infty$ such that

$$C^{-1} Q'(f \circ \Psi, f \circ \Psi) \leq Q(f, f) \leq C Q'(f \circ \Psi, f \circ \Psi) \; \forall f \in \mathcal{D}(Q)$$

and

$$C^{-1} Q(f \circ \phi, f \circ \phi) \leq Q'(f, f) \leq C Q(f \circ \phi, f \circ \phi) \; \forall f \in \mathcal{D}(Q').$$

We now modify questions 1 and 2 above.

**Question 1'** Is $Q$ equivalent to $\mathcal{E}^\nabla$ or, equivalently, is $Q$ equivalent to $Q^\nabla$?

**Question 2′** Let $\nabla^{(1)}$ and $\nabla^{(2)}$ be two covariant derivatives on $M$. Are the Dirichlet forms $\mathcal{E}^1 \equiv \mathcal{E}^{\nabla^{(1)}}$ and $\mathcal{E}^2 \equiv \mathcal{E}^{\nabla^{(2)}}$ equivalent?

The answer to both questions for $M = SO(n)$ (or more generally a compact Lie group) with $\nabla = \nabla^L$ or $\nabla = \nabla^R$, $\nabla^{(1)} = \nabla^L$ and $\nabla^{(2)} = \nabla^R$ is now yes. The following theorem is essentially Theorem 3.14 in Gross [12].

**Theorem 5.2** *Suppose that $M = G$ is a compact Lie group (ex. $G = SO(n)$), $Q^L \equiv Q^{\nabla^L}$, $Q^R \equiv Q^{\nabla^R}$, $\mathcal{E}^L \equiv \mathcal{E}^{\nabla^L}$, and $\mathcal{E}^R \equiv \mathcal{E}^{\nabla^R}$. Then $Q$, $Q^L$, $Q^R$, $\mathcal{E}^L$, and $\mathcal{E}^R$ are all equivalent. Moreover:*

$$\mathcal{E}^L(F \circ b^R, F \circ b^R) = Q(F, F) = \mathcal{E}^R(F \circ b^L, F \circ b^L),$$

*where $b^L \equiv b^{\nabla^L}$ and $b^R \equiv b^{\nabla^R}$.*

Unfortunately, as we will discuss below, the results of Theorem 4.4 seem to be essentially restricted to the case that $G$ is a compact Lie group. To investigate this lack of equivalence of forms, let us attempt to find an invertible measurable map $b : W(M) \to W$ such that (i) $b_*\nu \equiv \nu \circ b^{-1}$ is equivalent to $\mu$, (ii) $b$ is adapted, and (iii) $Q(f, f)$ is comparable to $\mathcal{E}^\nabla(f \circ b, f \circ b)$ for all $f \in \mathcal{D}(Q)$. (Note: it would be preferable to drop condition (ii) above, but at the present time I do not know how to handle the anticipatory case.) With the above assumptions, $b = \phi \circ b^\nabla$ for some adapted map $\phi : W \to W$ such that $\phi_*\mu$ and $\mu$ are equivalent. The following "structure" theorem is proved in Driver [5], see Theorem 2.1.

**Theorem 5.3 (Structure Theorem)** *Let $\phi : W \to W$ be an adapted map such that $\phi_*\mu$ is equivalent to $\mu$ and there is an adapted map $\phi^{-1} : W \to W$ such that $\phi \circ \phi^{-1}$ and $\phi^{-1} \circ \phi$ are both equal to the identity map $\mu-a.s.$ Then there exists an $O(d) \times \mathbb{R}^d -$valued predictable process $(\bar{O}, \bar{\alpha})$ on $W$ such that*

$$\phi(\omega) = \int \bar{O}(\omega)d\omega + \int \bar{\alpha}(\omega)ds, \tag{5.1}$$

*and $\int_0^1 |\bar{\alpha}_{s'}|^2 ds' < \infty \;\mu-a.s.$*

Because of this theorem, we learn that $b = \phi \circ b^\nabla$ may be written as

$$b = \int O db^\nabla + \int \alpha ds, \tag{5.2}$$

where $O \equiv \bar{O} \circ b^\nabla$ and $\alpha \equiv \bar{\alpha} \circ b^\nabla$. So our goal is to choose a predictable $O(d) \times \mathbb{R}^d -$valued process $(O, \alpha)$ on $W(M)$ such that $Q(f, f)$ is comparable to $\mathcal{E}^\nabla(f \circ b, f \circ b)$.

**Notation 5.4**  For $a, b \in T_o M$ and an isometry $u : T_o M \to T_m M$, let

$$\Omega_u \langle a, b \rangle \equiv u^{-1} R^\nabla \langle ua, ub \rangle u \in End(T_o M)$$

and

$$\Theta_u \langle a, b \rangle \equiv u^{-1} T^\nabla \langle ua, ub \rangle \in T_o M.$$

**Lemma 5.5**  *Let $h \in H$, $X^h$ be as in (3.3), then*

$$X^h b^\nabla \equiv \left. \frac{d}{dt} \right|_0 b^\nabla \circ e^{tX^h} = \int C^h \delta b^\nabla + h,$$

*where*

$$C_s^h \equiv \int_0^s \Omega_{u(s')} \langle h(s'), \delta b^\nabla(s') \rangle + \Theta_{u(s')} \langle h(s'), \cdot \rangle,$$

*and $u(s) \equiv P_s^\nabla$.*

**Proof**  See Theorem 5.1 in [4].  Q.E.D.

Assuming sufficient regularity on $b$ (i.e. on the "kernels" $O$ and $\alpha$) it is possible to compute $X^h b$ as:

$$
\begin{aligned}
X^h b &\equiv \left. \frac{d}{dt} \right|_0 b \circ e^{tX^h} \\
&= \int (X^h O) db^\nabla + \int O d(X^h b^\nabla) + \int (X^h \alpha) ds \\
&= \int (X^h O + O C^h) db^\nabla \quad \textbf{mod } H, \\
&= \int (X^h O + O C^h) O^{-1} db \quad \textbf{mod } H, \\
&= A(b)h + N(b)h,
\end{aligned}
$$

where

$$A(b)h \equiv \int (X^h O + O C^h) O^{-1} db \tag{5.3}$$

and $N(b) : H \to H$ is a bounded random linear operator on $H$. Working somewhat informally we have

$$\mathcal{E}^\nabla(f \circ b, f \circ b) = \sum_{h \in S} \int_{W(M)} |X^h(f \circ b)|^2 d\nu$$

$$= \sum_{h \in S} \int_{W(M)} (Df \circ b, X^h b)_H^2 d\nu$$

$$= \sum_{h \in S} \int_{W(M)} (Df \circ b, (A(b) + N(b))h)_H^2 d\nu$$

$$= \int_{W(M)} |(A(b) + N(b))^* Df \circ b|_H^2 d\nu.$$

If this last expression is to be finite for all $f \in \mathcal{D}(Q)$, it seems likely that $A(b)^*$ and hence $A(b)$ must be a bounded linear operator on $H$. But this is only the case when $A(b) \equiv 0$, since otherwise $A(b)h$ is expressed as a stochastic integral relative to $b$. Since $b$ is a Brownian motion on $(W(M), \tilde{\nu})$ where $\tilde{\nu}$ is a measure which is equivalent to $\nu$, it follows that $A(b)h$ is not in $H$ a.s. except when $A(b)h \equiv 0$.

For the reasons described above, we will try to choose $(O, \alpha)$ such that $A(b) \equiv 0$. By (5.3) $A(b) \equiv 0$ is equivalent to:

$$X^h O + O C^h \equiv 0 \ \forall h \in H. \tag{5.4}$$

In general a system of first order partial differential equations as in (5.4) will not have a solution unless the Frobenius integrability condition is satisfied. This condition is determined by formally requiring the equality of mixed partial derivatives of a supposed solution to (5.4). The next theorem describes the integrability conditions for (5.4). To avoid technical difficulties, I will state the theorem with $W(M)$ replaced by the Hilbert manifold $H(M)$ of finite energy paths in $W(M)$.

**Theorem 5.6** *The differential equations in* (5.4) *are locally solvable on* $H(M)$ *iff* $R^{\tilde{\nabla}} \equiv 0$, *where* $R^{\tilde{\nabla}}$ *is the curvature tensor of the covariant derivative* $\tilde{\nabla}$ *on* $M$ *defined by*

$$\tilde{\nabla}_X Y = \nabla_X Y - T^\nabla \langle X, Y \rangle. \tag{5.5}$$

*Moreover, if* $R^{\tilde{\nabla}} \equiv 0$ *and* $b \equiv b^{\tilde{\nabla}}$ *then* $\mathcal{E}^\nabla(f \circ b, f \circ b) = Q(f, f)$ *for all* $f \in \mathcal{D}(Q)$.

**Remark 5.7** In the case that $M = G = SO(n)$ as in Example 2.1, then $\tilde{\nabla}^L = \nabla^R$, $\tilde{\nabla}^R = \nabla^L$, and $R^{\tilde{\nabla}^L} = R^{\nabla^R} \equiv 0$ and $R^{\tilde{\nabla}^R} = R^{\nabla^L} \equiv 0$. Thus Theorem 5.6 explains "why" Theorem 5.2 is valid.

**Proof** (I will sketch only the proof of the first statement in Theorem 5.6. A detailed proof will appear in future work.) Let $O(M)$ be the orthogonal frame bundle over $M$. (The fiber $O_m(M)$ of $O(M)$ over $m$ is the set of isometries $u$ from $T_oM$ to $T_mM$.) Let $\tilde{\omega}$ be the connection 1-form on $O(M)$ corresponding to $\tilde{\nabla}$. More explicitly, for any path $u(s) \in O(M)$,

$$\tilde{\omega}\langle u'(s)\rangle \equiv u(s)^{-1}\tilde{\nabla}u(s)/ds.$$

Also let $C\langle\cdot\rangle$ denote the $\mathrm{L}^2([0,1], so(n))$ – valued 1-form on $H^1(M)$ determined by $C\langle X^h\rangle = C^h$ for all $h \in H$, where $so(n) \equiv T_I SO(n)$ is the set of all real $n \times n$ skew symmetric matrices. (Recall that $C^h$ was defined in Lemma 5.5.)

We now have the following facts:

1. $C = -(P^\nabla)^*\tilde{\omega}$, where $(P^\nabla)^*\tilde{\omega}$ denotes the differential form on $H^1(M)$ found by "pulling-back" $\tilde{\omega}$ by $P^\nabla$.

2. The integrability condition for (5.4) is equivalent to the statement that $-C$ has "zero curvature." That is

$$d(-C) + (-C) \wedge (-C) \equiv 0$$

Using these two facts, we learn that the integrability condition for (5.4) is equivalent to

$$0 = (P^\nabla)^*(d\tilde{\omega} + \tilde{\omega} \wedge \tilde{\omega}) \equiv \tilde{\Omega}\langle P^\nabla_*\cdot, P^\nabla_*\cdot\rangle, \tag{5.6}$$

where $\tilde{\Omega} \equiv d\tilde{\omega} + \tilde{\omega} \wedge \tilde{\omega}$ is the equivariant form of the curvature tensor $R^{\tilde{\nabla}}$. Let $\pi : O(M) \to M$ denote the natural projection map and $e_s : W(M) \to M$ be the evaluation map: $e_s(\sigma) = \sigma(s)$. Since $\pi \circ P^\nabla_s = e_s$, it easily follows that $(P^\nabla_s)_*$ maps $T_\sigma H^1(M)$ onto the "horizontal" vectors in $T_{P^\nabla_s(\sigma)}O(M)$. Since $\tilde{\Omega}$ annihilates "vertical" vectors it follows from (5.6) that $\tilde{\Omega} \equiv 0$ or equivalently $R^{\tilde{\nabla}} \equiv 0$.  Q.E.D.

The integrability condition in Theorem 5.6 is very restrictive as the next theorem indicates. For the statement of the theorem it is necessary to recall the notion of a Killing vector field.

**Definition 5.8** A *Killing vector field* on $M$ is a vector field $X$ on $M$ such that the flow $e^{tX}$ preserves the Riemannian metric $\langle\cdot,\cdot\rangle$. That is $\langle e^{tX}_* v, e^{tX}_* v\rangle = \langle v, v\rangle$ for all $v \in TM$.

**Theorem 5.9** *Assume that $M$ is a simply connected Riemannian manifold with Riemannian metric $\langle\cdot,\cdot\rangle$. Then there exists a covariant derivative $\nabla$ on $TM$ such* *that $R^{\tilde{\nabla}} \equiv 0$ ($\tilde{\nabla}$ as in (5.5)) iff there exists a global orthonormal frame of Killing vector fields $\{X_i\}_{i=1}^d$ on M. In particular, M is a homogeneous space with a trivial tangent bundle.*

**Example 5.10** Suppose that $G = SO(n)$ and $\{A_i\}_{i=1}^d$ ($d = n(n-1)/2$) is an orthonormal basis for $so(n) \equiv T_I SO(n)$. Set $X_i^L(g) \equiv gA_i$ and $X_i^R(g) \equiv A_i g$ for all $g \in G$. Then $\{X_i^L\}_{i=1}^d$ and $\{X_i^R\}_{i=1}^d$ are two orthonormal frames of Killing vector fields on $G$.

This example and its minor generalization to Lie groups of compact type are the only examples that I know of manifolds which admit an orthonormal basis of Killing vector-fields. It is well known that the generic Riemannian manifold does not admit **any** Killing vector fields, see Bochner [3]. In conclusion, it seems that the Dirichlet forms constructed in (3.4) are typically non-equivalent to the usual Dirichlet form $Q$ on $W$ (at least if the map $b : W(M) \to W$ is required to be adapted.)

**Acknowledgements** This research was partially supported by NSF Grant No. DMS 92-23177.

# References


[1] Albeverio, S. and Hoegh-Krohn, R.: The energy representation of Sobolev Lie groups; *Compositio Math.* **36** (1978) 37–52

[2] Bismut, J. -M.: *Large Deviations and the Malliavin Calculus.* Birkhäuser, Boston· Basel·Stuttgart, 1984

[3] Bochner, S.: Vector fields and Ricci curvature; *Bull. of A.M.S.* **Series 2, 52** (1946) 776–797

[4] Driver, B. K.: A Cameron-Martin type quasi-invariance theorem for Brownian motion on a compact Riemannian manifold; *J. of Funct. Anal.* **110** (1992) 272–376

[5] Driver, B. K.: Towards calculus and geometry on path spaces; to appear in the proceeding of the Summer Research Institute on Stochastic Analysis, held at Cornell Univ. 1993

[6] Driver, B. K.: The non-comparability of "left" and "right" Dirichlet forms on the Wiener space of a compact Lie group; in preparation

[7] Driver, B. K. and Röckner, M.: Construction of diffusions on path and loop spaces of compact Riemannian manifolds; *C. R. Acad. of Sci. Paris* **t.315, Série I** (1992) 603–608

[8] Eells, J. and Elworthy, K. D.; Wiener integration on certain manifolds; in: *Problems in Non-Linear Analysis*, G. Prodi. (ed.) 67–94. Centro Internazionale Matematico Estivo, IV Ciclo. Tome: Edizioni Cremonese (1971)

[9] Fang, S. and Malliavin, P.; Stochastic calculus on Riemannian manifolds; *J. of Funct. Anal.*, (1993) to appear.

[10] Frenkel, I. B.: Orbital theory for affine Lie algebras; *Invent. Math.* **77** (1984) 301–352

[11] Getzler, E.: Dirichlet forms on loop space; *Bull. Sc. Math.* **2 serie, 113** (1989) 151–174

[12] Gross, L.: Uniqueness of ground states for Schrödinger operators over loop groups; *J. of Funct. Anal.* **112** (1993) 373–441

[13] Hsu, E. P.: Quasiinvariance of the Wiener measure and integration by parts in the path space over a compact Riemannian manifold; *Preprint* (1992)

[14] Hsu, E. P.: Gradient flows on path spaces; talk given at the Summer Research Institute on Stochastic Analysis, 1993

[15] Leandre, R.: Integration by parts formulas and rotationally invariant Sobolev calculus on free loop spaces; *Preprint* (1992)

[16] Malliavin, P.: *Geometrie Differentielle Stochastique.* Pressesde l' Universite de Montreal, 1978

[17] Malliavin, P.: Stochastic calculus of variation and hypoelliptic operators; Proc. Int. Symp. S.D.E. Kyoto, (1976) 195–264. Wiley and Sons, New York, 1978

[18] Malliavin, P.: Naturality of quasi-invariance of some measures; in:*Proc. of the Lisbonne Conference*, A. B. Cruzerio (ed.), Birkhäuser, 1991

[19] Malliavin, M. -P. and Malliavin, P.: Integration on loop groups. I. Quasi invariant measures, Quasi invariant integration on loop groups; *J. of Funct. Anal.* **93** (1990) 207–237

[20] Shigekawa, I.: Transformations of the Brownian motion on the Lie group; in *Stochastic Analysis: proceedings of the Taniguchi International Symposium on Stochastic Analysis*, 409–422, K. Itô (ed.), North-Holland, Amsterdam/New York, 1984

[21] Shigekawa, I.: Transformations of the Brownian motion on the Riemannian symmetric space; *Z. Wahr. verw. Geb.* **65** (1984) 493–522

Bruce K. Driver
Department of Mathematics
University of California, San Diego
9500 Gilman Drive, Dept. 0112
La Jolla, CA 92093-0112, U.S.A.
*E-mail*: driver@euclid.ucsd.edu

T. FUNAKI

# Low temperature limit and separation of phases for Ginzburg-Landau stochastic equation

We investigate a problem of singular perturbation for a stochastic partial differential equation (SPDE) of Ginzburg-Landau type on $\mathbf{R}$. After the motivation from physics is explained at rather heuristic level in Section 1, the result is formulated precisely in Section 2. Then, the outline of the proof is given in Section 3. A complete proof can be found in [13]. Some related known results for both non-random and stochastic models are briefly reviewed in Section 4.

## 1. Problem and Motivation

Motivated by the investigation related to the so-called kinetic theory of phase transitions, we discuss a problem of singular perturbation for an SPDE of Ginzburg-Landau type:

$$\frac{\partial u}{\partial t} = \Delta u - \epsilon^{-1}U'(u) + \epsilon^{\gamma}a(x)\dot{w}_t(x), \quad t > 0, \; x \in \mathbf{R}, \qquad (1.1)$$

where $\Delta = d^2/dx^2$, $U(u) = (u^2 - 1)^2/4$ is a potential of double-well type, $a(x) \in C_0^2(\mathbf{R})$, $\gamma > 19/4$ is a constant and $\dot{w}_t(x)$ is a 2-parameter white noise. The (boundary) conditions $u(\pm\infty) = \pm 1$ at infinity are imposed on (1.1). Our main interest is to study the asymptotic behavior of the solution $u = u^{\epsilon}(t, x)$ of (1.1) as the parameter $\epsilon$, which indicates the temperature of the system, tends to 0. It is shown that after a proper scaling in time a phase separation point $\xi_t$ is formed for each time $t$ in the sense that $u^{\epsilon}(t, x)$ converges as $\epsilon \downarrow 0$ to $+1$ for $x > \xi_t$ and $-1$ for $x < \xi_t$; notice that $\pm 1$ are two bottoms (in other words, two stable phases) of the potential $U$ of equal depth. A stochastic differential equation (SDE) which describes the random motion of $\xi_t$ is also derived. Before formulating the result precisely, we explain in this section the motivation of this problem from a rather heuristic point of view.

(a) *static theory*: Let a configuration $u = \{u(x) \in \mathbf{R}, x \in \mathbf{R}^d\}$ of scalar field over $\mathbf{R}^d$ be given and let an energy $\mathcal{H}^{\epsilon}(u)$ be associated with $u$ by

$$\mathcal{H}^{\epsilon}(u) = \int_{\mathbf{R}^d} \left\{ \frac{1}{2}|\nabla u|^2(x) + \epsilon^{-1}U(u(x)) \right\} dx, \quad \nabla u = \left( \frac{\partial u}{\partial x_1}, \cdots, \frac{\partial u}{\partial x_d} \right). \qquad (1.2)$$

88

Then, the corresponding statistical mechanical system can be determined by the so-called Gibbs state which is described by a formal probability measure

$$\mu^\epsilon(du) = Z_\epsilon^{-1} e^{-\mathcal{H}^\epsilon(u)} \mathcal{D}(u). \tag{1.3}$$

Here, $\mathcal{D}(u) = \Pi_{x \in \mathbf{R}^d} du(x)$ is a "flat measure" (Feynman measure) on the space of all configurations and $Z_\epsilon$ is a "normalizing constant". Now, let the temperature $\epsilon$ tend to 0. Then, since $\mathcal{H}^\epsilon(u)$ diverges for such $u$ that $u(x) \neq \pm 1$ for some $x$, one would expect that $u(x)$ converges to $+1$ or $-1$ for each $x \in \mathbf{R}^d$ under the distribution $\mu^\epsilon$. We denote $x = (x_1, \underline{x}) \in \mathbf{R} \times \mathbf{R}^{d-1}$ and assume $\lim_{x_1 \to \pm\infty} u(x) = \pm 1$. Then, the phase separation hypersurface (interface) $\{x_1 = f(\underline{x})\}$ would be formed in such a way that $u(x)$ converges to $+1$ if $x_1 > f(\underline{x})$ and $-1$ if $x_1 < f(\underline{x})$. Since $f(\underline{x})$ is still random, the main interest is to determine the limit distribution of $f(\underline{x})$ under $\mu^\epsilon$. Diehl-Kroll-Wagner [8] gave an answer by computing an effective Hamiltonian $\mathcal{H}_{DH}(f)$ for $f(\underline{x})$. They integrated out the fluctuation by applying perturbation technique and finally derived

$$\mathcal{H}_{DH}(f) = \text{const} \int_{\mathbf{R}^{d-1}} \{1 + |\nabla_{\underline{x}} f|^2\}^{1/2} \, d\underline{x}. \tag{1.4}$$

Especially, the limit distribution of $f(\underline{x})$ would be given by the Gibbs state associated with $\mathcal{H}_{DH}(f)$. Notice that the integral in the right hand side of (1.4) gives the total area of the interface.

(b) *kinetic theory*: The corresponding kinetic theory was discussed by Kawasaki-Ohta [17]. One can introduce the Ginzburg-Landau stochastic equation:

$$\frac{\partial u}{\partial t} = -D\mathcal{H}^\epsilon(x, u) + \dot{w}_t(x), \quad t > 0, \; x \in \mathbf{R}^d, \tag{1.5}$$

as a dynamics associated with the Hamiltonian $\mathcal{H}^\epsilon$, where $D\mathcal{H}^\epsilon(x, u) = -\Delta u + \epsilon^{-1} U'(u(x))$ denotes the functional derivative of $\mathcal{H}^\epsilon$, $\Delta u = \sum_i \partial^2 u / \partial x_i^2$ and $\dot{w}_t(x)$ is the space-time white noise. In fact, at least formally, the equation (1.5) has the Gibbs state $\mu^\epsilon$ as its invariant measure (equilibrium state); mathematical theory for the relationship between the SPDE's of the type of (1.5) and the Gibbs states are developed in some detail in [12]. The first term in the right hand side of (1.5) determines a gradient flow which forces the solution $u$ toward lower energy level, while the second term provides a fluctuation around it. Kawasaki-Ohta studied the asymptotic behavior of the solution $u = u^\epsilon(t, x)$ of (1.5) as $\epsilon \downarrow 0$. They suggest that the interface between two phases $\pm 1$ is formed in the limit and moves governed by a stochastic equation with drift proportional to the mean curvature having additional fluctuation term. This conjecture is quite natural, since the functional derivative $-D\mathcal{H}_{DH}(\underline{x}, f)$ of $-\mathcal{H}_{DH}(f)$ is proportional to the mean curvature of the surface $\{x_1 = f(\underline{x})\}$ at this point.

Unfortunately, it is impossible to derive randomized mean curvature motion from (1.5), since this SPDE has mathematical meaning only when the spatial dimension $d = 1$ and in this case curvature plays no rule. Instead, we are interested in the form of the fluctuation term appearing in the limit stochastic equation. In addition, we multiply the noise by the factor $\epsilon^\gamma$ as in the SPDE (1.1), which makes the drift term dominant the noise term; the corresponding Hamiltonian $\mathcal{H}^\epsilon(u)$ is slightly modified into $\epsilon^{-2\gamma}\mathcal{H}^\epsilon(u)$ by this change. We consider general situation that the strength $a(x)$ of the noise varies with $x$, although we need to assume from a technical reason that the noise disappears outside of a certain compact region.

## 2. Result

A function $m$ defined by $m(y) = \tanh(y/\sqrt{2}), y \in \mathbf{R}$, is a critical point of the functional $\mathcal{H}^1(u)$ (i.e., $\mathcal{H}^\epsilon(u)$ with $\epsilon = 1$) satisfying the conditions $m(\pm\infty) = \pm 1$ and increasing in $y$. In fact, one can easily see that $m$ satisfies an ordinary differential equation $\Delta m = U'(m)$. We set $m_\xi^\epsilon(x) = m(\epsilon^{-1/2}(x - \xi)), x, \xi \in \mathbf{R}$. Then, $\{m_\xi^\epsilon\}_{\xi \in \mathbf{R}}$ are all critical points of $\mathcal{H}^\epsilon(u)$. Let us introduce the scaling in time to the solution $u = u^\epsilon(t, x)$ of the SPDE (1.1) by setting

$$\bar{u}^\epsilon(t, x) = u^\epsilon(\epsilon^{-1/2-2\gamma}t, x). \tag{2.1}$$

We assume $u^\epsilon(0, x) = m_\xi^\epsilon(x)$ for some $\xi \in \mathbf{R}$ for the initial data.

**Theorem** (i) *There exists an $\mathbf{R}$-valued process $\xi_t^\epsilon(\omega)$ such that*

$$\lim_{\epsilon \downarrow 0} P\left\{ \sup_{0 \le t \le T} \|\bar{u}^\epsilon(t, \cdot) - \chi_{\xi_t^\epsilon}\|_{L^2} > \delta \right\} = 0 \tag{2.2}$$

*for every $\delta > 0$ and $T > 0$, where $L^2 = L^2(\mathbf{R})$ and $\chi_\xi(x) = +1$ for $x > \xi$ and $-1$ for $x < \xi$.*

(ii) *The distribution on the space $C([0, T], \mathbf{R})$ of $\xi_t^\epsilon$ converges weakly to that of a solution $\xi_t$ of the following SDE starting at $\xi$ as $\epsilon \downarrow 0$.*

$$d\xi_t = \alpha_1 a(\xi_t)dB_t + \alpha_2 a(\xi_t)a'(\xi_t)dt, \tag{2.3}$$

*where $B_t$ is a standard Brownian motion,*

$$\begin{cases} \alpha_1 = \dfrac{1}{||\nabla m||_{L^2}} \left( = (9/8)^{1/4} \right), \qquad \nabla m = dm/dy, \\[4mm] \alpha_2 = \dfrac{1}{||\nabla m||_{L^2}^2} \displaystyle\int_0^\infty dt \int_{\mathbf{R}^2} y p_t(y,z)^2 U'''(m(z)) \nabla m(z) \, dy dz, \end{cases} \tag{2.4}$$

*and $p_t(y,z)$ denotes the fundamental solution of the PDE $\partial u/\partial t = \Delta u - U''(m)u$.*

The reason why (2.1) gives the proper time-scaling can be explained intuitively as follows: The drift term of the SPDE (1.1) is multiplied by $\epsilon^{-1/2-2\gamma}$ after this scaling, while the noise term is multiplied by $(\epsilon^{-1/2-2\gamma})^{1/2}$ so that it becomes $\epsilon^{-1/4} a(x) \dot{w}_t(x)$ in the sense of distribution. Then, the drift pushes the solution $\bar{u}$ strongly toward the set of minimizers $\{m_\xi^\epsilon\}$ of $\mathcal{H}^\epsilon(u)$ and only the fluctuations of $\xi$-variable survive in the limit. Since the typical width of the interface in the configuration $m_\xi^\epsilon$ is of order $\epsilon^{1/2}$, only the part of the white noise coming from this region becomes effective in the limit and consequently the noise term $\epsilon^{-1/4} a(x) \dot{w}_t(x)$ becomes a quantity of order 1.

**Remark** The assertion (2.2) holds for $||\bar{u}^\epsilon(t,\cdot) - m_{\xi_t^\epsilon}^\epsilon||_{L^\infty(\mathbf{R})}$ in place of $||\bar{u}^\epsilon(t,\cdot) - \chi_{\xi_t^\epsilon}||_{L^2}$.

## 3. Outline of the Proof

It is convenient to introduce a further spatial scaling to the SPDE (1.1):

$$v \equiv v^\epsilon(t,y) := \bar{u}^\epsilon(t, \epsilon^{1/2}y), \quad t > 0, \ y \in \mathbf{R}.$$

Notice that $x$ represents macroscopic spatial variable, while $y = \epsilon^{-1/2}x$ represents microscopic spatial variable. Then, since $\dot{w}_{\epsilon^{-1/2-2\gamma}t}(\epsilon^{1/2}y) = \epsilon^{-1/2-\gamma}\dot{w}_t(y)$ in the sense of distribution, $v$ satisfies an SPDE

$$\frac{\partial v}{\partial t} = \epsilon^{-3/2-2\gamma}\{\Delta v - U'(v)\} + \epsilon^{-1/2} a(\epsilon^{1/2}y)\dot{w}_t(y), \quad t > 0, \ y \in \mathbf{R}. \tag{3.1}$$

Note that the term inside curly brackets is the functional derivative of $-\mathcal{H}(v)$ defined by

$$\mathcal{H}(v) := \mathcal{H}^1(v) - \frac{2\sqrt{2}}{3} = \int_{\mathbf{R}} \left\{ \frac{1}{2}|\nabla v|^2(y) + U(v(y)) \right\} dy - \frac{2\sqrt{2}}{3}. \tag{3.2}$$

The last constant is subtracted to normalize $\mathcal{H}^1$ as $\inf_{v:v(\pm\infty)=\pm 1} \mathcal{H}(v) = 0$. The infimum is attained on the set $M = \{m_\eta; \eta \in \mathbf{R}\}$, where $m_\eta(y) = m(y - \eta)$. The proof of Theorem consists of two main steps.

*Step 1. Lyapunov type argument.* Here we prove that the solution $v = v_t^\epsilon$ of (3.1) stays in a neighborhood of $M$ with high probability as $\epsilon \downarrow 0$ by means of the strong drift $-\epsilon^{-3/2-2\gamma} D\mathcal{H}(\cdot, v)$ toward $M$.

For this purpose, we first introduce a nice coordinate in a neighborhood of $M$. We denote $\mathcal{X} := L^2 + m = \{v = s + m; s \in L^2\}$ and $\mathrm{dist}(v, M) := \inf_{\eta \in \mathbf{R}} \|v - m_\eta\|_{L^2}$ for $v \in \mathcal{X}$. Then, the infimum is attained at unique $\eta$ if $v$ is sufficiently close to $M$; more precisely, if $v \in \mathcal{X}$ satisfies $\mathrm{dist}(v, M) \leq \beta_1$ for some $\beta_1 > 0$. We denote $\eta$ by $\eta(v)$ and $s(v) := v - m_{\eta(v)}$. The pair $(\eta(v), s(v)) \in \mathbf{R} \times L^2$ is sometimes called by the Fermi coordinates of $v$ and satisfies $\langle s(v), \nabla m_{\eta(v)} \rangle = 0$, where $\langle \cdot, \cdot \rangle$ denotes the inner product of the space $L^2 = L^2(\mathbf{R})$.

Secondly, we notice that the Hessian of $\mathcal{H}$ is non-degenerate on $M$ toward the normal direction to $M$. Namely, the second functional derivative $D^2\mathcal{H}$ at $m \in M$ given by

$$D^2\mathcal{H}(m) = -\Delta + U''(m), \tag{3.3}$$

is a self-adjoint operator on $L^2$ and its spectrum consists of $\{0, 3/2\} \cup [2, \infty)$. Two eigenvalues $0$ and $3/2$ are both simple and the eigenspace corresponding to $0$ is spanned by $\nabla m$, which can be regarded as a tangent vector to $M$ at $m$.

This property of $D^2\mathcal{H}(m)$ combined with the translation-invariance of $\mathcal{H}$ proves the following two bounds with the help of Sobolev's imbedding theorem and Gårding type inequality:

$$\begin{cases} c_1 \|s(v)\|_{H^1}^2 \leq \mathcal{H}(v) \leq c_2 \|s(v)\|_{H^1}^2, \\ c_1 \|s(v)\|_{H^2}^2 \leq \|D\mathcal{H}(\cdot, v)\|_{L^2}^2 \leq c_2 \|s(v)\|_{H^2}^2, \end{cases} \tag{3.4}$$

for all $v \in H^2 + m$ satisfying $\|s(v)\|_{H^1} \leq \beta_2$ for some $c_1, c_2, \beta_2 > 0$, where $H^n = H^n(\mathbf{R}), n = 1, 2$, stand for the Sobolev spaces equipped with the usual norms $\|s\|_{H^n}^2 = \sum_{k=0}^n \|d^k s/dy^k\|_{L^2}^2$. In particular, (3.4) implies

$$\mathcal{H}(v) \leq C \|D\mathcal{H}(\cdot, v)\|_{L^2}^2, \tag{3.5}$$

if $\|s(v)\|_{H^1} \leq \beta_2$ for some $C > 0$.

The main idea in the present step is to use the functional $\mathcal{H}$ as a Lyapunov function for $v_t^\epsilon$. For simplicity, if there is no noise term in the SPDE (3.1), then we have

$$\frac{\partial}{\partial t} \mathcal{H}(v_t^\epsilon) = -\epsilon^{-3/2-2\gamma} \|D\mathcal{H}(\cdot, v_t^\epsilon)\|_{L^2}^2$$

$$\leq -C^{-1} \epsilon^{-3/2-2\gamma} \mathcal{H}(v_t^\epsilon) \tag{3.6}$$

from the estimate (3.5) and this shows that $\mathcal{H}(v_t^\epsilon)$ decreases rapidly to 0 as $\epsilon \downarrow 0$ if $v_0^\epsilon$ is in a neighborhood of $M$. Unfortunately, this idea does not work directly, since the solution $v_t^\epsilon(x)$ of the SPDE (3.1) is not differentiable in $x$ (but $\alpha$-Hölder continuous in $x$ a.s. for all $\alpha < 1/2$) and therefore $\mathcal{H}(v_t^\epsilon)$ has no meaning. To overcome this inconvenience, we introduce a smooth approximation $(v_t^\epsilon)^\delta$ by acting Friedrichs' mollifier of support size $\delta$ to $v_t^\epsilon$ and compute $\{\mathcal{H}((v_t^\epsilon)^\delta)\}^p, p \geq 1$, by using Itô's formula. Then, choosing $\delta = \delta(\epsilon)$ in a proper manner, we can finally prove

$$\lim_{\epsilon \downarrow 0} P\{ \sup_{0 \leq t \leq T} \mathrm{dist}(v_t^\epsilon, M) \leq \epsilon^{1/20 + \gamma/5 - \kappa}\} = 1, \quad T > 0 \tag{3.7}$$

for arbitrary $\kappa > 0$ if $\gamma \geq 1$; a similar assertion holds true by replacing the distance in the space $L^2(\mathbf{R})$ with that of $L^\infty(\mathbf{R})$. These estimates give a rate of convergence of $v_t^\epsilon$ to $M$.

The following uniform bound, which is shown with the help of the maximum principle for PDE's and by relying on the assumption $a \in C_0(\mathbf{R})$ in an essential way, is used frequently in the proof:

$$\lim_{\epsilon \downarrow 0} P\left\{ \sup_{0 \leq t \leq T, y \in \mathbf{R}} |v_t^\epsilon(y)| \leq 1 + \delta \right\} = 1, \qquad T, \delta > 0. \tag{3.8}$$

This means that the solution $v_t^\epsilon$ of the SPDE (3.1) stays in a neighborhood of $[-1, 1]$, the region containing two phases $\pm 1$, with high probability as $\epsilon \downarrow 0$.

*Step 2. Identification of the limit.* To complete the proof of Theorem, we apply the method which was originally employed by Kazenberger [16] for a finite dimensional problem. Let us consider a PDE

$$\frac{\partial v_t}{\partial t} = \Delta v_t - V'(v_t), \qquad t > 0, \ y \in \mathbf{R}; \qquad v_0 = v, \tag{3.9}$$

where $V(v)$ is a mild potential obtained by cutting off $U(v)$ for $|v| \geq 2$. From (3.8), this cut-off does not harm our argument. We denote $v_t$ by $v_t(\cdot; v)$ to indicate its initial data. Then, it can be proved that $\lim_{t \to \infty} v_t(\cdot; v)$ exists in $M$ for $v \in \mathcal{X}_{\beta_3} := \{v \in \mathcal{X}; \mathrm{dist}(v, M) \leq \beta_3\}$ for some $\beta_3 > 0$. We denote the limit by $m_{\zeta(v)}$.

*Properties of $\zeta(v)$:* (i) The function $\zeta(v)$ is twice Fréchet differentiable in $v \in \mathcal{X}_{\beta_3}$. We denote the kernel functions for integral representations of its first and second Fréchet derivatives by $D\zeta(y, v)$ and $D^2\zeta(y_1, y_2, v)$, respectively.
(ii) $D\zeta(\cdot, v) \in \cap_{\alpha < 2} H^\alpha$ and $\langle D\zeta(\cdot, v), \Delta v - U'(v) \rangle = 0$ for every $v \in \mathcal{X}_{\beta_3} \cap (H^\delta + m)$

with some $\delta > 0$ satisfying $\|v\|_{L^\infty} \leq 2$, where $H^\alpha = H^\alpha(\mathbf{R}), \alpha \geq 0$, stand for the Sobolev spaces.

(iii) $\qquad \|D\zeta(\cdot, v) - D\zeta(\cdot, m_{\zeta(v)})\|_{L^2} \leq C\sqrt{\operatorname{dist}(v, M)}, \qquad C > 0, \ v \in \mathcal{X}_{\beta_3}.$

(iv) $\qquad \left| \int_{\mathbf{R}} \{y - \zeta(v)\}^p \{D^2\zeta(y, y, v) - D^2\zeta(y, y, m_{\zeta(v)})\} \, dy \right|$

$$\leq C_p \sqrt{\operatorname{dist}(v, M)}, \quad C_p > 0, \ p \geq 0, \ v \in \mathcal{X}_{\beta_3}.$$

(v) $\qquad \sup_{v \in \mathcal{X}_{\beta_3}} \int_{\mathbf{R}} \{y - \zeta(v)\}^2 |D^2\zeta(y, y, v)| \, dy < \infty.$

(vi) $\qquad \int_{\mathbf{R}} D^2\zeta(y, y, m_\eta) \, dy = 0 \quad$ for every $\eta \in \mathbf{R}.$

The proof of these six properties of $\zeta(v)$ is rather involved. The identity in (ii) is shown roughly from

$$0 = \frac{d}{dt}\zeta(v_t) = \langle D\zeta(\cdot, v_t), \Delta v_t - V'(v_t)\rangle, \quad t > 0,$$

by noting that $\zeta(v_t)$ is constant along the solution $v_t$ of (3.9). We use the interpolation theory for the proof of $D\zeta(\cdot, v) \in \cap_{\alpha<2} H^\alpha$. To show the bounds (iv) and (v), we are forced to introduce weighted $L^2$-spaces $L^2(\mathbf{R}, e^{2\lambda|y|}dy), \lambda > 0$, and investigate certain semigroups on these spaces by means of perturbation technique. The centering condition (vi) follows from the reflection and shift-invariance of the dynamics defined by (3.9). The details are omitted.

Now, we return to the proof of Theorem. Set $\xi_t^\epsilon := \epsilon^{1/2}\zeta(v_t^\epsilon)$ and apply Itô's formula to obtain

$$\xi_t^\epsilon = \xi_0^\epsilon + \mathfrak{m}_t^\epsilon + \frac{1}{2}\int_0^t b_s^\epsilon \, ds$$

from (3.1), where

$$\mathfrak{m}_t^\epsilon = \int_0^t \langle a(\epsilon^{1/2}\cdot)D\zeta(\cdot, v_s^\epsilon), dw_s\rangle,$$

$$b_t^\epsilon = \epsilon^{-1/2}\int_{\mathbf{R}} a^2(\epsilon^{1/2}y)D^2\zeta(y, y, v_t^\epsilon) \, dy.$$

We have employed the property (ii), from which a diverging term disappears. This is the main advantage of introducing the function $\zeta(v)$ and the trick essentially due to Katzenberger.

The limit of the martingale term can be computed by using (iii) and (3.7) as follows:

$$\frac{d}{dt}[\mathrm{m}^\epsilon]_t = \int_{\mathbf{R}} a^2(\epsilon^{1/2}y)D\zeta(y, v_t^\epsilon)^2\, dy$$

$$= \int_{\mathbf{R}} a^2(\epsilon^{1/2}y)D\zeta(y, m_{\zeta(v_t^\epsilon)})^2\, dy + o(1)$$

$$= a^2(\xi_t^\epsilon)\int_{\mathbf{R}} D\zeta(y, m)^2\, dy + o(\epsilon^{1/2}) + o(1) = a^2(\xi_t^\epsilon)\alpha_1^2 + o(1),$$

where $[\mathrm{m}^\epsilon]_t$ denotes the quadratic variational process of $\mathrm{m}_t^\epsilon$. We have used an identity

$$D\zeta(y, m_\eta) = -\nabla m_\eta(y)/\|\nabla m\|_{L^2}^2.$$

On the other hand, by the Taylor's expansion for the term $a^2(\epsilon^{1/2}y)$,

$$b_t^\epsilon = \epsilon^{-1/2} a^2(\xi_t^\epsilon)\int_{\mathbf{R}} D^2\zeta(y, y, v_t^\epsilon)\, dy$$

$$+ (a^2)'(\xi_t^\epsilon)\int_{\mathbf{R}} (y - \zeta(v_t^\epsilon))D^2\zeta(y, y, v_t^\epsilon)\, dy$$

$$+ \int_{\mathbf{R}} O\left(\epsilon^{1/2}(y - \zeta(v_t^\epsilon))^2\right)D^2\zeta(y, y, v_t^\epsilon)\, dy$$

$$=: b_{t,1}^\epsilon + b_{t,2}^\epsilon + b_{t,3}^\epsilon.$$

From (iv) with $p = 0$, (vi) and (3.7), we obtain

$$|b_{t,1}^\epsilon| \leq \text{const } \epsilon^{-1/2}(\epsilon^{1/20+\gamma/5-\kappa})^{1/2} \longrightarrow 0,$$

if $\gamma > 19/4$. On the other hand, from (iv) with $p = 1$,

$$\left|b_{t,2}^\epsilon - (a^2)'(\xi_t^\epsilon)\int_{\mathbf{R}} yD^2\zeta(y, y, m)\, dy\right| \longrightarrow 0.$$

Finally we get $b_{t,3}^\epsilon \to 0$ from (v). Deriving a concrete formula for $D^2\zeta(y, y, m)$, these computations are summarized into $b_t^\epsilon = \alpha_2(a^2)'(\xi_t^\epsilon) + o(1)$ as $\epsilon \downarrow 0$. Therefore, showing the tightness of $\{\xi_t^\epsilon\}_{0<\epsilon<1}$, we see that the distribution of $\xi_t^\epsilon$ converges weakly to that of the solution $\xi_t$ of the SDE (2.3) as $\epsilon \downarrow 0$. Then, the conclusion of Theorem follows.

## 4. Concluding Remarks

Finally in this section we review briefly some related results.

(a) *Deterministic equations*: There are several results on the asymptotic behavior of the solution of PDE of the form (1.1) without noise.

(a-1) 1-*dimensional case*: Fife-Hsiao [11] considered a PDE

$$\frac{\partial u}{\partial t} = \epsilon \Delta u - \epsilon^{-1} U'(x, u), \quad t > 0, \ x \in \mathbf{R}, \tag{4.1}$$

with a potential $U = U(x, u)$ varying with $x$ and of double-well type for each $x$ but not necessarily of equal depth, where $U' = \partial U/\partial u$. For simplicity, assuming that $\pm 1$ are bottoms of $U(x, \cdot)$ for all $x$, they proved that $u_t(x)$ converges to $\chi_{\xi_t}(x)$ as $\epsilon \downarrow 0$ and $\xi_t$ moves with velocity $c(\xi_t)$, where $c(\xi)$ denotes the speed of travelling wave solution for $\partial v/\partial t = \Delta v - U'(\xi, v), y \in \mathbf{R}$. Carr-Pego [2] treated the case where many transition layers appear.

(a-2) *Multi-dimensional case*: de Mottoni-Schatzman [7] derived the mean curvature motion from

$$\frac{\partial u}{\partial t} = \Delta u - \epsilon^{-2} U'(u), \quad t > 0, \ x \in \mathbf{R}^d, \tag{4.2}$$

where $U$ is the double-well potential of equal depth. The time scale of (4.2) is much longer than (4.1). Chen [3] has some further results. These results are valid as long as the interface remains smooth. Evans-Soner-Souganidis [9] succeeded to derive the mean curvature motion for all time, which means that their results are applicable also after the appearance of geometric singularities. The mean curvature motion with singularities was covered by the results due to Evans-Spruck [10] and Chen-Giga-Goto [4]. They introduced the notion of the "generalized solution". The book [15] by Gurtin gives a nice introduction to the theory of interfacial evolution.

(b) *Stochastic models*: There are only a few results on stochastic dynamics. Spohn [19] gave rather physical arguments on the derivation of mean curvature motion from Ginzburg-Landau stochastic lattice model and spin flip model of Glauber's type. De Masi-Orlandi-Presutti-Triolo [5, 6] derived the mean curvature motion after taking two stages of scaling limits from the Glauber evolution with Kac potentials. Brassesco-De Masi-Presutti [1] considered

$$\frac{\partial u}{\partial t} = \epsilon^{-1}\{\Delta u - U'(u)\} + \dot{w}_t(y), \quad t > 0, \ y \in [-\epsilon^{-1}, \epsilon^{-1}], \tag{4.3}$$

by imposing Neumann boundary conditions at both edges and proved that $u = u_t^\epsilon$ converges to $m_{\eta_t}$ in distribution's sense, where $\eta_t = \eta_0 + DB_t$ with diffusion coefficient $D > 0$ and a Brownian motion $B_t$. Comparing to (1.1) and (3.1), the equation (4.3) is of microscopic level. Hence, the limit keeps the shape $m_\eta$ and the sharp transition $\chi_\eta$ does not arise. Mueller-Sowers [18] derived a result on random travelling waves for an SPDE

$$\frac{\partial u}{\partial t} = \Delta u + (u - u^2) + \lambda\sqrt{u(1-u)}\dot{w}_t(y), \quad t > 0,\ y \in \mathbf{R}. \qquad (4.4)$$

Here, $\lambda > 0$ is small but fixed and notice that $u = 1$ is stable while $u = 0$ is unstable for the equation without noise. Funaki-Nagai [14] gave a similar result to [16] for finite dimensional diffusions.

## References


[1] Brassesco, S., De Masi, A., and Presutti, E.: Brownian fluctuations of the instanton in the d=1 Ginzburg-Landau equation with noise; *Preprint* (1994)

[2] Carr, J. and Pego, R. L.: Metastable patterns in solutions of $u_t = \epsilon^2 u_{xx} - f(u)$; *Commun. Pure Appl. Math.* **XLII** (1989) 523–576

[3] Chen, X.: Generation and propagation of interfaces in reaction diffusion systems; *J. Diff. Equations* **96** (1992) 116–141

[4] Chen, Y. -G., Giga, Y., and Goto, S.: Uniqueness and existence of viscosity solutions of generalized mean curvature flow equation; *J. Diff. Geom.* **33** (1991) 749–786

[5] De Masi, A., Orlandi, E., Presutti, E., and Triolo, L.: Glauber evolution with Kac potentials I. Mesoscopic and macroscopic limits, interface dynamics; *Preprint* (1992)

[6] De Masi, A., Orlandi, E., Presutti, E., and Triolo, L.: Motion by curvature by scaling non local evolution equations; *J. Statis. Phys.* **73** (1993) 543–570

[7] de Mottoni, P. and Schatzman, M.: Geometrical evolution of developed interfaces; *Preprint* (1989)

[8] Diehl, H. W., Kroll, D. M., and Wagner, H.: The interface in a Ginsburg-Landau-Wilson model: Derivation of the drumhead model in the low-temperature limit; *Z. Physik B* **36** (1980) 329–333

[9] Evans, L. C., Soner, H. M., and Souganidis, P. E.: Phase transitions and generalized motion by mean curvature; *Commun. Pure Appl. Math.* **XLV** (1992) 1097–1123

[10] Evans, L. C. and Spruck, J.: Motion of level sets by mean curvature I; *J. Diff. Geom.* **33** (1991) 635–681

[11] Fife, P. C. and Hsiao, L.: The generation and propagation of internal layers; *Nonlinear Analysis* **12** (1988) 19–41

[12] Funaki, T.: The reversible measures of multi-dimensional Ginzburg-Landau type continuum model; *Osaka J. Math.* **28** (1991) 463–494

[13] Funaki, T.: Scaling limit for a stochastic PDE and the separation of phases, *Preprint* (1993)

[14] Funaki, T. and Nagai, H.: Degenerative convergence of diffusion process toward a submanifold by strong drift; *Stochastics* **44** (1993) 1–25

[15] Gurtin, M. E.: *Thermomechanics of evolving phase boundaries in the plane.* Oxford, 1993

[16] Katzenberger, G. S.: Solutions of a stochastic differential equation forced onto a manifold by a large drift; *Ann. Prob.* **19** (1991) 1587–1628

[17] Kawasaki, K. and Ohta, T.: Kinetic drumhead model of interface I; *Prog. Theoret. Phys.* **67** (1982) 147–163

[18] Mueller, C. and Sowers, R.: Random travelling waves for the KPP equation with noise; *Preprint* (1993)

[19] Spohn, H.: Interface motion in models with stochastic dynamics; *J. Statis. Phys.* **71** (1993) 1081–1132

Tadahisa Funaki
Department of Mathematics
School of Science
Nagoya University
Nagoya 464-01, JAPAN
*E-mail*: funaki@math.nagoya-u.ac.jp

L. GROSS

# Harmonic analysis for the heat kernel measure on compact homogeneous spaces

## 1. Introduction

The earliest efforts to construct a viable theory of infinite dimensional nonlinear manifolds which support a reasonable integration theory seem to be those of H.-H. Kuo [20, 21, 22] and J. Eells and D. Elworthy [7, 8, 9]. Their manifolds were modeled on abstract Wiener spaces. The central technical issue consisted in showing that on overlapping charts the Wiener measure class was preserved by the transition maps. The degree of generality of these manifolds depended therefore on the state of the art for nonlinear transformations of Wiener measure, an art which has had a long history since the initial paper of Cameron and Martin [2] in 1949 and has advanced greatly since that time, [1, 3, 4, 5, 10, 17, 20, 23, 24, 25, 26, 27, 30]. Among the most interesting infinite dimensional manifolds are the loop spaces $C(S^1; M)$, wherein M is a finite dimensional Riemannian manifold. Although these special kinds of manifolds do not seem to be included among those defined in the early pioneering works of Kuo, Eells and Elworthy the study of analysis over these loop spaces has received impetus in the past five years from the techniques developed partly in constructive quantum field theory and partly in the context of Malliavin's approach to hypoellipticity. For further history and references see the introduction to [11].

In [11] this author studied the question of uniqueness of lowest eigenstates for a naturally occuring Schrödinger operator over certain based loop spaces. The manifold M was taken to be a finite dimensional Lie group G of compact type. The word "based" means that one is actually studying those functions in $C([0, 1]; G)$ which take 0 and 1 into the identity element of G. The long range goal of this work is to prove a theorem of Hodge-de Rham type over this infinite dimensional manifold. See [13] for a survey of these problems.

A byproduct of the work in [11] was a theorem which may be regarded as a noncommutative extension of the Wiener-Ito-Kakutani-Segal isomorphism [18, 19, 28, 29, 31]. One replaces Gauss measure over $R^n$ by heat kernel measure over a compact Lie group. This theorem will be stated in detail in Section 2. In this note we will extend this theorem further, replacing the compact group by a homogeneous space G/K.

## 2. Background and Preliminaries

Denote by G a connected, simply connected compact Lie group and by $\mathcal{G}$ its Lie algebra. We will identify $\mathcal{G}$ with the tangent space to G at the identity element e. $\langle\,,\rangle$ will denote throughout a fixed Ad G invariant inner product on $\mathcal{G}$. Write $|\xi|$ for $\langle\xi,\xi\rangle^{1/2}$.

Denote by $\mathcal{G}^{\otimes n}$ the n-fold tensor product of $\mathcal{G}$ with itself and write $\langle\,,\rangle_n$ for the inner product on $\mathcal{G}^{\otimes n}$ with associated cross norm $|\cdot|_n$. Denote by T the space of algebraic tensors over $\mathcal{G}$. For an element $\beta$ in T define

$$\|\beta\|_T^2 = \sum_{k=0}^{\infty}(k!)|\beta_k|_k^2, \qquad \beta = \sum_{k=0}^{\infty}\beta_k, \qquad \beta_k \in \mathcal{G}^{\otimes k}.$$

Write $\bar{T}$ for the completion of T in this norm. T is an algebra with respect to the operation of tensor product. Let $J$ be the two sided ideal in T generated by the elements of the form

$$\xi \otimes \eta - \eta \otimes \xi - [\xi,\eta], \qquad \xi,\eta \in \mathcal{G}.$$

Denote by $J^{\perp}$ the orthogonal complement of $J$ in $\bar{T}$. One sees easily that the zero rank tensor

$$\omega = (1,0,0,\ldots)$$

is in $J^{\perp}$.

Define the operator of right multiplication in T by

$$R_\xi\beta = \beta \otimes \xi.$$

Write $R_\xi^*$ for the adjoint of $R_\xi$ on $\bar{T}$. Since $R_\xi J \subset J, J^{\perp}$ is invariant under the (unbounded, densely defined) operator $R_\xi^*$. Let

$$A_\xi = R_\xi^*|J^{\perp}.$$

It is shown in [11, Section 6] that $A_\xi$ is densely defined in $J^{\perp}$.

Let $e_1, \ldots, e_r$ be an orthonormal basis of $\mathcal{G}$. For any $\xi$ in $\mathcal{G}$ denote by $\tilde{\xi}$ the corresponding right invariant vector field on G. Then the Laplacian on G may be defined by

$$\Delta = \sum_{j=1}^{r} (\tilde{\xi}_j)^2.$$

It is well known and easy to verify that $\Delta$ is independent of the choice of O.N. basis of $\mathcal{G}$ and that as an operator on $C^\infty(G)$ it has a self-adjoint closure in $L^2(G, dx)$ which we also denote by $\Delta$. $dx$ refers here to Haar measure on G. The semigroup $e^{t\Delta/2}$ is given by convolution on G with a strictly positive $C^\infty$ function $p_t : G \to R$ such that $\int_G p_t(x)\, dx = 1$.

The present paper is an extension of the following theorem, which was proved in [11].

**Theorem 2.1** [11] *There exists a unique linear isometry $U$ from $L^2(G, p_1(x)dx)$ onto $J^\perp$ such that*

(a) $U1 = \omega$.

(b) $U\tilde{\xi}U^{-1} = -A_\xi, \quad \xi \in \mathcal{G}$.

*Here $\tilde{\xi}$ is to be regarded as the closure of $\tilde{\xi}\,\big|_{C^\infty(G)}$ in $L^2(G, p_1(x)dx)$.*

**Remark 2.2** The proof of the preceding theorem given in [11] is highly probabilistic. A much simpler and largely analytical proof has recently been given by O. Hijab [16]. Moreover the remaining probabilistic portion of the proof has now also been replaced by an analytic proof by B. Driver [6]. Further properties of this isomorphism have been discussed in [12]. A survey of the infinite dimensional problems which have led to this finite dimensional theorem is given in [13].

## 3. Extension to Compact Homogeneous Spaces

Let K be a connected closed subgroup of G and denote by $\mathcal{K}$ its Lie subalgebra of $\mathcal{G}$. Let M $=$G/K be the space of left cosets aK. It is well known how to give M a Riemannian structure which is invariant under the left action of G [14, 15]. We will give in the Appendix a brief review for the reader's convenience. The following proposition summarizes what will be needed here. Write $\pi : G \to M$ for the quotient space map and $dm := \pi_* dx$ for the measure on M induced by Haar measure. Denote by $L^2(G, p_1(x)dx)^K$ the space of right K invariant functions on G which are square integrable with respect to the measure $p_1(x)dx$. A similar notation will be used for other right K invariant function spaces.

**Proposition 3.1** *There is a unique Riemannian metric on $M$ which is invariant under the left action of $G$ and for which the linear map*

$$d\pi_e\big|_{\mathcal{K}^\perp} : \mathcal{K}^\perp \to T_{eK}M \tag{3.1}$$

*is an isometry. If $\triangle_M$ is the associated Laplacian on $M$ then the heat kernel $q_t(y,z)$ on $M$ is defined by*

$$(e^{t\triangle_M/2}h)(y) = \int_M q_t(y,z)h(z)\,dm(z)$$

*for $h$ in $C(M)$. Let*

$$d\nu(y) := q_1(y,eK)dm(y)$$

*denote the heat kernel measure for the pointed Riemannian manifold $(eK, M)$ at time $t = 1$. Then the pullback map $S : h \mapsto h \circ \pi$ is an isometry from $L^2(M,\nu)$ onto $L^2(G, p_1(x)dx)^K$ satisfying $S1 = 1$ and $S^{-1}\tilde{\xi}S = d\pi\,\tilde{\xi}$.*

A proof will be given in the appendix.

Denote now by $\mathcal{K}T$ the right ideal generated by $\mathcal{K}$ in the algebra T and by $J + \mathcal{K}T$ the linear span of $J$ and $\mathcal{K}T$.

**Theorem 3.2** *The linear isometry $U$ of Theorem 2.1 takes $L^2(G, p_1(x)\,dx)^K$ onto $(J + \mathcal{K}T)^\perp$. The restriction of $U$ to $L^2(G, p_1(x)dx)^K$ is the unique surjective linear isometry*

$$W : L^2(G, p_1(x)dx)^K \to (J + \mathcal{K}T)^\perp$$

*such that*

    (1) $W1 = \omega$.

    (2) $W\tilde{\xi}W^{-1} = -A_\xi$    *for all $\xi$ in $\mathcal{G}$.*

**Theorem 3.3** *There exists a unique surjective linear isometry*

$$V : L^2(M,\nu) \to (J + \mathcal{K}T)^\perp$$

*such that*

    (1) $V1 = \omega$.

    (2) $V(d\pi\tilde{\xi})V^{-1} = -A_\xi$    *for all $\xi$ in $\mathcal{G}$.*

**Notation** If $\alpha = \xi_1 \otimes \cdots \otimes \xi_n$ is in T define $\alpha' = \tilde{\xi}_n \cdots \tilde{\xi}_1$ and extend this map linearly to T.

**Lemma 3.4** *For any function $f$ in $L^2(G, p_1(x)\,dx)$ and $\alpha \in T$, we have*

$$\langle Uf, \alpha \rangle = \int_G f(x)\alpha' p_1(x)\,dx. \tag{3.2}$$

*In particular $\{p_1(x)^{-1}(\alpha' p_1)(x) : \alpha \in T\}$ is dense in $L^2(G, p_1(x)\,dx)$.*

**Proof** It suffices to prove (3.2) in case $f$ is in $C^\infty(G)$ since both sides are continuous in $L^2$ norm. For such f we have

$$\begin{aligned}
\langle Uf, \xi_1 \otimes \cdots \otimes \xi_n \rangle &= \langle Uf, R_{\xi_n} \cdots R_{\xi_1}\omega \rangle \\
&= \langle A_{\xi_1} \cdots A_{\xi_n} Uf, \omega \rangle \\
&= (-1)^n \int_G \tilde{\xi}_1 \cdots \tilde{\xi}_n f(x) p_1(x)\,dx \\
&= \int_G f(x) \tilde{\xi}_n \cdots \tilde{\xi}_1 p_1(x)\,dx
\end{aligned}$$

which proves the Lemma. Note that $f$ and $Uf$ are in the domains of the unbounded operators which have been applied to them by condition (b) of Theorem 2.1 and because $f$ is smooth.

**Lemma 3.5** *Let $\phi$ be in $C^\infty(G)$ and assume $\phi$ is central. That is, $\phi(ax) = \phi(xa)$ for all $x$ and $a$ in $G$. Denote by $\hat{\eta}$ the <u>left</u> invariant vector field extending the element $\eta$ in $\mathcal{G}$. Then, for any $\alpha \in T$,*

$$(\eta^{\otimes n} \otimes \alpha)'\phi = (\hat{\eta})^n \alpha'\phi \tag{3.3}$$

**Proof** If $\xi_1, \ldots, \xi_k$ are in $\mathcal{G}$, then

$$\tilde{\xi}_k \cdots \tilde{\xi}_1 \tilde{\eta}^n \phi(x) = \partial^{n+k}\phi(\,\exp\,(t_0\eta)\exp(t_1\xi_1)\cdots\exp(t_k\xi_k)x)/\partial t_0^n \partial t_1 \cdots \partial t_k$$

with all partial derivatives evaluated at $t_0 = t_1 = \cdots = t_k = 0$. Since $\phi$ is central the factor $\exp(t_0\eta)$ may be commuted to the far right of the argument of $\phi$, proving the Lemma.

**Proof of Theorem 3.2** If f is in $C^\infty(G)^K$ then for any $\eta$ in $\mathcal{K}$ and $\alpha$ in T we have

$$\begin{aligned}
\langle Uf, \eta \otimes \alpha \rangle &= \int_G f(x)\hat{\eta}\alpha' p_1(x)\,dx \\
&= -\int_G (\hat{\eta}f)(x)\alpha' p_1(x)\,dx \\
&= 0
\end{aligned}$$

because $p_1$ is central and $f(xe^{t\eta})$ is constant in $t$. Since such functions $f$ are dense in $L^2(G, p_1(x)dx)^K$ it follows that $\langle Uf, \eta \otimes \alpha \rangle = 0$ for all $f$ in $L^2(G, p_1(x)dx)^K$. So U maps $L^2(G, p_1(x)dx)$ into $J^\perp \cap (\mathcal{K}T)^\perp$ which equals $(J + \mathcal{K}T)^\perp$.

To prove surjectivity observe that we already know that U maps $L^2(G, p_1(x)dx)$ onto $J^\perp$ by Theorem 2.1. Suppose then that $f$ is in $L^2(G, p_1(x)dx)$ and that $Uf$ is orthogonal to $\mathcal{K}T$. It suffices to show that $f$ is equal almost everywhere to a right K invariant function. Now if $\eta$ is in $\mathcal{K}$ and $\alpha$ is in T then $(\alpha' p_1)(xe^\eta)$ is analytic in $\eta$, hence is given by the power series

$$(\alpha' p_1)(xe^\eta) = \sum_{n=0}^{\infty} (1/n!)\hat{\eta}^n (\alpha' p_1)(x)$$

which for each $\eta$ converges uniformly in x. For all $n \geq 1$, by Lemmas 3.4 and 3.5,

$$\int_G f(x)\hat{\eta}^n(\alpha' p_1)(x)\, dx = \int_G f(x)(\eta^{\otimes n} \otimes \alpha)' p_1(x)\, dx$$
$$= \langle Uf, \eta^{\otimes n} \otimes \alpha \rangle = 0.$$

Hence

$$\int_G f(x)(\alpha' p_1)(xe^\eta)\, dx = \int_G f(x)\alpha' p_1(x)\, dx \qquad \text{for all } \eta \in \mathcal{K}.$$

Thus

$$\int_G (f(xe^{-\eta}) - f(x))(\alpha' p_1(x))\, dx = 0 \qquad \text{for all } \eta \in \mathcal{K} \text{ and all } \alpha \in T.$$

By the density statement of Lemma 3.4 we therefore have $f(xe^{-\eta}) = f(x)$ a.e.. Since K is connected every element of K is a finite product of elements of the form $e^{-\eta}, \eta \in \mathcal{K}$. Thus for each element $a$ in K

$$f(xa) = f(x) \qquad a.e.$$

Since $p_1$ is bounded away from zero on G, $f$ is in $L^2(G, dx)$. Hence for any smooth function $u$ on G the convolution $\int_G u(xy^{-1})f(y)\, dy$ is a smooth right K invariant function on G by the preceding equation and therefore is measurable with respect to the $\sigma$-field $\mathcal{S}$ of right K invariant Borel sets in G. Choosing a sequence of smooth functions $u$ converging to a $\delta$ function we obtain $f$ as an $L^2(dx)$ limit on the one hand while we obtain an $\mathcal{S}$ measurable function as an $L^2(dx|\mathcal{S})$ limit on the other hand. So $f$ is equal to a right K invariant funtion almost everywhere.

To prove the uniqueness of W observe that the identity (3.2) was derived using only the properties (a) and (b) of Theorem 2.1 and therefore holds with U replaced by any map W satisfying conditions (1) and (2) of Theorem 3.2 provided that $f$ is in $L^2(G, p_1(x)\, dx)^K$. Thus for such a function $f$ we have $\langle Wf, \alpha \rangle = \langle Uf, \alpha \rangle$ for all $\alpha \in T$. Hence $Wf = Uf$.

**Proof of Theorem 3.3** By Proposition 3.1 the map $V := WS$ satisfies the requirements of Theorem 3.3. Conversely if V satisfies the conditions of Theorem 3.3 then $W := VS^{-1}$ satisfies the conditions of Theorem 3.2, yielding uniqueness of V.

**Remark 3.6** It was pointed out in [12, Remark 3.6] that there are some special functions on G which are analogs of the Hermite polynomials. For example if $(\tilde{\xi})^*$ is computed in the Hilbert space $L^2(G, p_1(x)dx)$ then $(\tilde{\xi}^*)^n 1$ is a typical such function and is a Hermite polynomial in case $G = R$. O. Hijab [16] has shown that these functions can be expressed in terms of the heat kernel $p_1(x)$ by the same kind of formula that applies to ordinary Hermite polynomials and that the isometry of Theorem 2.1 can be expressed neatly in terms of them. For the quotient space G/K a similar formulation holds as follows.

Let $q(x) = \int_K p_1(xk)\, dk$. Then $q$ is right K invariant and so is $\alpha' q$ for any $\alpha$ in T. Define

$$H_\alpha(x) = q(x)^{-1}(\alpha' q)(x) \tag{3.4}$$

Since $q$ is bounded away from zero $H_\alpha$ is a right K invariant $C^\infty$ function on G.

**Corollary 3.7** *Let P be the projection of $\bar{T}$ onto $(J + \mathcal{K}T)^\perp$. Then for any element $\alpha$ in $T$*

$$W H_\alpha = P\alpha. \tag{3.5}$$

**Proof** If the function $f$ in equation (3.2) is right K invariant we may replace its argument by $xk^{-1}$ for any element $k$ in K, or equivalently, by right translation invariance of Haar measure, we may replace $(\alpha' p_1)(x)$ in (3.2) by $(\alpha' p_1)(xk)$. Now integrate the resulting equation over K with respect to Haar measure on K and observe that $\int_K (\alpha' p_1)(xk)\, dk = (\alpha' q)(x)$ because $\alpha'$ commutes with right translations. Therefore the equation (3.2) yields

$$\langle Uf, \alpha \rangle = \int_G f(x) H_\alpha(x) q(x)\, dx \tag{3.6}$$

if $f$ is in $L^2(G, p_1(x)dx)^K$. But $Uf$ is in the range of P and all functions in the integrand of (3.6) are measurable with respect to the $\sigma$ field $\mathcal{S}$ of right K invariant

Borel sets in G. Hence the last equation can be written

$$\langle Wf, P\alpha \rangle = \langle f, H_\alpha \rangle_{L^2(G,\mathcal{S},q\,dx)}. \tag{3.7}$$

Since W is an orthogonal transformation from $L^2(G,\mathcal{S},q\,dx)$ onto the range of P we have $W^{-1}P\alpha = H_\alpha$ which proves (3.5).

Finally we note that Corollary 3.7 has its natural formulation over M. By equations (A13) and (A14) the push-down of $q$ to M is the heat kernel on M measured from the point eK and $\nu$ is the push-down of the measure $q(x)dx$. The right K invariant functions $H_\alpha$ define functions on M which we may also write as $H_\alpha$. Then in the notation of Theorem 3.3 we have proven $VH_\alpha = P\alpha$.

## Appendix

We will discuss the Riemannian structure on M and prove Proposition 3.1. All of this material is standard [14, 15].

Let $\mathcal{M} = \mathcal{K}^\perp$, the orthogonal complement of $\mathcal{K}$ in $\mathcal{G}$. Write $\mathcal{M}_a = (Ada)\mathcal{M}$ for $a \in G$. Since $(Adk)\mathcal{K} = \mathcal{K}$ for $k \in \mathcal{K}$ and the inner product on $\mathcal{G}$ is $AdG$ invariant we also have $(Adk)\mathcal{M} = \mathcal{M}$. Hence

$$\mathcal{M}_{ak} = \mathcal{M}_a \qquad a \in G,\ k \in K. \tag{A1}$$

Thus $\mathcal{M}_a$ depends only on the coset $aK$.

Next observe that if $f \in C^\infty(G), \xi \in \mathcal{G}$ and $a \in G$ then, defining $L_a x = ax$, we have

$$(dL_a|_x \tilde{\xi})f = ((Ada\,\xi)\check{}f)(ax) \tag{A2}$$

because

$$\begin{aligned}
((dL_a\tilde{\xi})f)(x) &= (\tilde{\xi}(f \circ L_a))(x) \\
&= (d/dt)f(a(\exp(t\xi)x)|_{t=0} \\
&= (d/dt)f((\exp(tAda)\xi)ax)|_{t=0}
\end{aligned}$$

which is (A2). Now if $a \in G$, $h \in C^\infty(M)$ and $f = h \circ \pi$ then a similar computation shows that $(d\pi_a\tilde{\xi})h = (\tilde{\xi}f)(a) = (d/dt)f(a\exp(tAda^{-1}\xi))|_{t=0}$ which is zero by right K invariance of $f$ if $\xi \in Ada\,\mathcal{K}$. Thus by a dimension argument $d\pi_a$ is an isomorphism from $\mathcal{M}_a$ onto $T_{aK}M$. We henceforth identify these two spaces and take over the inner product on $\mathcal{M}_a$ to $T_{aK}M$. Thus

$$|d\pi_a\tilde{\xi}|_{aK} = |\xi| \qquad \text{if } \xi \in \mathcal{M}_a. \tag{A3}$$

G acts on M on the left by $\tau(a)xK := axK$ for $a$ and $x$ in G. That is, $\tau(a)\pi = \pi L_a$. We will show that $d\tau(a)$ preserves the Riemannian metric defined by (A3). To see this apply (A2) with $f = h \circ \pi$. Then

$$(d\tau(a)d\pi_x\tilde{\xi})h = (d(\tau(a)\pi)_x\tilde{\xi})h = (d(\pi \circ L_a)_x\tilde{\xi})h$$
$$= d\pi_{ax}(dL_a)|_x\tilde{\xi})h = (dL_a|_x\tilde{\xi})f$$
$$= (Ada\,\xi)\,\widetilde{\phantom{x}}|_{ax}f = d\pi_{ax}(Ada\,\xi)\widetilde{\phantom{x}}h$$

Thus

$$d\tau(a)d\pi|_x\tilde{\xi} = d\pi|_{ax}(Ada\,\xi)\widetilde{\phantom{x}}. \tag{A4}$$

Hence if $\xi \in \mathcal{M}_x$ then $d\tau(a)d\pi|_x\tilde{\xi}$ "comes from" $Ada\,\xi$ which is clearly in $\mathcal{M}_{ax}$ and norms are therefore preserved. This establishes the first assertion of Proposition 3.1.

We may now define the gradient and Laplacians over G and M. If $f$ is in $C^\infty(G)$ then define $\nabla f(x)$ to be that element of $\mathcal{G}$ such that

$$\langle \nabla f(x), \xi \rangle = (\tilde{\xi}f)(x) \qquad \text{for all } \xi \text{ in } \mathcal{G}. \tag{A5}$$

Then in the notation of Section 2 we have $\nabla f(x) = \sum_{j=1}^r e_j(\tilde{e}_jf)(x)$ and the Laplacian given in Section 2 is equally well given by the equation

$$\int_G (\Delta f(x))g(x)\,dx = -\int_G \langle \nabla f(x), \nabla g(x) \rangle\,dx \qquad f,g \in C^\infty(G). \tag{A6}$$

Similarly, if h is in $C^\infty(M)$ then $\nabla h(aK)$ may be defined as the unique element of $\mathcal{M}_a$ such that

$$\langle \nabla h(aK), \xi \rangle = ((d\pi\tilde{\xi})h)(aK) \qquad \text{for all } \xi \text{ in } \mathcal{G} \tag{A7}$$

since $d\pi_a\tilde{\xi} = 0$ if $\xi \perp \mathcal{M}_a$. Equivalently

$$\langle \nabla h(aK), \xi \rangle = (\tilde{\xi}(h \circ \pi))(a) \qquad \text{for all } \xi \text{ in } \mathcal{G}. \tag{A8}$$

Thus

$$(\nabla h)(\pi(a)) = (\nabla(h \circ \pi))(a). \tag{A9}$$

Now writing dk for normalized Haar measure on K and recalling $dm(y) = \pi_* dx$ we have $\int_M h(y)\,dm(y) = \int_G (h \circ \pi)(x)\,dx$. $\Delta_M$ is defined by

$$\int_M (\Delta_M h_1(y))h_2(y)\,dm(y) = -\int_M \langle (\nabla h_1)(y), \nabla h_2(y) \rangle\,dm(y). \tag{A10}$$

Let $f_i = h_i \circ \pi$. Then

$$\int_G (\triangle_M h_1)(\pi(x)) h_2(\pi(x)) \, dx$$

$$= \int_M (\triangle_M h_1)(y) h_2(y) \, dm(y) = -\int_M \langle \nabla h_1(y), \nabla h_2(y) \rangle \, dm(y)$$

$$= -\int_G \langle \nabla f_1(x), \nabla f_2(x) \rangle \, dx = \int_G (\triangle f_1)(x) f_2(x) \, dx. \qquad (A11)$$

Since the right invariant vector fields $\tilde{\xi}$ commute with right translations $\triangle f_1$ is right K invariant. Hence we may remove the integral to get

$$(\triangle_M h) \circ \pi = \triangle(h \circ \pi) \qquad (A12)$$

Next let us consider the heat kernels on M and G. Let

$$q_t(x, y) = \int_K p_t(x k y^{-1}) \, dk \qquad x, y \in G \qquad (A13)$$

It is straight forward to verify that

(1) $q_t(x, y)$ is right K invariant in $x$ and also in $y$.
(2) $\int_G q_t(x, y) \, dy = 1$ for each $t > 0$.
(3) $\int_G q_t(x, y) f(y) \, dy \to f(x)$ as $t \downarrow 0$ if $f$ is in $C(G)^K$.
(4) $\partial q_t(x, y)/\partial t = (1/2) \triangle_x q_t(x, y)$ for $t > 0$ and each $y$ in G.

By (1), (4) and (A12) $q_t(\cdot, y)$ defines a solution to the heat equation on M for $\triangle_M$ and therefore by (2) and (3) it is the heat kernel on M for $\triangle_M$.

To prove the last assertion of Proposition 3.1 observe that if $h$ is in $L^2(M, \nu)$ and $f = h \circ \pi$ then of course f and $|f|^2$ are right K invariant and

$$\int_G |f(x)|^2 p_1(x) \, dx = \int_G |f(xk^{-1})|^2 p_1(x) \, dx$$

$$= \int_G |f(x)|^2 p_1(xk) \, dx \qquad \text{for all } k \text{ in K.}$$

Integrating this equality over K gives

$$\int_G |f(x)|^2 p_1(x) \, dx = \int_G |f(x)|^2 q_1(x, e) \, dx$$

$$= \int_M |h(y)|^2 \, d\nu(y). \qquad (A14)$$

108

Thus the map $S : h \mapsto h \circ \pi$ is an isometry. It is clearly onto $L^2(G, p_1(x)dx)^K$ and $S1 = 1$. The equation $\tilde{\xi}S = S \, d\pi \, \tilde{\xi}$ is the definition of $d\pi$ and the equation $S^{-1}\tilde{\xi}S = d\pi \, \tilde{\xi}$ holds for the closures of these operators also. This completes the proof of Proposition 3.1.

**Acknowledgements**  This work was partially supported by NSF Grant DMS-9200774.

**References**

[1] Albeverio, S. and Hoegh–Krohn, R.: The energy representation of Sobolev Lie groups; *Compositio Math.* **36** (1978) 37–52

[2] Cameron, R. H. and Martin, W. T.: The transformation of Wiener integrals by nonlinear transformations; *Trans. Amer. Math. Soc.* **66** (1949) 253–283

[3] Cruzeiro, A. B.: Équations différentielles sur l'espace de Wiener et formules de Cameron-Martin non linéaires; *J. Funct. Anal.* **54** (1983) 206–227

[4] Driver, B.: A Cameron-Martin type quasi-invariance theorem for Brownian motion on a compact Riemannian manifold; *J. of Funct. Anal.* **110** (1992) 272–376

[5] Driver, B.: A Cameron-Martin type quasi-invariance theorem for pinned Brownian motion on a compact Riemannian manifold; *Trans. Amer. Math. Soc.*, in press (1993)

[6] Driver, B.: private communication

[7] Eells, J.: Integration on Banach manifolds; *Proc. 13th Biennial Seminar of the Canadian Mathematical Congress, Halifax* (1971) 41–49

[8] Eells, J. and Elworthy, K. D.: Wiener integration on certain manifolds; *C.I.M.E.* IV, *Some Problems in Non-Linear Analysis* Edizioni Cremonese, Rome, 67–94 (1971)

[9] Elworthy, K. D.: Stochastic differential equations on manifolds; *London Math. Soc. Lecture Note Series* **70**, Cambridge Univ. Press (1982)

[10] Gross, L.: Integration and nonlinear transformations in Hilbert space; *Trans. Amer. Math. Soc.* **94** (1960) 404–440

[11] Gross, L.: Uniqueness of ground states for Schrödinger operators over loop groups; *J. of Funct. Anal.* **112** (1993) 373–441

[12] Gross, L.: The homogeneous chaos over compact Lie groups; in: *Stochastic Processes, A Festschrift in Honour of Gopinath Kallianpur* (1993) 117–123, S. Cambanis et al. (eds.), Springer-Verlag, New York, Inc.

[13] Gross, L.: Analysis on loop groups; *NATO ASI* (1993) to appear

[14] Helgason, S.: *Differential Geometry and Symmetric Spaces*, Academic Press, San Diego (1962)

[15] Helgason, S.: *Groups and Geometric Analysis*, Academic Press, San Diego (1984)

[16] Hijab, O.: Hermite functions on compact Lie Groups; *J. of Funct. Anal.*, to appear (1993)

[17] Hsu, E. P.: Quasiinvariance of the Wiener measure and integration by parts in the path space over a compact Riemannian manifold; *J. of Funct. Anal.*, to appear (1993)

[18] Itô, K. : Multiple Wiener integral; *J. Math. Soc. of Japan* **3** (1951) 157–169

[19] Kakutani, S.: Spectrum of the flow of Brownian motion; *Proc. Nat. Acad. Sci. U.S.A.* **36** (1950) 319–323

[20] Kuo, H. -H.: Integration theory on infinite-dimensional manifolds; *Trans. Amer. Math. Soc.* **159** (1971) 57–78

[21] Kuo, H. -H.: Diffusion and Brownian motion on infinite-dimensional manifolds; *Trans. Amer. Math. Soc.* **169** (1972) 439–457

[22] Kuo, H. -H.: Ornstein-Uhlenbeck process on a Riemann-Wiener manifold; *Proc. Symp. on Stoch. Diff. Eqs., Kyoto*, K. Itô (ed.) (1976) 187–193

[23] Kusuoka, S.: The nonlinear transformation of Gaussian measure on Banach space and its absolute continuity (I); *J. Fac. Sci. Univ. of Tokyo, Sec. IA* **29** (1982) 567–598

[24] Kusuoka, S.: The nonlinear transformation of Gaussian measure on Banach space and its absolute continuity (II); *J. Fac. Sci. Univ. of Tokyo Sec. IA*, **30** (1983) 199–220

[25] Kusuoka, S.: Analysis on Wiener spaces I. Nonlinear maps; *J. of Funct. Anal.*, **98** (1991) 122–168

[26] Malliavin, M. -P. and Malliavin, P.: Integration on loop groups I. Quasi invariant measures; *J. of Funct. Anal.* **93** (1990) 207–237

[27] Ramer, R. : On nonlinear transformations of Gaussian measures; *J. of Funct. Anal.* **15** (1974) 166–187

[28] Segal, I. E.: Tensor algebras over Hilbert spaces; *Trans. Amer. Math. Soc.* **81** (1956) 106–134

[29] Segal, I. E.: Distributions in Hilbert space and canonical systems of operators; *Trans. Amer. Math. Soc.* **88** (1958) 12–41

[30] Shigekawa, I.: Transformations of the Brownian motion on the Lie group; in: *Proceedings, Taniguchi International Symposium on Stochastic Analysis* (1984) 409–422, Kiyosi Ito (ed.), Katata and Kyoto, 1982, North-Holland, Amsterdam

[31] Wiener, N.: The homogeneous chaos; *Amer. J. Math.* **60** (1938) 897–936

Leonard Gross
White Hall
Department of Mathematics
Cornell University
Ithaca, NY 14853, U.S.A.
*E-mail*: gross@math.cornell.edu

T. HIDA

# Some recent results in white noise analysis

We discuss, as an application of white noise analysis, operators of exponential type $\exp\left[\int_{-\infty}^{t} F(t-u)\partial_u^* \, du\right]$ in terms of the creations $\partial_u^*$. Causality in time and the canonical property for representation of a Gaussian process are always taken into account.

## 1. Introduction

White noise analysis has undergone a vast development in the past two decades. A systematic and basic theory has been established (see e.g., [2]). We are now in search of a new direction of development of the analysis.

The purpose of this short note is to propose an application that can be carried out within the framework of white noise analysis. This application involves generalized stationary stochastic processes which enjoy essentially infinite dimensional and nonlinear character. The present paper discusses such processes which can be expressed in white noise analysis in terms of a class of operators that have arisen in molecular biology (see [4]). We will see the symmetry as well as asymmetry in time for those processes in question.

## 2. Operators of Exponential Type

We shall deal with operators which are functions of annihilation and creation operators denoted by $\partial_t$ and $\partial_t^*$, respectively.

Set

$$A(t) = \int_{-\infty}^{t} F(t-u)\partial_u \, du, \quad t \in \mathbf{R},$$

where $F(u)$ is defined for $u \geq 0$ and is a continuous $L^2$-function. Its formal adjoint $A(t)^*$ is given by

$$A(t)^* = \int_{-\infty}^{t} F(t-u)\partial_u^* \, du. \tag{2.1}$$

Such an operator has been suggested by the paper by F. Oosawa [4]. As is easily seen, the domain $D$ of the $A(t)$ is taken to be sufficiently large in $(S)^*$ and is independent of $t$.

Take $\varphi$ in the domain $D$ and set

$$X(t) = A(t)^*\varphi, \quad t \in \mathbf{R}.$$

Then we are given a stochastic process living in $(\mathcal{S})^*$. The $X(t)$, $t \in \mathbf{R}$, span a subspace of $(\mathcal{S})^*$ with unit multiplicity and the $A(t)^*$ raises the degree by one.

We then come to an operator of exponential type with exponent $A(t)$. Namely, we set

$$E(t) = \exp[A(t)], \quad t \in \mathbf{R}.$$

It is almost obvious that the adjoint operator of $E(t)$ can be expressed in the form

$$E(t)^* = \exp[A(t)^*].$$

Take $\psi$ to be an exponential function

$$\psi(x) = \exp\left[\langle x, \eta \rangle - \|\eta\|^2/2\right], \quad \eta \in E.$$

Then, direct computation shows

$$\langle E(t)^*\varphi, \psi \rangle_\mu = \langle \varphi, E(t)\psi \rangle_\mu$$
$$= \exp\left[(F * \eta)(t)\right] \cdot \langle \varphi, \psi \rangle_\mu, \tag{2.2}$$

where $\mu$ is the white noise measure.

We can rephrase (2.2) as a formula below that is often used later.

**Proposition 1** *The $S$-transform of $E(t)^*\varphi$ is given by*

$$(SE(t)^*\varphi)(\xi) = \exp\left[(F * \eta)(t)\right] \cdot (S\varphi)(\xi).$$

As for causality in time, we observe an interesting result as follows. Let $S_t$ be the (time) shift, which describes development of the time. The family $\{S_t; t \in \mathbf{R}\}$ forms a continuous one-parameter group of operators that act on the basic nuclear space $E$ in such a way that

$$(S_t\xi)(u) = \xi(u - t), \quad t \in \mathbf{R}, \ \xi \in E.$$

Then $\{T_t\}$, with $T_t = S_t^*$, is the *flow of Brownian motion*. An operator $U_t$ acting on $(\mathcal{S})^*$ is now defined by

$$U_t\varphi(x) = \varphi(T_t x).$$

If $\varphi$ is restricted to $(L^2)$, then $\{U_t, t \in \mathbf{R}\}$ forms a continuous one parameter group of *unitary* operators.

**Lemma** *The following relationships hold :*

$$U_t \partial_u = \partial_{u+t} U_t, \tag{2.3}$$

$$U_t \partial_u^* = \partial_{u-t}^* U_t.$$

Now we set

$$Y(t) = U_t \varphi, \ t \in \mathbf{R}.$$

It can be called a *generalized stationary stochastic process* that shares the time shift with the white noise. In fact, $U_t$ represents the propagation of random functions, formed from white noise, as time $t$ goes by.

We obtain the expectation $\langle Y(t) \rangle$ in the sense that

$$\langle Y(t) \rangle \equiv \langle Y(t), 1 \rangle.$$

It is easy to see that

$$\langle Y(t) \rangle = c, \quad c \text{ being a constant.}$$

The following proposition gives an interesting connection between the group $\{U_t\}$ and the operator $E(t)$ (and also its adjoint $E(t)^*$).

**Proposition 2** *The following relationships hold:*

$$E(\tau + t)U_t = U_t E(\tau), \tag{2.4}$$

$$E(\tau)^* U_t \varphi = U_t E(\tau + t)^* \varphi. \tag{2.5}$$

**Proof** The equality (2.4) comes easily from the formula (2.3) and the fact that $U_t A(s)^n = A(s+t)^n U_t$. The formula (2.5) follows easily from (2.4). $\square$

We further come to a system $\{Z(t), t \in \mathbf{R}\}$ of $(\mathcal{S})^*$-functionals given by

$$Z(t) = E(t)^* U_t \varphi, \ \varphi \in (\mathcal{S})^*. \tag{2.6}$$

It is easy to prove from (2.5) that

$$U(t)Z(s) = U_t E(s)^* U_s \varphi = E(s+t)^* U_t U_s \varphi$$

$$= E(t+s)^*U_{t+s}\varphi = Z(s+t).$$

Namely, we obtain the following

**Theorem** *Let $Z(t)$ be given by (2.6). Then the collection $\{Z(t), t \in \mathbf{R}\}$ forms a generalized stationary stochastic process.*

It can be proved that by taking the time derivative of $Z(t)$ we can recover the white noise. Namely,

**Proposition 3** *Let $\varphi$ be in $(L^2)$. Then the singular part of $dZ(t)/dt$ determines the original white noise and hence the differential operator $\partial_t$.*

## 3. Symmetry and Asymmetry

Returning to the formula (2.1), we take an operator with the canonical kernel $F_c$ which defines $A_c(t)^*$. But if a non-canonical kernel $F_{nc}$ is given, we are also given an operator $A_{nc}(t)^*$. With a choice of $F_{nc}$ the following two Gaussian processes $X_c(t)$ and $X_{nc}(t)$ are proved to be the same Gaussian process:

$$X_c(t) = A_c(t)^*1,$$

$$X_{nc}(t) = A_{nc}(t)^*1.$$

To fix the idea we assume that the above processes enjoy the finite multiple, say $N$, Markov property. With this assumption we find a unique non-canonical kernel $F_{nc}(t - u)$ which is a Goursat kernel of order $N$.

Now set

$$E_c(t) = \exp[A_c(t)]$$

and

$$E_{nc}(t) = \exp[A_{nc}(t)].$$

As is easily seen, $E_c(t)^*1$ and $E_{nc}(t)^*1$ are the same stochastic process. For each $\varphi$ in $(\mathcal{S})^*$ we can form generalized stationary stochastic processes

$$\{Z_c(t) = E_c(t)^*U_t\varphi; t \in \mathbf{R}\}$$

and

$$\{Z_{nc}(t) = E_{nc}(t)^*U_t\varphi; t \in \mathbf{R}\}.$$

They are in general not the same stochastic process, depending on the choice of $\varphi$.

114

It is noted that the non-canonical kernel $F_{nc}$ can be chosen among others so that $F_{nc}(u - t)$ is the backward canonical kernel (see [1, Chapter V, §6]). Also note that $|F_c(0)| = |F_{nc}(0)|$. Furthermore, on can see [1] that the white noise giving the backward canonical representation is a generalized linear functional of $x$ in $(\mathcal{S})^*$.

Now take $Z_c(t)$ and apply the differential operator $\partial_s$, $s < t$. From Proposition 3 we can recover the operator $\partial_s$ from $Z_c(s)$ so that we have

$$\partial_s Z(t) = F_c(t - s)Z(t) + E(t)^* U_t \partial_{s-t}\varphi.$$

If, in particular, $\varphi$ is chosen to be an exponential function of the form $:\exp[\langle x, \eta \rangle]:$, then we have

$$\partial_s Z(t) = \{F_c(t - s) + \eta(s - t)\}Z(t), \tag{3.1}$$

which is a result coming from exponential type operator $E(t)$.

On the other hand, the non-canonical kernel defines the same Gaussian process in the same way as the canonical kernel does. More precisely, we can discuss a counterpart of (3.1) in the following manner.

Keep the white noise as before. Take the non-canonical kernel $F_{nc}(t - u)$ and form the backward canonical kernel $F_{nc}(u - t)$, $u \geq t$, to define

$$A^+(t) = \int_t^\infty F_{nc}(u - t)\partial_u \, du$$

and

$$E^+(t) = \exp[A^+(t)].$$

A generalized stationary stochastic process $Z^+(t)$ can be defined in a similar manner for $Z(t)$ and we have for an exponential function $\varphi$, the same as before, i.e.,

$$\partial_s Z^+(t) = \{F_{nc}(s - t) + \eta(t - s)\}Z^+(t).$$

We now observe a symmetry and an asymmetry of a random phenomenon expressible as $Z(t)$ as time $t$ propagates.

**Acknowledgements**   The author thanks Professors H. Kunita and H. -H. Kuo for their invitation to the U.S.-Japan Bilateral Seminar on "Stochastic Analysis on Infinite Dimensional Spaces" held at Baton Rouge in January 1994. He is also grateful to Professor K. Saitô for helping him to prepare this manuscript.

## References

[1] Hida, T. and Hitsuda, M.: *Gaussian Processes*. Translations of Mathematical Monographs, Vol. 120, American Mathematical Society, 1993

[2] Hida, T., Kuo, H. -H., Potthoff, J., and Streit, L.: *White Noise: An Infinite Dimensional Calculus*. Kluwer Academic Publishers, 1993

[3] Lévy, P.: *Problèmes Concrets d'Analyse Fonctionnelle*. Gauthier-Villars, Paris, 1951

[4] Oosawa, F., Tsuchiya, M., and Kubori, T.: Fluctuation in living cells: effect of field fluctuation and asymmetry of fluctuations; in:*Proc. Stochastic Method in Biology, Nagoya, Lecture Notes in Biomathematics* **70** (1985) 158–170, Springer-Verlag

Takeyuki Hida
Department of Mathematics
Meijo University
Nagoya 468, JAPAN
*E-mail*: f43925g@nucc.cc.nagoya-u.ac.jp

T. KAZUMI & I. SHIGEKAWA

# Differential calculus on a submanifold of an abstract Wiener space, I. Covariant derivative

## 1. Introduction

After Airault-Malliavin [3] introduced the notion of a submanifold of the Wiener space, it receives much attention recently. Infinite dimensional manifolds were already studied in geometry. But on the Wiener space, one crucial point is that we are given a measure on it. It enables us to develop the integration theory on a submanifold. The measure is given as a conditional probability and the conditional probability is realized by means of positive generalized function on the Wiener space. We can deal with it in the language of the Malliavin calculus. Further, on the path space, the conditional probability is a pinned measure. For example, Getzler [8] considered a based loop group, which can be identified with a submanifold of the Wiener space through the Itô map. Then Airault-Van Biesen [5], Airault [2] and Van Biesen [19] developed a calculus on a submanifold systematically. In this paper we attempt to rearrange their works in the setting of abstract Wiener space.

On the other hand, we have a long range goal to establish the de Rham theory on a submanifold of the Wiener space, or a loop space. So differential forms are also our main object. The other purpose of this paper is dealing with differential forms.

Let $(B, H, \mu)$ be an abstract Wiener space. A submanifold $S$ is defined as an inverse image:

$$S = \{x \in B; F(x) = a\}$$

where $a \in \mathbf{R}^d$ and $F$ is an $\mathbf{R}^d$-valued non-degenerate smooth function in the sense of Malliavin. We already have the derivative on an abstract Wiener space, which is well-developed in the Malliavin calculus. So the Malliavin calculus is a fundamental tool throughout the paper. By using the derivative on an abstract Wiener space, we can define a covariant derivative on a submanifold as in the finite dimensional case. By using the covariant derivative, we define the Lie bracket, the Lie derivative, the Riemannian curvature, etc. We also give some formulas concerning them. For example, we can prove the Bianchi identity, Cartan identity, etc. similarly in the finite dimensional case. Many proofs are well-known in the finite dimensional case. We only have to see that they work in the framework of the Malliavin calculus.

The organization of this paper is as follows. In Section 2, after quick survey of the Malliavin calculus, we introduce a submanifold of the Wiener space and define a gradient operator and its dual operator on a submanifold. We develop a differential calculus in Section 3. We define a covariant derivative on a submanifold. Everything is defined through the covariant derivative. We also discuss differential forms.

## 2. Submanifold of an Abstract Wiener Space

### 2A. Malliavin calculus

Let us review the Malliavin calculus quickly. Let $(B, H, \mu)$ be an abstract Wiener space, i.e., $B$ is a real separable Banach space and $\mu$ is a Gaussian measure with a reproducing kernel Hilbert space $H$. The following Ornstein-Uhlenbeck semigroup is fundamental in the Malliavin calculus.

$$T_t f(x) = \int_B f(e^{-t}x + \sqrt{1 - e^{-2t}}y)\mu(dy). \tag{2.1}$$

$\{T_t\}$ is a strongly continuous contraction semigroup on $L^p(B, \mu)$ for $p \geq 1$. The generator of $\{T_t\}$ is called an *Ornstein-Uhlenbeck operator* and is denoted by $L$. To define the Sobolev space, we recall the following gamma transformation: for $r \geq 0$,

$$(1 - L)^{-r/2} = \frac{1}{\Gamma(r/2)} \int_0^\infty t^{(r/2)-1} e^{-t} T_t dt.$$

Then the *Sobolev space* $W^{r,p}$ can be defined as follows:

$$W^{r,p} := \mathrm{Ran}((1 - L)^{-r/2}) = (1 - L)^{-r/2}(L^p(B, \mu)).$$

We define the norm $\| \cdot \|_{r,p}$ as follows:

$$\|v\|_{r,p} := \|u\|_p \quad \text{for} \quad v = (1 - L)^{-r/2}u,$$

where $\| \cdot \|_p$ is the $L^p$-norm. Then $(W^{r,p}, \| \cdot \|_{r,p})$ forms a Banach space and we denote its dual space by $W^{-r,q}$, where $q$ is the conjugate exponent of $p$: $\frac{1}{p} + \frac{1}{q} = 1$. We set

$$W^{\infty,\infty-} = \bigcap_{\substack{r \geq 0 \\ p > 1}} W^{r,p}, \quad W^{-\infty,1+} = \bigcup_{\substack{r \geq 0 \\ q > 1}} W^{-r,q}.$$

So $W^{\infty,\infty-}$ is a space of smooth functions and $W^{-\infty,1+}$ is a space of generalized functions.

Similarly, we can define Hilbert space valued smooth functions. To do this, take a real separable Hilbert space $K$ and note that the semigroup (2.1) can be defined for $u \in L^p(B, \mu; K)$ where $L^p(B, \mu; K)$ is the space of all $p$-th integrable $K$-valued functions. We denote it by the same symbol $\{T_t\}$. It is easy to see that $\{T_t\}$ is a strongly continuous contraction semigroup. We also denote its generator by $L$. Then we can define a $K$-valued Sobolev space $W^{r,p}(K)$ as

$$W^{r,p}(K) := \mathrm{Ran}((1 - L)^{-r/2}) = (1 - L)^{-r/2}(L^p(B, \mu; K))$$

and the norm $\| \cdot \|_{r,p}$ is defined as

$$\|v\|_{r,p} := \|u\|_p \quad \text{for} \quad v = (1 - L)^{-r/2}u.$$

In the connection to Sobolev spaces $W^{r,p}$, we give a definition of the $(r, p)$-capacity. The $(r, p)$-capacity $C_{r,p}$ is defined as follows: for an open set $G \subseteq B$,

$$C_{r,p}(G) := \inf\{\|u\|_{r,p}^p \,;\, u \in W^{r,p}, \, u \geq 1 \quad \mu\text{-a.e. on } G\}$$

and for an arbitrary set $A \subseteq B$,

$$C_{r,p}(A) := \inf\{C_{r,p}(G) \,;\, G \text{ is open and } G \supseteq A\}.$$

Since $W^{r,p} \cap C_b(B)$ is dense in $W^{r,p}$, any function $f \in W^{r,p}$ admits a quasi-continuous modification. Here a function $f$ on $B$ is said to be *quasi-continuous* if for any $\epsilon > 0$, there exists a closed set $F$ such that $C_{r,p}(B \setminus F) \leq \epsilon$ and the restriction $f|_F$ is continuous on $F$. This is valid for Hilbert valued functions $W^{r,p}(K)$.

Concerning the quasi-continuity, we remark the following general fact: if $f$ and $g$ are quasi-continuous and $f = g$ a.e., then $f = g$ q.e. For the proof, see [7]. We will use this fact frequently in the sequel. The following lemma can be obtained from this fact.

**Lemma 2.1** *If $f, 1/f \in W^{r,p}$, then $C_{r,p}(\tilde{f} = 0) = 0$ where $\tilde{f}$ is a quasi-continuous modification of $f$.*

**Proof** Let $(1/f)^{\sim}$ be a quasi-continuous modification of $1/f$. Then, $\tilde{f} \cdot (1/f)^{\sim}$ is quasi-continuous and $\tilde{f} \cdot (1/f)^{\sim} = f \cdot (1/f) = 1$ $\mu$-a.e. Hence, by using the above fact, we have $\tilde{f} \cdot (1/f)^{\sim} = 1$ q.e. This completes the proof. $\square$

## 2B. Submanifolds

Let us take $F = (F^1, \ldots, F^d) \in W^{\infty, \infty -}(\mathbf{R}^d)$, i.e., $F^i \in W^{\infty, \infty -}$, $i = 1, \ldots, d$. Set $\sigma^{ij} = (DF^i | DF^j)_{H^*}$, where $DF^i$ is the $H$-derivative of $F^i$. $\sigma = (\sigma^{ij})$ is called the *Malliavin covariance matrix*. We assume that $F$ is non-degenerate in the sense of Malliavin, i.e., $\det \sigma > 0$ a.e. and $(\det \sigma)^{-1} \in L^{\infty -}(\mu) = \bigcap_{p \geq 1} L^p(\mu)$. Now by Lemma 2.1, we have $\det \sigma > 0$ q.e. We denote the inverse matrix of $(\sigma^{ij})$ by $(\gamma_{ij})$, that is, of course, defined q.e.

By the Riesz theorem, the dual space $H^*$ is isomorphic to $H$. We denote the isomorphism by $\sharp: H^* \to H$ and its inverse by $\flat: H \to H^*$.

We define 1-forms $\omega^i$ and vector fields $X_i$, $i = 1, 2, \ldots, d$, by

$$\omega^i = DF^i, \quad X_i = \gamma_{ij}(DF^j)^\sharp.$$

Here, following the custom, we abbreviate the summation sign for repeated indices appearing once at the top and once at the bottom. In the sequel, we use this convention without mentioning. Then the followings can be easily obtained:

$$(\omega^i | \omega^j)_{H^*} = \sigma^{ij},$$
$$(X_i | X_j)_H = \gamma_{ij}, \qquad (2.2)$$
$$\omega^i(X_j) = \delta^i_j, \qquad (2.3)$$

where $\delta^i_j$ is the Kronecker delta. For example, let us see (2.2).

$$(X_i, X_j)_H = \gamma_{ik}\gamma_{jl}((DF^k)^\sharp | (DF^l)^\sharp)_H = \gamma_{ik}\gamma_{jl}\sigma^{kl} = \gamma_{ij}.$$

The others are similar.

Take an $a \in \mathbf{R}^d$ and fix it. Set

$$S = \{x \in B; F(x) = a\}.$$

Note that $S$ is defined up to quasi-every equivalence. We regard $S$ as a submanifold of the Wiener space. For $x \in S$, define projections $Q_x$ and $P_x$ in $H$ as follows:

$$Q_x h = \gamma_{ij}(x)DF^i_x(h)(DF^j_x)^\sharp = \omega^i(h)X_i, \qquad P_x h = h - Q_x h.$$

Similarly we have for $h^* \in H^*$,

$$Q_x^* h^* = \gamma_{ij}(x)\,(h^* | DF^i_x)_{H^*} DF^j_x = h^*(X_j)\omega^j, \qquad P_x^* h^* = h^* - Q_x^* h^*.$$

By (2.3) it is easy to see that $Q_x$ is symmetric and idempotent, i.e., $Q_x^2 = Q_x$. So $Q_x$ and $P_x$ are projections. Using these operators, we define the *tangent bundle* $T(S)_x$ and the *normal bundle* $N(S)$ as follows:

$$N(S)_x := Q_x(H) = \text{ linear span of } \{X_j; j = 1, \ldots, d\},$$
$$T(S)_x := P_x(H) = \{h \in H; DF^i(h) = 0,\ i = 1, \ldots, d\}.$$

The inner product in $H$ induces an inner product in $T(S)_x$ for all $x \in S$. We suppose that $T(S)$ is equipped with a Riemannian metric by this inner product. We denote it by $g$. Since we have gotten the tangent space, we can define the *tangential differentiation*. For $f \in W^{\infty,\infty-}(K)$,

$$D_F f(x)(h) = Df(x)(P_x h).$$

We denote the set of all Hilbert-Schmidt class operators from $H$ to $K$ by $\mathcal{L}_{(2)}(H; K)$. Then, from the definition, it follows that $D_F f \in W^{\infty,\infty-}(\mathcal{L}_{(2)}(H; K))$.

## 2C. Area measure on a submanifold

Following Airault-Malliavin [3], we introduce the area measure on a submanifold $S$. Set

$$m = \sqrt{\det \sigma}\, \delta_a(F),$$

where $\delta_a(F)$ is the composition of the Dirac measure $\delta_a$ and the non-degenerate Wiener function $F$ in the sense of Watanabe (see [20]). Since the right hand side is a positive generalized function, it defines a measure on $B$ (see, e.g., [18] or [3]). Further we note that for sufficiently large $r$ and $p$,

$$m(A) = 0 \quad \text{if} \quad C_{r,p}(A) = 0.$$

In the whole paper, we assume that $r$ and $p$ are taken to satisfy the above condition.

The necessity of the term $\sqrt{\det \sigma}$ can be seen in the following heuristic argument. Let $\varphi \colon \mathbf{R}^d \to \mathbf{R}^d$ be a diffeomorphism and set $G = \varphi \circ F$. Then, noting $\delta_{\varphi(a)} \circ \varphi = |\det \varphi'|^{-1}\delta_a$, $\det \varphi' = \det\left(\frac{\partial \varphi^i}{\partial \xi^j}\right)$ and $\det \sigma_G = (\det \varphi')^2 \det \sigma$, $\sigma_G = ((DG^i, DG^j)_{H^*})$, we have

$$\sqrt{\det \sigma_G}\, \delta_{\varphi(a)}(G) = |\det \varphi'|\sqrt{\det \sigma}\,|\det \varphi'|^{-1}\delta_a(F) = \sqrt{\det \sigma}\,\delta_a(F).$$

We take $m$ as a reference measure.

We can regard $m$ as a measure on $S$ by the following lemma:

**Lemma 2.2** $m(\tilde{F} \neq a) = 0$, where $\tilde{F}$ is a quasi-continuous modification of $F$.

**Proof** For any $\epsilon > 0$, take $\varphi \in C^\infty(\mathbf{R}^d)$ such that $0 \leq \varphi \leq 1$ and

$$\varphi = \begin{cases} 1 & \text{on } \{\xi; |\xi - a| \geq \epsilon\}, \\ 0 & \text{on } \{\xi; |\xi - a| \leq \epsilon/2\}. \end{cases}$$

Let $\{f_n\} \subseteq C_0^\infty(\mathbf{R}^d)$ be a sequence tending to $\delta_a$ in the sense of distribution. We may assume that $\mathrm{supp} f_n \subseteq \{\xi; |\xi - a| \leq 1/n\}$. Then we have

$$\int_B \varphi(\tilde{F}) dm = \langle \varphi(F) \sqrt{\det \sigma}, \delta_a(F) \rangle = \lim_{n \to \infty} \langle \varphi(F) \sqrt{\det \sigma}, f_n(F) \rangle.$$

We may assume that $\mathrm{supp} f_n \subseteq \{\xi; |\xi - a| \leq 1/n\}$. Now it is easy to see that

$$\int_B \varphi(\tilde{F}) dm = 0.$$

Since $\epsilon$ is arbitrary, we can get the desired result. $\square$

In the following, we discuss duality with respect to the measure $m$. Unfortunately, we already have the measure $\mu$. To avoid the confusion, $*$ indicates exclusively the dual operation with respect to $m$ in the sequel.

Let us calculate the dual operator of $D_F$. We denote the dual operator of $D$ with respect to $\mu$ by $\delta$, i.e., for $u \in W^{\infty,\infty-}(K)$, $v \in W^{\infty,\infty-}(\mathcal{L}_{(2)}(H;K))$,

$$\int_B (Du|v)_{HS} \, d\mu = \int_B (u|\delta v)_K \, d\mu,$$

where $(\cdot, \cdot)_{HS}$ denotes the Hilbert-Schmidt inner product in $\mathcal{L}_{(2)}(H;K)$. The Ornstein Uhlenbeck operator $L$ can be written as $L = -\delta D$.

We denotes the interior product by $i$:

$$i(h)v := v(h) \quad \text{for} \quad h \in H.$$

For $v \in W^{\infty,\infty-}(\mathcal{L}_{(2)}(H;K))$, set

$$D_F^* v = \delta(v \circ P) - \frac{1}{2} v((D_F \log \det \sigma)^\sharp)$$

$$= \delta v - \delta(v \circ Q) - \frac{1}{2} i((D_F \log \det \sigma)^\sharp) v.$$

Let us see that the above $D_F^*$ is actually the dual operator:

**Lemma 2.3** *For $u \in W^{\infty,\infty-}(K)$, $v \in W^{\infty,\infty-}(\mathcal{L}_{(2)}(H;K))$, it holds that*

$$\int_S (D_F u | v)_{HS}\, dm = \int_S (u | D_F^* v)_K\, dm. \qquad (2.4)$$

**Proof** Take any $f \in C_0^\infty(\mathbf{R}^d)$, then

$$\int_{\mathbf{R}^d} f(b)\langle (D_F u|v)_{HS}\sqrt{\det \sigma},\, \delta_b(F)\rangle db = \int_B (f \circ F)(D_F u|v)_{HS}\,\sqrt{\det \sigma}\, d\mu$$

$$= \int_B (Du|(f \circ F)\sqrt{\det \sigma}\,(v \circ P))_{HS}\, d\mu$$

$$= \int_B (u|\delta\{(f \circ F)\sqrt{\det \sigma}\,(v \circ P)\})_K\, d\mu.$$

Now noting

$$\delta((f \circ F)\sqrt{\det \sigma}\,(v \circ P))$$
$$= -(\partial_j f \circ F)\sqrt{\det \sigma}\,v(P(DF^j)^\sharp) - (f \circ F)v(P(D\sqrt{\det \sigma})^\sharp)$$
$$\quad + (f \circ F)\sqrt{\det \sigma}\,\delta(v \circ P)$$
$$= -\frac{1}{2}(f \circ F)\sqrt{\det \sigma}\,v((D_F \log \det \sigma)^\sharp) + (f \circ F)\sqrt{\det \sigma}\,\delta(v \circ P) \quad (\because P(DF^j)^\sharp = 0)$$
$$= (f \circ F)\sqrt{\det \sigma}\,D_F^* v,$$

we have

$$\int_{\mathbf{R}^d} f(b)\langle (D_F u|v)_{HS}\sqrt{\det \sigma},\, \delta_b(F)\rangle db = \int_B (f \circ F)(u|D_F^* v)_K \sqrt{\det \sigma}\, d\mu$$

$$= \int_{\mathbf{R}^d} f(b)\langle (u|D_F^* v)_K \sqrt{\det \sigma},\, \delta_b(F)\rangle db.$$

Since $f$ is arbitrary, we have (2.4). $\quad\square$

Let us calculate $D_F$ more explicitly. To do this, we calculate $\delta Q$ with $Q$ being the projection to the normal space. So $Q \in W^{\infty,\infty-}(H^*\otimes H)$ and $\delta Q \in W^{\infty,\infty-}(H)$. We denote the trace in the normal direction by $\mathrm{tr}_{N(S)}$, i.e., for $K \in H^* \otimes H^*$

$$\mathrm{tr}_{N(S)} K = \sigma^{jk} K(X_j, X_k) = \gamma_{jk}(K|\omega^j \otimes \omega^k).$$

**Lemma 2.4** *It holds that*

$$P\delta Q = -\frac{1}{2}\gamma_{jk}P(D\sigma^{jk})^\sharp, \qquad (2.5)$$
$$Q\delta Q = -(LF^j - \mathrm{tr}_{N(S)}D^2 F^j)X_j. \qquad (2.6)$$

**Proof** From the definition, we have

$$\delta Q = \delta(\gamma_{jk} DF^j \otimes (DF^k)^\sharp)$$
$$= -(D\gamma_{jk}|DF^j)(DF^k)^\sharp + \gamma_{jk}(\delta DF^j)(DF^k)^\sharp - \gamma_{jk} D^2 F^k(DF^j, \cdot)$$
$$= -(D\gamma_{jk}|DF^j)(DF^k)^\sharp - LF^j X_j - \gamma_{jk} D^2 F^k(DF^j, \cdot).$$

On the other hand,

$$- (D\gamma_{jk}|DF^j)(DF^k)^\sharp$$
$$= \gamma_{jl}\gamma_{mk}(D\sigma^{lm}|DF^j)(DF^k)^\sharp$$
$$= \gamma_{jl}(D(DF^l|DF^m)|DF^j)X_m$$
$$= \gamma_{jl}(D^2 F^l|DF^j \otimes DF^m)X_m + \gamma_{jl}(D^2 F^m|DF^j \otimes DF^l)X_m$$
$$= \frac{1}{2}\gamma_{jl}(D(DF^l|DF^j)|DF^m)X_m + \gamma_{jl}(D^2 F^m|DF^j \otimes DF^l)X_m$$
$$= \frac{1}{2}\gamma_{jl}(D\sigma^{lj}|DF^m)X_m + \gamma_{jl}(D^2 F^m|DF^j \otimes DF^l)X_m$$
$$= \frac{1}{2}\gamma_{jl}Q(D\sigma^{lj})^\sharp + \mathrm{tr}_{N(S)} D^2 F^m X_m$$

and

$$-\gamma_{jk}D^2 F^k(DF^j, \cdot) = -\frac{1}{2}\gamma_{jk}D(DF^j|DF^k) = -\frac{1}{2}\gamma_{jk}D\sigma^{jk}.$$

Thus

$$\delta Q = \mathrm{tr}_{N(S)} D^2 F^m X_m - LF^j X_j - \frac{1}{2}\gamma_{jl}P(D\sigma^{lj})^\sharp.$$

Noting $X_j \in N(S)$, we have (2.5) and (2.6). $\square$

For later use, we remark that $\mathrm{tr}_{N(S)}$ also can be written as

$$\mathrm{tr}_{N(S)}K = \sum_\lambda K(Q\varphi_\lambda, Q\varphi_\lambda)$$

where $\{\varphi_\lambda\}$ is a c.o.n.s. for $H$.

Now we can give an explicit form of $D_F^*$.

**Proposition 2.5** *For $v \in \mathcal{W}(S \to H^* \otimes K)$, it holds that*

$$\delta(v \circ Q) + \frac{1}{2}i((D_F \log \det \sigma)^\sharp)v = -\mathrm{tr}_{N(S)}Dv + i(Q\delta Q)v. \tag{2.7}$$

*Moreover,* $D_F^* v = \delta v + \mathrm{tr}_{N(S)}Dv - i(Q\delta Q)v.$

**Proof** It is enough to prove (2.7). First we note that for a matrix $a$

$$\frac{d}{dt}\det a = (\det a) \cdot \mathrm{tr}\left(a^{-1}\frac{da}{dt}\right).$$

Using this identity, we have

$$D\log\det\sigma = (\det\sigma)^{-1}D\det\sigma = (\det\sigma)^{-1}(\det\sigma)\mathrm{tr}(\sigma^{-1}D\sigma) = \gamma_{jk}D\sigma^{jk}.$$

Thus

$$D_F\log\det\sigma = \gamma_{jk}P^*D\sigma^{jk}.$$

Combining this with Lemma 2.4 we have

$$\delta(v\circ Q) + \frac{1}{2}i((D_F\log\det\sigma)^\sharp)v = v(\delta Q) - (Dv|\hat{Q}) + \frac{1}{2}i(\gamma_{jk}P(D\sigma^{jk})^\sharp)v$$
$$= -(Dv|\hat{Q}) + i(Q\delta Q)v,$$

where $\hat{Q} \in W^{\infty,\infty-}(H^* \otimes H^*)$ is defined by

$$\hat{Q} = \gamma_{jk}DF^j \otimes DF^k.$$

Hence

$$(Dv|\hat{Q}) = \gamma_{jk}(Dv|DF^j \otimes DF^k) = \mathrm{tr}_{N(S)}Dv.$$

Now we arrive at (2.7).    □

By Lemma 2.3, we easily have that if $u = w$ $m$-a.e., then $D_F u = D_F w$ $m$-a.e. Here after, we identify two functions $u$ and $w$ if $u = w$ $m$-a.e. By this equivalence relation, we denote the equivalence class of $W^{\infty,\infty-}(K)$ by $\mathcal{W}(S \to K)$. Hence $D_F$ is a linear operator from $\mathcal{W}(S \to K)$ to $\mathcal{W}(S \to \mathcal{L}_{(2)}(H;K))$. Similarly, $D_F^*$ is a linear operator from $\mathcal{W}(S \to \mathcal{L}_{(2)}(H;K))$ to $\mathcal{W}(S \to K)$.

## 2D. Tensor fields

We have defined the tangent bundle. Then we can define tensor bundles as usual. First note that the cotangent space at $x$ can be defined as follows:

$$T^*(S)_x := (T(S)_x)^* = P_x^*(H^*).$$

Using this, we define for $(k,l) \in \mathbf{Z}_+ \times \mathbf{Z}_+$

$$T_l^k(S)_x := \overbrace{T(S)_x \otimes \cdots \otimes T(S)_x}^{k} \otimes \overbrace{T^*(S)_x \otimes \cdots \otimes T^*(S)_x}^{l}.$$

Set

$$T_l^k(S) := \bigcup_{x \in S} T_l^k(S)_x$$

and we call $T_l^k(S)$ the *tensor bundle* of type $(k, l)$.

To proceed to the tensor fields, we define $H_l^k$ by

$$H_l^k := \overbrace{H \otimes \cdots \otimes H}^{k} \otimes \overbrace{H^* \otimes \cdots \otimes H^*}^{l}$$

and set

$$\pi_x := \overbrace{P_x \otimes \cdots \otimes P_x}^{k} \otimes \overbrace{P_x^* \otimes \cdots \otimes P_x^*}^{l}.$$

$\pi_x$ depends on $(k, l)$, but we do not specify it to avoid the complexity. Note that $\pi_x$ is a projection operator on $H_l^k$ and $\mathrm{Ran}(\pi_x) = T_l^k(S)_x$. Now we can define the space of *smooth tensor fields* $\Gamma(T_l^k(S))$ to be a set of all $u \in \mathcal{W}(S \to H_l^k)$ such that $\pi_x u(x) = u(x)$, i.e., $u(x) \in T_l^k(S)_x$ for all $x \in S$.

Similarly we can define the space of *smooth sections* of $N(S)$ that we denote by $\Gamma(N(S))$. We also denote the set of all vector fields by $\mathfrak{X}(S)$ in place of $\Gamma(T(S))$.

## 3. Covariant Derivative

### 3A. Covariant derivative and second fundamental form

We define the *covariant derivative* on the submanifold $S$ as follows: for $X, Y \in \mathfrak{X}(S)$,

$$\nabla_X Y = P(D_F Y(X)).$$

More precisely,

$$(\nabla_X Y)_x = P_x (D_F Y)_x (X_x).$$

Here $(D_F Y)_x \in H^* \otimes H$. Further, the *second fundamental form* $\alpha$ is defined by

$$\alpha(X, Y) = Q(D_F Y(X)) = D_F Y(X) - P(D_F Y(X)).$$

**Theorem 3.1** *The second fundamental form $\alpha$ is written in the following form:*

$$\alpha(X, Y) = -D^2 F^j(X, Y) X_j.$$

*In particular, $\alpha$ is symmetric.*

**Proof** It is enough to prove that

$$D_F Y(X) + D^2 F^j(X, Y) X_j \in \mathrm{Ker}(\omega^k) \quad k = 1, 2, \ldots, d.$$

First note that

$$D_F F^j = 0 \quad \text{on} \quad B.$$

Hence

$$0 = D_F F^j(Y) = DF^j(PY) = DF^j(Y) \quad \text{on} \quad S.$$

Further

$$0 = D_F(DF^j(Y))(X) = D(DF^j(Y))(PX) = D(DF^j(Y))(X)$$
$$= D^2 F^j(X, Y) + DF^j(DY(X)) = D^2 F^j(X, Y) + DF^j(D_F Y(X)).$$

Therefore,

$$\omega^k(D_F Y(X) + D^2 F^j(X, Y) X_j) = DF^k(D_F Y(X)) + D^2 F^k(X, Y) = 0$$

as desired.  $\square$

**Remark 3.1** For convention, we assume that the second fundamental form vanishes in the normal direction. So we had better to write $\alpha(\cdot, \cdot) = -D^2 F^j(P\cdot, P\cdot) X_j$.

We can define the *covariant differentiation* of the normal bundle $N(S)$. For $X \in \mathfrak{X}(S)$ and $\nu \in \Gamma(N(S))$,

$$\nabla_X^\perp \xi = Q(D_F \nu(X)).$$

Further, for $X \in \mathfrak{X}(S)$ and $\nu \in \Gamma(N(S))$, set

$$\beta(X, \nu) = P(D_F \nu(X)) = D_F \nu(X) - \nabla_X^\perp \nu.$$

Then $\nabla^\perp$ defined the *connection* on the normal bundle $N(S)$.

**Lemma 3.2** *For $X, Y \in \mathfrak{X}(S)$ and $\nu \in \Gamma(N(S))$, it holds that*

$$(\alpha(X, Y)|\nu)_H + (Y|\beta(X, \nu))_H = 0.$$

**Proof** Noting that $(Y|\nu)_H = 0$, we have

$$0 = (D_F(Y|\nu)_H)(X)$$
$$= (D_F Y(X)|\nu) + (Y|D_F \nu(X))_H$$
$$= (\alpha(X, Y)|\nu)_H + (Y|\beta(X, \nu))_H$$

which completes the proof. $\square$

In the connection to the second fundamental form, we calculate $DQ$.

**Proposition 3.3** *We have the following identities:*

$$DQ(P\cdot, P\cdot) = -\alpha(\cdot, \cdot), \tag{3.1}$$

$$QDQ(\cdot, \cdot) = DQ(\cdot, P\cdot), \tag{3.2}$$

$$QDQ(\cdot, Q\cdot) = 0. \tag{3.3}$$

*In particular, $DQ(Ph, Pk)$ is symmetric in $h$ and $k$. Further for $\xi \in \Gamma(T^*(S))$, it holds that*

$$D\xi(P\cdot, Qh) = (\alpha(\cdot, \xi^\sharp)|h) \quad \text{for } h \in H^*. \tag{3.4}$$

**Proof** We first prove (3.1). For $X \in \mathfrak{X}(S)$, it holds that $QX = 0$ on $S$. Hence

$$\begin{aligned}
0 &= D_F(QX) \\
&= D_F Q(\cdot, X) + Q(D_F X(\cdot)) \\
&= DQ(P\cdot, X) + Q(DX(P\cdot)) \\
&= DQ(P\cdot, X) + \alpha(P\cdot, X).
\end{aligned}$$

Since $X$ is arbitrary, we have (3.1).

To get (3.2), we note $Q(Q\cdot) = Q(\cdot)$. By differentiating both hands, we have

$$DQ(\cdot, Q(\cdot)) + Q(DQ(\cdot, \cdot)) = DQ(\cdot, \cdot).$$

Hence

$$QDQ(\cdot, \cdot) = DQ(\cdot, \cdot) - DQ(\cdot, Q\cdot) = DQ(\cdot, P\cdot),$$

which is (3.2). By substituting $Q\cdot$ into the second variable, we easily get (3.3).

Lastly, from the definition of the second fundamental form, it holds that for $X \in \mathfrak{X}(S)$,

$$\alpha(\cdot, X) = QDX(P\cdot).$$

Setting $X = \xi^\sharp$ for $\xi \in \Gamma(T^*(S))$, we have for $h \in H^*$

$$(\alpha(\cdot, \xi^\sharp)|h) = (QD\xi^\sharp(P\cdot)|h) = D\xi(P\cdot, Qh)$$

which shows (3.4). This completes the proof. $\square$

## 3B. Lie bracket

For $X, Y \in \mathfrak{X}(S)$ we define *Lie bracket* $[X, Y]$ by

$$[X, Y] = \nabla_X Y - \nabla_Y X. \tag{3.5}$$

Let us see that the definition of Lie bracket is consistent with the usual one.

**Theorem 3.4**  *For $X, Y \in \mathfrak{X}(S)$ and $f \in \mathcal{W}(S \to \mathbf{R})$, we have*

$$[X, Y]f = X(Yf) - Y(Xf).$$

**Proof**  First we note that

$$
\begin{aligned}
X(Yf) &= D(Yf)(X) \\
&= D(Df(Y))(X) \\
&= D^2 f(X, Y) + Df(DY(X)) \\
&= D^2 f(X, Y) + Df(\nabla_X Y) + Df(\alpha(X, Y)) \\
&= D^2 f(X, Y) + (\nabla_X Y)f + Df(\alpha(X, Y)).
\end{aligned}
$$

Using the symmetry of $D^2 f$ and $\alpha$, we have

$$
\begin{aligned}
X(Yf) - Y(Xf) &= D^2 f(X, Y) + (\nabla_X Y)f + Df(\alpha(X, Y)) \\
&\quad - D^2 f(Y, X) - (\nabla_Y X)f - Df(\alpha(Y, X)) \\
&= (\nabla_X Y)f - (\nabla_Y X)f.
\end{aligned}
$$

This completes the proof.  $\square$

By this characterization of Lie bracket, we easily see the Jacobi identity:

$$[[X, Y], Z] + [[Y, Z], X] + [[Z, X], Y] = 0.$$

## 3C. Curvature

Now we can define the *curvature tensor*. For $X, Y \in \mathfrak{X}(S)$, we define $R(X, Y) \in \Gamma(T_1^1(S))$ by

$$R(X, Y) := [\nabla_X, \nabla_Y] - \nabla_{[X, Y]} = \nabla_X \nabla_Y - \nabla_Y \nabla_X - \nabla_{[X, Y]}.$$

Hence $R \in \Gamma(T_3^1(S))$. But we have a Riemannian metric, we may regard $R$ as an element of $\Gamma(T_4^0(S))$ as follows:

$$R(X, Y, Z, W) = (R(X, Y)Z | W)_H.$$

**Theorem 3.5** *The curvature $R$ can be written as follows:*

$$R(X, Y, Z, W) = -(\alpha(X, Z)|\alpha(Y, W))_H + (\alpha(Y, Z)|\alpha(X, W))_H. \qquad (3.6)$$

**Proof** First, we note

$$D_F Z(Y) = \nabla_Y Z + \alpha(Y, Z).$$

Since $\nabla_Y Z$ is tangential and $\alpha(Y, Z)$ is normal, we have further

$$D_F(D_F Z(Y))(X) = \nabla_X \nabla_Y Z + \alpha(X, \nabla_Y Z) + \nabla_X^\perp \alpha(Y, Z) + \beta(X, \alpha(Y, Z)).$$

Interchanging $X$ and $Y$, we have

$$D_F(D_F Z(X))(Y) = \nabla_Y \nabla_X Z + \alpha(Y, \nabla_X Z) + \nabla_Y^\perp \alpha(X, Z) + \beta(Y, \alpha(X, Z)).$$

On the other hand,

$$\begin{aligned}
D_F&(D_F Z(Y))(X) - D_F(D_F Z(X))(Y) \\
&= D(DZ(Y))(X) - D(DZ(X))(Y) \\
&= D^2 Z(X, Y) + DZ(DY(X)) - D^2 Z(Y, X) - DZ(DX(Y)) \\
&= DZ(D_F Y(X) - D_F X(Y)) \\
&= DZ(\nabla_X Y + \alpha(X, Y) - \nabla_Y X - \alpha(Y, X)) \\
&= DZ([X, Y]) \\
&= \nabla_{[X,Y]} Z + \alpha([X, Y], Z).
\end{aligned}$$

Thus we have

$$\begin{aligned}
\nabla_{[X,Y]} Z + \alpha([X, Y], Z) &= \nabla_X \nabla_Y Z + \alpha(X, \nabla_Y Z) + \nabla_X^\perp \alpha(Y, Z) + \beta(X, \alpha(Y, Z)) \\
&\quad - \nabla_Y \nabla_X Z - \alpha(Y, \nabla_X Z) - \nabla_Y^\perp \alpha(X, Z) - \beta(Y, \alpha(X, Z)).
\end{aligned}$$

Comparing the tangential component, we get

$$\begin{aligned}
R(X, Y)Z &= \nabla_X \nabla_Y Z - \nabla_Y \nabla_X Z - \nabla_{[X,Y]} Z \\
&= -\beta(X, \alpha(Y, Z)) + \beta(Y, \alpha(X, Z)).
\end{aligned}$$

Now using Lemma 3.2, we obtain

$$\begin{aligned}
R(X, Y, Z, W) &= -(W|\beta(X, \alpha(Y, Z)))_H + (W|\beta(Y, \alpha(X, Z)))_H \\
&= (\alpha(X, W)|\alpha(Y, Z))_H - (\alpha(Y, W)|\alpha(X, Z))_H,
\end{aligned}$$

which completes the proof. $\quad\square$

Now we can see that $R \in \Gamma(T_4^0(S))$. The following identities can be deduced from the expression (3.6). We can give also a direct proof from the definition in the same manner as the finite dimensional case (see, e.g., [11]).

**Theorem 3.6** *We have*

$$R(X,Y) = -R(Y,X), \qquad \mathfrak{S}R(X,Y)Z = 0,$$

$$R(X,Y,Z,W) = -R(X,Y,W,Z), \qquad R(X,Y,Z,W) = R(Z,W,X,Y).$$

*Here $\mathfrak{S}$ denotes the cyclic sum with respect to $X$, $Y$ and $Z$.*

### 3D. Covariant derivative of tensor fields

We can define the *covariant derivative of tensor fields* as follows: for $u \in \Gamma(T_l^k(S))$,

$$\nabla u := \pi D_F u.$$

This means that the covariant derivative $\nabla : \Gamma(T_l^k(S)) \to \Gamma(T_{l+1}^k(S))$ is the composite of the following mappings:

$$D_F : \mathcal{W}(S \to H_l^k) \longrightarrow \mathcal{W}(S \to H_{l+1}^k)$$

$$\pi : \mathcal{W}(S \to H_{l+1}^k) \longrightarrow \Gamma(T_{l+1}^k(S)).$$

Similarly, we can define the *covariant derivative along the tangent vector:* for $u \in \Gamma(T_l^k(S))$ and $X \in \mathfrak{X}(S)$,

$$\nabla_X u := \nabla u(X) = \pi(D_F u(X)).$$

The following derivation property is obvious from the definition: for $u \in \Gamma(T_l^k(S))$, $v \in \Gamma(T_n^m(S))$ and $X \in \mathfrak{X}(S)$

$$\nabla_X(u \otimes v) = \nabla_X u \otimes v + u \otimes \nabla_X v.$$

Moreover, this operation is commutative with the contraction, e.g., for $Y \in \mathfrak{X}(S)$ and $\omega \in \Gamma(T^*(S))$

$$\nabla_X(\omega(Y)) = X(\omega(YT(S))) = (\nabla_X \omega)(Y) + \omega(\nabla_X Y).$$

In fact, noting $PY = Y$ and $P^*\omega = \omega$, we have

$$X(\omega(Y)) = D_F(\omega(Y))(X)$$

$$= (D_F\omega(X))(Y) + (\omega(D_F Y(X))$$

$$= (P^*D_F\omega(X))(Y) + \omega(PD_F Y(X))$$

$$= \nabla_X\omega(Y) + \omega(\nabla_X Y).$$

In general, for $u \in \Gamma(T_l^k(S))$, $X, Y_1, \ldots Y_l \in \mathfrak{X}(S)$, it holds that

$$(\nabla_X u)(Y_1, \ldots, Y_l) := \nabla u(X, Y_1, \ldots, Y_l)$$

$$= \nabla_X(u(Y_1, \ldots, Y_l)) - \sum_{j=1}^{l} u(Y_1, \ldots, \nabla_X Y_j, \ldots, Y_l). \tag{3.7}$$

A natural Riemannian metric in $T(S)$ is just a restriction of the inner product of $H$. We already used this metric. For a while we specify it by $g$. Clearly $g \in \Gamma(T_2^0(S))$.

**Theorem 3.7**  *The Riemannian metric $g$ is parallel, i.e., $\nabla g = 0$.*

**Proof**  It is enough to prove that $\nabla X g = 0$ for any $X \in \mathfrak{X}(S)$. To see this, for $Y, Z \in \mathfrak{X}(S)$, we note that $\alpha(X, Y) \perp Z$ and $\alpha(X, Z) \perp Y$ and thereby we have

$$\begin{aligned}
\nabla_X g(Y, Z) &= X(g(Y, Z)) - g(\nabla_X Y, Z) - g(Y, \nabla_X Z) \\
&= D_F((Y|Z)_H)(X) - (\nabla_X Y|Z)_H - (Y|\nabla_X Z)_H \\
&= (D_F Y(X)|Z)_H + (Y|D_F Z(X))_H - (\nabla_X Y|Z)_H - (Y|\nabla_X Z)_H \\
&= (\nabla_X Y + \alpha(X, Y)|Z)_H + (Y|\nabla_X Z + \alpha(X, Z))_H \\
&\quad - (\nabla_X Y|Z)_H - (Y|\nabla_X Z)_H \\
&= 0.
\end{aligned}$$

This completes the proof.  $\square$

By the above theorem and (3.5), we may call the connection $\nabla$ the *Levi-Civita connection*. In the sequel, the Riemannian metric will be denoted by $(\ |\ )$ instead of $g$ for simplicity. We can show the following second Bianchi identity as well (see, e.g., [11]).

**Theorem 3.8**  *The following identity holds: $\mathfrak{S} \nabla_X R(Y, Z) = 0$.*

For $u \in \Gamma(T_l^k(S))$, we can define inductively $\nabla^2 u$, $\nabla^3 u$, $\ldots$. For example, the second covariant derivative $\nabla^2 u$ belongs to $\Gamma(T_{l+2}^k(S))$. In general, $\nabla^2 u(X, Y)$ is not symmetric in $X$ and $Y$. We set

$$R^{T_l^k(S)}(X, Y)u = \nabla^2 u(X, Y) - \nabla^2 u(Y, X).$$

$R^{T_l^k(S)}$ is called a curvature of the vector bundle $T_l^k(S)$. Using (3,22), we easily have

$$\nabla^2 u(X, Y) = \nabla_X \nabla_Y u - \nabla_{\nabla_X Y} u.$$

132

Therefore, we can write $R^{T_l^k(S)}$ as follows:

$$R^{T_l^k(S)}(X,Y)u = [\nabla_X, \nabla_Y]u - \nabla_{[X,Y]}u.$$

Thus the Riemannian curvature is the curvature of the tangent bundle $T(S)$.

For $\eta \in T^*(S) = T_1(S)$, the curvature is given by

$$R^{T^*(S)}(X,Y)\eta = \{R(X,Y)\eta^\sharp\}^\flat,$$

i.e.,

$$R^{T^*(S)}(X,Y) = \flat \circ R(X,Y) \circ \sharp.$$

To see this, for $X$, $Y$, $Z \in \mathfrak{X}(S)$ and $\eta \in \Gamma(T_1^0(S))$, note that

$$\nabla_Y\langle \eta, Z\rangle = \langle \nabla_Y\eta, Z\rangle + \langle \eta, \nabla_Y Z\rangle,$$

where $\langle\ ,\ \rangle$ denotes a natural paring. Further we have

$$\nabla_X\nabla_Y\langle \eta, Z\rangle = \langle \nabla_X\nabla_Y\eta, Z\rangle + \langle \nabla_Y\eta, \nabla_X Z\rangle + \langle \nabla_X\eta, \nabla_Y Z\rangle + \langle \eta, \nabla_X\nabla_Y Z\rangle.$$

Interchanging $X$ and $Y$, we easily get

$$\begin{aligned}
0 &= \nabla_X\nabla_Y\langle \eta, Z\rangle - \nabla_Y\nabla_X\langle \eta, Z\rangle - \nabla_{[X,Y]}\langle \eta, Z\rangle \\
&= \langle R^{T^*(S)}(X,Y)\eta, Z\rangle + \langle \eta, R(X,Y)Z\rangle \\
&= \langle R^{T^*(S)}(X,Y)\eta, Z\rangle + R(X,Y,Z,\eta^\sharp) \\
&= \langle R^{T^*(S)}(X,Y)\eta, Z\rangle - R(X,Y,\eta^\sharp,Z) \\
&= \langle R^{T^*(S)}(X,Y)\eta, Z\rangle - (R(X,Y)\eta^\sharp|Z) \\
&= \langle R^{T^*(S)}(X,Y)\eta, Z\rangle - \langle \{R(X,Y)\eta^\sharp\}^\flat, Z\rangle.
\end{aligned}$$

Further we have for $\eta_1$, $\eta_2, \ldots, \eta_n$,

$$R^{T_n(S)}(X,Y)(\eta_1 \otimes \cdots \otimes \eta_n) = \sum_{j=1}^n \eta_1 \otimes \cdots \otimes R^{T_1(S)}(X,Y)\eta_j \otimes \cdots \otimes \eta_n.$$

## 3E. Lie derivative

For $X \in \mathfrak{X}(S)$, define a homomorphism $S_X : T(S) \to T(S)$ by

$$S_X Y = -\nabla_Y X,$$

i.e., $S_X = -P\,DX\,P$. Let $S_X^* : T^*(S) \to T^*(S)$ be the dual homomorphism of $S_X$. It can be written as $S_X^* = -P^*(DX)^*P^*$. Now we can define the homomorphism $d\Gamma(S_X) : T_l^k(S) \to T_l^k(S)$ such that

$$d\Gamma(S_X)(Y_1 \otimes \cdots \otimes Y_k \otimes \eta_1 \otimes \cdots \otimes \eta_l)$$
$$= \sum_{i=1}^{k} Y_1 \otimes \cdots \otimes S_X Y_i \otimes \cdots Y_k \otimes \eta_1 \otimes \cdots \otimes \eta_l$$
$$+ \sum_{j=1}^{l} Y_1 \otimes \cdots \otimes Y_k \otimes \eta_1 \otimes \cdots \otimes (-S_X^*)\eta_j \otimes \cdots \otimes \eta_l,$$

where $Y_1, \ldots, Y_k \in \mathfrak{X}(S)$ and $\eta_1, \ldots, \eta_l \in T^*(S)$. $d\Gamma(S_X)$ is called the *derivation extension*. We set $d\Gamma(S_X) = 0$ on $\Gamma(T_0^0(S)) = \mathcal{F}(S)$. Then it is easy to see that $d\Gamma(S_X)$ commutes with the contraction, e.g.,

$$d\Gamma(S_X)(\omega(X)) = 0 = (-S_X^*\omega)(X) + \omega(S_X X).$$

Now the *Lie differentiation* can be defined as follows:

$$L_X := \nabla_X + d\Gamma(S_X).$$

Then, for $X, Y \in \mathfrak{X}(S)$,

$$L_X Y = \nabla_X Y - \nabla_Y X = [X, Y].$$

Further $L_X$ satisfies the derivation property and commutes with the contraction.

**Proposition 3.9** *Let $X \in \mathfrak{X}(S)$ and $u \in \Gamma(T_n^0(S))$. Then for all $Y_1, \ldots, Y_n \in \mathfrak{X}(S)$:*

$$L_X u(Y_1, \ldots, Y_n) = X(u(Y_1, \ldots, Y_n)) - \sum_{j=1}^{n} u(Y_1, \ldots, [X, Y_j], \ldots, Y_n)$$

**Proof** Note that

$$(\nabla_X u)(Y_1, \ldots, Y_n) = X(u(Y_1, \ldots, Y_n)) - \sum_{j=1}^{n} u(Y_1, \ldots, \nabla_X Y_j, \ldots, Y_n)$$

and

$$(d\Gamma(S_X)u)(Y_1, \ldots, Y_n) = \sum_{j=1}^{n} u(Y_1, \ldots, -S_X Y_j, \ldots, Y_n)$$
$$= \sum_{j=1}^{n} u(Y_1, \ldots, \nabla_{Y_j} X, \ldots, Y_n).$$

Hence, by (3.1),

$$(L_X u)(Y_1,\ldots,Y_n) = X(u(Y_1,\ldots,Y_n)) - \sum_{j=1}^{n} u(Y_1,\ldots,\nabla_X Y_j - \nabla_{Y_j} X,\ldots,Y_n)$$

$$= X(u(Y_1,\ldots,Y_n)) - \sum_{j=1}^{n} u(Y_1,\ldots,[X,Y_j],\ldots,Y_n)$$

which completes the proof. $\square$

We introduce the interior product: for $X \in \mathfrak{X}(S)$, the map $i(X)\colon \Gamma(T^0_{n+1}(S)) \to \Gamma(T^0_n(S))$ is defined by

$$i(X)u(Y_1,\ldots,Y_n) = u(X,Y_1,\ldots,Y_n)$$

for $Y_1,\ldots,Y_n \in \mathfrak{X}(S)$. Then we can show the following identities by standard argument (see, e.g., [16]):

**Proposition 3.10** *For $X,Y \in \mathfrak{X}(S)$ and $u \in \Gamma(T^0_n(S))$, we have*

$$i([X,Y])u = [L_X,i(Y)]u := L_X i(Y)u - i(Y)L_X u$$

$$L_{[X,Y]}u = [L_X,L_Y]u := L_X L_Y u - L_Y L_X u$$

$$i(\nabla_X Y)u = [\nabla_X,i(Y)]u := \nabla_X i(Y)u - i(Y)\nabla_X u.$$

## 3F. Differential forms

For a tensor field $\omega \in \Gamma(T^0_p(S))$, we can define the *alternation mapping* $A_p$ as

$$(A_p\omega)(Y_1,Y_2,\ldots,Y_p) = \frac{1}{p!} \sum_{\sigma \in \mathfrak{S}_p} \mathrm{sgn}\sigma\, \omega(Y_{\sigma(1)},Y_{\sigma(2)},\ldots,Y_{\sigma(p)}),$$

where $\mathfrak{S}_p$ denotes the set of all permutations of $\{1,2,\ldots,p\}$ and sgn denotes the signature of the permutation. If $A_p\omega = \omega$, we call it a *differential form of degree* $p$ or a *$p$-form* for short. We denote the set of all $p$-forms by $\Omega^p(S)$. The exterior product of $\omega \in \Omega^p(S)$ and $\eta \in \Omega^q(S)$ can be defined as

$$\omega \wedge \eta = \frac{(p+q)!}{p!q!} A_{p+q}(\omega \otimes \eta).$$

Then the followings are easily seen: for $\omega \in \Omega^p(S)$, $\eta \in \Omega^q(S)$ and $\xi \in \Omega^r(S)$,

$$\omega \wedge \eta = (-1)^{pq}\eta \wedge \omega, \qquad \omega \wedge (\eta \wedge \xi) = (\omega \wedge \eta) \wedge \xi.$$

Now we can define the *exterior derivation* $d: \Omega^p(S) \to \Omega^{p+1}(S)$ as follows: for $\omega \in \Omega^p(S)$,

$$d\omega = (p+1)A_{p+1}\nabla\omega.$$

By (3.7), we have

$$\nabla_X\omega(Y_1,\ldots,Y_p) = X(\omega(Y_1,\ldots,Y_p)) - \sum_{j=1}^{p}\omega(Y_1,\ldots,\nabla_X Y_j,\ldots,Y_p). \qquad (3.8)$$

By this expression, it is easy to see that $\nabla_X\omega \in \Omega^p(S)$. Combining this with the definition of exterior product $\wedge$, we have, for $\omega \in \Omega^p(S)$ and $\xi \in \Omega^q(S)$

$$\nabla_X(\omega \wedge \xi) = \nabla_X\omega \wedge \xi + \omega \wedge \nabla_X\xi.$$

Further

$$d\omega(Y_1,Y_2,\ldots,Y_{p+1}) = \sum_{k=1}^{p+1}(-1)^{k-1}\nabla_{Y_k}\omega(Y_1,\ldots\overset{k}{\vee}\ldots,Y_{p+1})$$

$$= \sum_{k=1}^{p+1}(-1)^{k-1}\nabla\omega(Y_k,Y_1,\ldots\overset{k}{\vee}\ldots,Y_{p+1}), \qquad (3.9)$$

where $\ldots\overset{k}{\vee}$ means that $Y_k$ is deleted. Now we can express the exterior differential without using the covariant derivative as follows:

**Proposition 3.11**  *For $\omega \in \Omega^p(S)$, $\eta \in \Omega^q(S)$,*

$$d\omega(Y_1,Y_2,\ldots,Y_{p+1}) = \sum_{k=1}^{p+1}(-1)^{k-1}Y_k(\omega(Y_1,\ldots\overset{k}{\vee}\ldots,Y_{p+1}))$$

$$+ \sum_{j<k}(-1)^{j+k}\omega([Y_j,Y_k],Y_1,\ldots\overset{j}{\vee}\ldots\overset{k}{\vee}\ldots,Y_{p+1}).$$

**Proof**  Combining (3.9) with (3.8), we have

$$d\omega(Y_1,Y_2,\ldots,Y_{p+1}) = \sum_{k=1}^{p+1}(-1)^{k-1}Y_k(\omega(Y_1,\ldots\overset{k}{\vee}\ldots,Y_{p+1}))$$

$$- \sum_{j\neq k}(-1)^{k-1}\omega(Y_1,\ldots,\nabla_{Y_k}Y_j,\ldots\overset{k}{\vee}\ldots,Y_{p+1})$$

$$= \sum_{k=1}^{p+1} (-1)^{k-1} Y_k(\omega(Y_1, \dots \overset{k}{\vee} \dots, Y_{p+1}))$$

$$- \sum_{j<k} (-1)^{k-1+j-1} \omega(\nabla_{Y_k} Y_j, Y_1, \dots \overset{j}{\vee} \dots \overset{k}{\vee} \dots, Y_{p+1})$$

$$- \sum_{j>k} (-1)^{k-1+j-2} \omega(\nabla_{Y_k} Y_j, Y_1, \dots \overset{k}{\vee} \dots \overset{j}{\vee} \dots, Y_{p+1})$$

$$= \sum_{k=1}^{p+1} (-1)^{k-1} Y_k(\omega(Y_1, \dots \overset{k}{\vee} \dots, Y_{p+1}))$$

$$+ \sum_{j<k} (-1)^{j+k} \omega([Y_j, Y_k], Y_1, \dots \overset{j}{\vee} \dots \overset{k}{\vee} \dots, Y_{p+1})$$

which completes the proof. $\square$

Next we show the following Cartan identity:

**Proposition 3.12** *For $\omega \in \Omega^p(S)$, $p \geq 0$, and $X \in \mathfrak{X}(S)$, it holds that*

$$L_X \omega = i(X) d\omega + d i(X)\omega.$$

*Here if $\omega \in \Omega^0(S) = \mathcal{F}(S)$, we define $i(X)\omega = 0$.*

**Proof** By Proposition 3.11, we have

$$i(X) d\omega(Y_1, \dots, Y_p) = d\omega(X, Y_1, \dots, Y_p)$$

$$= X(\omega(Y_1, \dots, Y_p)) + \sum_{k=1}^{p+1} (-1)^k Y_k(\omega(X, Y_1, \dots \overset{k}{\vee} \dots, Y_p))$$

$$+ \sum_{j=1}^{p} (-1)^k \omega([X, Y_k], Y_1, \dots \overset{k}{\vee} \dots, Y_p)$$

$$+ \sum_{j<k} (-1)^{j+k} \omega([Y_j, Y_k], X, Y_1, \dots \overset{j}{\vee} \dots \overset{k}{\vee} \dots, Y_{p+1}).$$

By virtue of Proposition 3.9, the sum of the first term and the third term equals $L_X \omega(Y_1, \dots, Y_p)$. Further the sum of the second term and the fourth term equals $-(di(\omega))(Y_1, \dots, Y_p)$ by Proposition 3.11. This completes the proof. $\square$

Using the Cartan identity, we can show the following theorem (for the proof, see [16]):

**Theorem 3.13** *We have $d^2 = 0$ and for $X \in \mathfrak{X}(S)$,*

$$L_X d = d L_X. \tag{3.10}$$

*Further we have, for $\omega \in \Omega^p(S)$ and $\xi \in \Omega^q(S)$*

$$L_X(\omega \wedge \xi) = L_X \omega \wedge \xi + \omega \wedge L_X \xi,$$

$$d(\omega \wedge \xi) = d\omega \wedge \xi + (-1)^p \omega \wedge d\xi.$$

The above theorem shows that $d$ and $L_X$ commute. But $d$ and $\nabla_X$ do not commute in general. We will see it in the case of 1-form.

**Proposition 3.14** *For $\eta \in \Omega^1(S)$, it holds that*

$$(\nabla_X d\eta - d\nabla_X \eta)(Y, Z) = -(\nabla_{\nabla_Y X}\eta)(Z) + (\nabla_{\nabla_Z X}\eta)(Y) + \eta(R(Y, Z)X). \tag{3.11}$$

**Proof** By (3.10),

$$\begin{aligned}
\nabla_X d\eta - d\nabla_X \eta &= (\nabla_X - L_X)d\eta + L_X d\eta - d(\nabla_X - L_X)\eta - dL_X \eta \\
&= (\nabla_X - L_X)d\eta - d(\nabla_X - L_X)\eta.
\end{aligned}$$

First we calculate $(\nabla_X - L_X)d\eta$.

$$\begin{aligned}
(\nabla_X - L_X)d\eta(Y, Z) &= -(d\Gamma(S_X)d\eta)(Y, Z) \\
&= d\eta(S_X Y, Z) + d\eta(Y, S_X Z) \\
&= d\eta(-\nabla_Y X, Z) + d\eta(Y, -\nabla_Z X) \\
&= -\nabla_Y X \eta(Z) + Z\eta(\nabla_Y X) + \eta([\nabla_Y X, Z]) \\
&\quad - Y\eta(\nabla_Z X) + \nabla_Z X \eta(Y) + \eta([Y, \nabla_Z X]).
\end{aligned}$$

Secondly,

$$\begin{aligned}
(d(\nabla_X - L_X)\eta)(Y, Z) &= -(d(d\Gamma(S_X)\eta))(Y, Z) \\
&= -Y(d\Gamma(S_X)\eta(Z)) + Z(d\Gamma(S_X)\eta(Y)) + d\Gamma(S_X)\eta([Y, Z]) \\
&= -Y\eta(-S_X Z) + Z\eta(-S_X Y) + \eta(-S_X[Y, Z]) \\
&= -Y\eta(\nabla_Z X) + Z\eta(\nabla_Y X) + \eta(\nabla_{[Y,Z]} X).
\end{aligned}$$

Therefore,

$$(\nabla_X d\eta - d\nabla_X \eta)(Y, Z)$$

$$= -\nabla_Y X(\eta(Z)) + Z\eta(\nabla_Y X) + \eta([\nabla_Y X, Z])$$

$$- Y\eta(\nabla_Z X) + \nabla_Z X\eta(Y) + \eta([Y, \nabla_Z X])$$

$$+ Y\eta(\nabla_Z X) - Z\eta(\nabla_Y X) - \eta(\nabla_{[Y,Z]} X)$$

$$= -(\nabla_{\nabla_Y X}\eta)(Z) - \eta(\nabla_{\nabla_Y X} Z) + \eta([\nabla_Y X, Z])$$

$$+ (\nabla_{\nabla_Z X}\eta)(Y) + \eta(\nabla_{\nabla_Z X} Y) + \eta([Y, \nabla_Z X]) - \eta(\nabla_{[Y,Z]} X)$$

$$= -(\nabla_{\nabla_Y X}\eta)(Z) + (\nabla_{\nabla_Z X}\eta)(Y)$$

$$+ \eta(-\nabla_{\nabla_Y X} Z + [\nabla_Y X, Z] + \nabla_{\nabla_Z X} Y + [Y, \nabla_Z X] - \nabla_{[Y,Z]} X).$$

On the other hand, by (3.5), we have

$$\nabla_{\nabla_Y X} Z - \nabla_Z \nabla_Y X = [\nabla_Y X, Z], \qquad \nabla_{\nabla_Z X} Y - \nabla_Y \nabla_Z X = [\nabla_Z X, Y].$$

Hence

$$-\nabla_{\nabla_Y X} Z + [\nabla_Y X, Z] + \nabla_{\nabla_Z X} Y + [Y, \nabla_Z X] - \nabla_{[Y,Z]} X$$

$$= -\nabla_Z \nabla_Y X + \nabla_Y \nabla_Z X - \nabla_{[Y,Z]} X$$

$$= R(Y, Z)X.$$

Eventually, we arrive at

$$(\nabla_X d\eta - d\nabla_X \eta)(Y, Z) = -(\nabla_{\nabla_Y X}\eta)(Z) + (\nabla_{\nabla_Z X}\eta)(Y) + \eta(R(Y, Z)X)$$

which is (3.11).　□

**Acknowledgments** The second author is very grateful to Professor H. -H. Kuo for his hospitality during the U.S.-Japan Bilateral Seminar in January 1994 at Baton Rouge.

## References

[1] Aida, S.: On the Ornstein-Uhlenbeck operators on Wiener-Riemannian manifolds; *J. Funct. Anal.* (to appear)

[2] Airault, H.: Differential calculus on finite codimensional submanifolds of the Wiener space—The divergence operator; *J. Funct. Anal.* **100** (1991) 291–316

[3] Airault, H. and Malliavin, P.: Intégration géometrique sur l'espace de Wiener; *Bull. Sci. Math.* **112** (1988) 3–52

[4] Airault, H. and Malliavin, P.: Integration on loop groups. II. Heat equation for the Wiener measure; *J. Funct. Anal.* **104** (1992) 71–109

[5] Airault, H. and Van Biesen, J.: Géometrie riemannienne en codimension finie sur l'espace de Wiener; *C. R. Acad Sci. Paris Série I* **311** (1990) 125–130

[6] Feyel, D. and de La Pradelle, A.: Capacités gaussiennes; *Ann. Inst. Fourier, Grenoble* **41** (1991) 49–76

[7] Fukushima, M. and Kaneko, H.: On $(r,p)$-capacities for general Markovian semigroups; in:*Infinite dimensional analysis and stochastic processes*, 41–47, S. Albeverio (ed.), Pitman, 1985

[8] Getzler, E.: Dirichlet forms on loop space; *Bull. Sci. Math.* **113** (1989) 151–174

[9] Kazumi, T.: De Rham complex in an abstract Wiener space; Master thesis (in Japanese), Kyoto University, 1989

[10] Kazumi, T. and Shigekawa, I.: Measures of finite $(r,p)$-energy and potentials on a separable metric space; Séninaire de Probabilités, XXVI, J. Azéma, P.A. Meyer, and M. Yor (eds.), *Lecture Notes in Math.* **1526** (1992) 415–444, Springer-Verlag

[11] Kobayashi, S. and Nomizu, K.: *Foundations of Differential Geometry.* Vol. I (1963), Vol. II (1969), Interscience Publishers, New York-London

[12] Kusuoka, S.: Analysis on Wiener space, I. Nonlinear maps; *J. Funct. Anal.* **98** (1991) 122–168

[13] Kusuoka, S.: Analysis on Wiener space, II. Differential forms, *J. Funct. Anal.* **103** (1992) 229–274

[14] Malliavin, M. -P. and Malliavin, P.: Integration on loop groups. I. Quasi Invariant measures; *J. Funct. Anal.* **93** (1990) 207–237

[15] Malliavin, M. -P. and Malliavin, P.: Integration on loop groups. III. Asymptotic Peter-Weyl orthogonality; *J. Funct. Anal.* **108** (1992) 13–46

[16] Matsushima, Y.: *Differentiable Manifolds.* Marcel Dekker, New York, 1972

[17] Shigekawa, I.: De Rham-Hodge-Kodaira's decomposition on an abstract Wiener space; *J. Math. Kyoto Univ.* **26** (1986) 191–202

[18] Sugita, H.: Positive generalized Wiener functions and potential theory over abstract Wiener spaces; *Osaka J. Math.* **25** (1988) 665–696

[19] Van Biesen, J.: The divergence on submanifolds of the Wiener space; *J. Funct. Anal.* **113** (1993) 426–461

[20] Watanabe, S.: *Lectures on Stochastic Differential Equations and Malliavin Calculus.* Tata Institute of Fundamental Research, Springer-Verlag, Berlin-Heidelberg-New York, 1984

Tetsuya Kazumi
Department of Mathematics
University of Osaka Prefecture
Sakai 591, JAPAN
*E-mail*: kazumi@mathsun.cias.osakafu-u.ac.jp

Ichiro Shigekawa
Department of Mathematics
Faculty of Science
Kyoto University
Kyoto 606-01, JAPAN
*E-mail*: ichiro@kusm.kyoto-u.ac.jp

A. KOHATSU-HIGA & P. PROTTER

# The Euler scheme for SDE's driven by semimartingales

## 1. Introduction

We consider here stochastic differential equations driven by semimartingales and the rate of convergence of the Euler scheme approximations to the solutions. It is already known [2] that Euler schemes converge, but a complete characterization of the rate of convergence is still open. We describe here the convergence rate for convergence in probability, in terms of the Lipschitz constant of the coefficients and more importantly the $L^2$ norm of the time changed increments of the driving semimartingales. In the important special case when the driving semimartingales are diffusions or Lévy processes, we obtain — under mild additional technical assumptions — the same rate of convergence as in the classical case of Brownian motion and Lebesgue measure as differentials.

For all unexplained notation and undefined terms we refer the reader to Protter [4]. A standard reference on rates of convergence for the "classical" case of stochastic differential equations driven by Brownian motion and Lebesgue measure is the recent book by Kloeden and Platen [1]. For a survey paper of recent results see, e.g., Talay [6].

We work only on the time interval $[0, 1]$, and in our interpretation of the Euler scheme we follow [2]: let $0 \equiv \tau_0^n < \tau_1^n < \ldots < \tau_k^n \equiv 1$ be stopping times and define

$$\eta_n(t) = \tau_k^n \quad \text{for } \tau_k^n \le t < \tau_{k+1}^n.$$

We of course assume $\lim_{n\to\infty} \sup_k \tau_{k+1}^n - \tau_k^n = 0$, a.s. We then have that the solution

$$X_t^n = X_0 + \int_0^t f(X_{\eta_n(s-)}^n)dY_s$$

represents the Euler scheme for the equation

$$X_t = X_0 + \int_0^t f(X_{s-})dY_s.$$

141

## 2. The Quasi-left Continuous Case

In this section, we assume the driving term of the stochastic differential equation is quasi–left continuous, which leads to a much cleaner theory. Quasi–left continuity (QLC) is often verified in practice: for example, Lévy processes are QLC and so are most strong Markov processes. (Any continuous process is *a fortiori* QLC.)

**Definition** A stochastic process $Y$ is *quasi-left continuous* (QLC) if all of its jump times are totally inaccessible stopping times.

See [4, p.99] for the definition of totally inaccessible stopping times. A canonical example is the jump times of a Poisson process, or more generally of a Lévy process.

Let $Y$ be a QLC special semimartingale with canonical decomposition $Y = M + V$, where $M$ is assumed to be locally square integrable and the finite variation term $V$ is predictable. Let

$$A_t = \langle M, M \rangle_t + \int_0^t |dV_s| + t, \tag{2.1}$$

where $\langle M, M \rangle$ denotes the compensator of the quadratic variation process $[M, M]$ of $M$. Then $A$ is adapted, continuous, and strictly increasing. Define

$$\gamma_u = \inf\{t > 0 : A_t > u\}. \tag{2.2}$$

Then $\gamma_u$ is a stopping time for each $u$, and $u \to \gamma_u$ is also continuous and strictly increasing, and $A_{\gamma_u} = \gamma_{A_u} = u$.

Let $F$ be a process Lipschitz coefficient with (non–random) Lipschitz constant $k$, and suppose further $F$ is bounded by a constant $c$. Let $X^i$ denote the unique solution of

$$X_t^i = U_t^i + \int_0^t F(X^i)_{s-} dY_s, \tag{2.3}$$

where $U^i$ are adapted, square integrable, càdlàg processes ($i = 1, 2$).

**Lemma 2.4** *With $X^i$ as in (2.3) we have*

$$E\left\{ \sup_{s \leq \gamma(u)} |X_s^1 - X_s^2|^2 \right\} \leq 3 e^{\beta(\alpha, k) u} E\left\{ \sup_{s \leq \gamma(u)} |U_1^2 - U_s^2|^2 \right\},$$

*where $\beta(a, k) = (12 \vee 3\alpha)k^2$, (where "$\vee$" denotes "max") if $\int_0^t |dV_s| \leq \alpha < \infty$ a.s.*

**Proof** Recall $Y = M + V$ is the canonical decomposition of $Y$. Then

$$|X_t^1 - X_t^2|^2 \le 3|U_t^1 - U_t^2|^2 + 3\left|\int_0^t F(X^1)_{s-} - F(X^2)_{s-}\,dM_{s-}\right|^2$$
$$+ 3\left|\int_0^t F(X^1)_{s-} - F(X^2)_{s-}\,dV_{s-}\right|^2.$$

Let $Z_t = \sup_{s \le t} |X_s^1 - X_s^2|^2$, and recalling that $\gamma(u)$ is a stopping time, use Doob's martingale quadratic inequality on the second term and Jensen's inequality on the third on the right side above to get ($k =$ Lipschitz constant for $F$):

$$E\{Z_{\gamma(u)}\} \le 3E\{\sup_{s \le \gamma(u)} |U_s^1 - U_s^2|^2\}$$
$$+ 12k^2 E\left\{\int_0^{\gamma(u)} |X_{s-}^1 - X_{s-}^2|^2 d\langle M, M\rangle_s\right\}$$
$$+ 3k^2 \alpha E\left\{\int_0^{\gamma(u)} |X_{s-}^1 - X_{s-}^2|\,|dV_s|\right\}$$
$$\le 3E\{\sup_{s \le \gamma_u} |U_s^1 - U_s^2|^2\} + \beta(\alpha, k)E\left\{\int_0^{\gamma(u)} Z_{s-}dA_s\right\}$$

where $A_t = \langle M, M\rangle_t + \int_0^t |dV_s| + t$, and $\beta(\alpha, k) = (12 \vee 3\alpha)k^2$. By Lebesgue's change of time lemma (cf., e.g., [5, p. 379]) and Fubini's theorem we obtain

$$E\{Z_{\gamma(u)}\} \le 3E\{\sup_{s \le \gamma_u} |U_s^1 - U_s^2|^2\} + \beta(\alpha, k)\int_0^u E\{Z_{\gamma(s-)}\}ds.$$

Since $u \to E[\sup_{s \le \gamma_u} |U_s^1 - U_s^2|^2\}$ is an increasing function of $u$, by a version of Gronwall's inequalities (cf., e.g., [3, p. 359]) we have the result. $\square$

**Theorem 2.5** *Let*

$$X_t = X_0 + \int_0^t F(X)_{s-}dY_s,$$
$$X_t^n = X_0 + \int_0^t F(X^n)_{\eta^n(s-)}dY_s.$$

*Assume $Y$ is a quasi–left continuous, special semimartingale, and that if $Y = M + V$ is its canonical decomposition, then $M$ is locally square integrable. Also assume $F$ is process Lipschitz and bounded. Let*

$$\rho_n(s) = E\{(Y_{\gamma(s-)} - Y_{\eta_n(\gamma(s-))})^2\},$$

$\bar{\rho}_n(t) = \int_0^t \rho_n(s)ds$, and $\bar{\rho}_n = \overline{\rho_n}(A_1)$. Then $X^n - X$ converges to $O$ in ucp at rate $\sqrt{\bar{\rho}_n}$.

**Proof** Assume first that $M$ is a square integrable martingale and $\int_0^t |dV_s| \leq \alpha$ a.s. we can rewrite

$$X_t^n = X_0 + \int_0^t (F(X^n)_{\eta_n(s-)} - F(X^n)_{s-})dY_s + \int_0^t F(X^n)_{s-}dY_s$$

$$= X_0 + U_t^n + \int_0^t F(X^n)_{s-}dY_s.$$

Using $(a+b)^2 \leq 2a^2 + 2b^2$, Doob's inequality, and the Lipschitz property of $F$:

$$E\{ \sup_{s \leq \gamma_u} |U_s^n|^2 \} \leq (8 \vee 2\alpha)k^2 E\left\{ \int_0^{\gamma_u} |X_{\eta_n(s-)}^n - X_{s-}^n|^2 dA_s \right\}$$

$$= (8 \vee 2\alpha)k^2 E\left\{ \int_0^{1v} F(X_{\eta_n(s-)}^n)^2 (Y_{s-} - Y_{\eta_n(s-)})^2 dA_s \right\},$$

and since $F$ is assumed bounded by a constant $c$:

$$\leq (8 \vee 2\alpha)k^2 c E\left\{ \int_0^{\gamma_u} (Y_s - Y_{\eta_n(s-)})^2 dA_s \right\}$$

$$= \gamma(\alpha, k, c) \int_0^u E\{(Y_{\gamma_s} - Y_{\eta_n(\gamma_s-)})^2\}ds,$$

where the last equality is by Lebesgue's change of time lemma and Fubini's theorem. Since the equation for $X$ has $U^0 \equiv 0$, Lemma 2.4 implies

$$E\{ \sup_{s \leq \gamma_u} |X_s - X_{\eta_n(s)}^n|^2 \} \leq 3e^{\beta(\alpha,k)u} \delta(\alpha, k, c) \int_0^u \rho_n(s)ds.$$

Since $\overline{\rho_n}(t) = \int_0^t \rho_n(s)ds$ and $A_1 \geq 1$ a.s., we have the result.

To remove the simplifying assumptions, choose $\varepsilon > 0$ and $T$ so large that $P(T < A_1) < \varepsilon$, but $M^T$ is a square integrable martingale and $\int_0^T |dV_s| \leq \alpha$ a.s., for some $\alpha$, however large. (Recall that $t \to \int_0^t |dV_s|$ is continuous, since $Y$ is QLC and $V$ is predictable.) Let $m_n$ be a sequence increasing to $\infty$. Then,

$$P\left( \sup_{s \leq 1} \frac{1}{m_n \sqrt{\bar{\rho}_n}} |X_s - X_s^n| > \delta \right)$$

$$\leq P\left( \frac{1}{m_n \sqrt{\bar{\rho}_n}} (X^T - (X^n)^T)_1^* > \delta \right) + P(T < A_1)$$

$$\leq \varepsilon + \frac{E\left\{ \sup_{s \leq 1} |X_s^T - (X_s^n)^T|^2 \right\}}{(m_n)^2 \bar{\rho}_n \delta^2}$$

$$\leq \varepsilon + \frac{3e^{\beta(\alpha,k)} \delta(\alpha, k, c) \bar{\rho}_n}{(m_n)^2 \bar{\rho}_n \delta^2},$$

where we have used the Chebyshev–Markov inequality. Since the $\overline{\rho_n}$'s cancel, and since $\varepsilon$ was arbitrary, we have

$$\frac{1}{m_n\sqrt{\overline{\rho_n}}}(X - X^n)_1^* \text{ tends to } 0$$

in probability, hence $X - X^n$ tends to $O$ in ucp at rate $\sqrt{\overline{\rho_n}}$. $\quad\square$

**Example 2.6** Let $Y$ be a Lévy process such that $E(Y_t^2) < \infty$ each $t$, $Y_0 = 0$. Then $Y$ has a canonical decomposition $Y_t = (Y_t - \xi t) + \xi t$, where $\xi$ is a constant. (For example, $Y$ having bounded jumps would more than suffice!) Then $A_t = \tau t$ for a constant $\tau > 1$, and $\gamma_u = \frac{1}{\tau}u$. One then has

$$\rho_n(s) = E\left\{\left(Y_{\gamma(s)} - Y_{\eta_n(\gamma(s-))}\right)^2\right\} = E\left\{\left(Y_{(\frac{s}{\tau})} - Y_{\eta_n(\frac{s}{\tau})}\right)^2\right\} \sim \frac{1}{n}$$

and therefore $\overline{\rho_n} \sim \frac{1}{n}$, which means that $X^n - X$ converges to $O$ in ucp at rate $\frac{1}{\sqrt{n}}$, which is the same rate for solutions of stochastic differential equations driven by Brownian motion and Lebesgue measure.

**Example 2.7** Let $Y$ be a diffusion that is the solution of a stochastic differential equation driven by standard Brownian motion and Lebesgue measure. Then since $B_t$ and $t$ have continuous paths, they are *a fortiori* QLC, and one easily checks that with mild assumptions on the coefficients that determine $Y$, that $X^n - X$ converges to $O$ in ucp again at rate $\frac{1}{\sqrt{n}}$.

We note that the same proofs can be used for systems of SDEs, which yield the same rate of convergence.

## 3. The General Case

When the jumps of a semimartingale are not necessarily totally inaccessible, then the process $A$ given in (2.1) need not be continuous, and the situation becomes more complicated. For simplicity we again assume we have one driving semimartingale $Y$ and one equation; our results extend trivially to several driving semimartingales and systems.

### 3.1 Assumptions on the driving semimartingale $Y$:

*We assume $Y$ is a special semimartingale with canonical decomposition $Y = M + V$, where $M$ is assumed to be locally square integrable; the finite variation term $V$ is predictable and also there exists a constant $\alpha$ such that $\int_0^1 |dV_s| \le \alpha$ a.s. Finally assume $Y$ has square–integrable jumps.*

We consider the equations

$$X_t = X_0 + \int_0^t f(X_{s-})dY_s,$$

$$X_t^n = X_0 + \int_0^t f(X_{\eta_n(s-)}^n)dY_s,$$

(3.2)

where $f$ is Lipschitz continuous with constant $k$ and *bounded*. ($\eta_n$ is defined in Section 1 and determines an Euler approximate scheme.)

Let $\xi$ be a constant such that $(12 \vee 3\alpha)k^2 < \frac{1}{\xi}$. Let

$$A_t = \langle M, M \rangle_t + \int_0^t |dV_s| + t.$$

(3.3)

Then $A$ is predictable, right continuous, and strictly increasing. Let $T_0 = 0$, and

$$T_{i+1} = \inf\{t > T_i : A_t - A_{T_i} \geq \xi\}.$$

(3.4)

Define

$$A_t^i = \langle M, M \rangle_t^{T^{i+1}-} - \langle M, M \rangle_t^{T^i} + \int_{t \wedge T^i}^{t \wedge T^{i+1}-} |dV_s| + (t \wedge T^{i+1} - t \wedge T^i),$$

where $Z^{T-} = Z_t 1_{(t<T)} + Z_{T-} 1_{(t \geq T)}$, for a càdlàg process $Z$ and random time $T > 0$. (Of course, $s \wedge t = \min(s,t)$.) Next define

$$\gamma^i(u) = \inf\{t > 0 : A_t^i > u\},$$

(3.5)

so that $\gamma^i$ is the right continuous inverse of $A^i$. Note that $\gamma^i(u)$ is a stopping time for each $u > 0$. Let $Y^i = Y^{T^{i+1}-} - Y^{T_i}$ and define also:

$$\rho_n^i(s) = E\{(Y_{\gamma_i(s)-}^i - Y_{\eta_n(\gamma_i(s)-)}^i)^2\},$$

(3.6)

and note that $\rho_n^i(s) < \infty$. Also,

$$\bar{\rho}_n^i(s) = \rho_n^i(s) + \int_0^s \rho_n^i(r)dr$$

and

$$\bar{\rho}_n = \sum_{i=1}^{[\alpha/\xi]+1} \bar{\rho}_n^i(\xi),$$

(3.7)

146

where $[\alpha/\xi]$ denotes the integer part of $\alpha/\xi$.

**Theorem 3.8** *Let $f$ be Lipschitz continuous and bounded and assume assumptions (3.1) hold. Let $X^n$, $X$ be solutions of (3.2). Then $X^n$ converges to $X$ uniformly in probability on $[0,1]$ at rate $\sqrt{\rho_n}$. That is, for any sequence $m_n$ increasing to $\infty$,*

$$\lim_{n\to\infty} \frac{1}{m_n\sqrt{\rho_n}} \sup_{t\leq 1} |X_t^n - X_t| = 0,$$

*with convergence in probability.*

Before proving Theorem (3.8) we establish three lemmas. With the same $f$ and $Y$, it is convenient to consider more general equations:

$$
\begin{aligned}
X_t^1 &= U_t^1 + \int_0^t f(X_{s-}^1)dY_s, \\
X_t^2 &= U_t^2 + \int_0^t f(X_{s-}^2)dY_s,
\end{aligned}
\tag{3.9}
$$

where $U^1$, $U^2$ are adapted, càdlàg processes.

**Lemma 3.10** *Assume $\sup_t \Delta A_t \leq \xi$, where $\beta(\alpha,k)\xi < 1$, and $\beta(\alpha,k) = (12\vee 3\alpha)k^2$. Let $Z_t = \sup_{s\leq t} |X_s^1 - X_s^2|^2$. Then*

$$E\{Z_{\gamma(u)}\} \leq \frac{3}{1-\beta(\alpha,k)\xi} \exp\left( \frac{\beta(\alpha,k)}{1-\beta(\alpha,k)\xi} u \right) E\{ \sup_{s\leq\gamma(u)} |U_s^1 - U_s^2|^2\}.$$

**Proof** Since

$$
|X_t^1 - X_t^2| \leq |3U_t^1 - U_t^2|^2 + 3\left| \int_0^t f(X_{s-}^1) - f(X_{s-}^2)dM_s \right|^2 \\
+ 3\left| \int_0^t f(X_{s-}^1) - f(X_{s-}^2)dV_s \right|^2,
$$

using Doob's quadratic martingale inequality and Jensen's inequality implies

$$E\{Z_{\gamma(u)}\} \leq 3E\{ \sup_{s\leq\gamma(u)} |U_s^1 - U_s^2|^2\} + \beta(\alpha,k)E\left\{ \int_0^{\gamma(u)} Z_{s-}dA_s \right\},$$

and by Lebesgue's change of time lemma and Fubini's theorem,

$$\leq 3E\left\{ \sup_{s\leq\gamma(u)} |U_s^1 - U_s^2|^2\right\} + \beta(\alpha,k)E\{(A_{\gamma(u)} - u)Z_{\gamma(u)-}\} + \beta(\alpha,k) \int_0^u E\{Z_{\gamma(s-)}\}ds.$$

However, $0 \leq A \circ \gamma(u) - u \leq \sup_t \Delta A_t \leq \xi$, whence

$$E\{Z_{\gamma(u)}\} \leq 3E\left\{ \sup_{s \leq \gamma(u)} |U_s^1 - U_s^2|^2 \right\} + \beta(\alpha,k)\xi E\{Z_{\gamma(u)-}\} + \beta(\alpha,k) \int_0^u E\{Z_{\gamma(s)-}\} ds,$$

and since $\beta(\alpha,k)\xi < 1$ and $u \to Z_{\gamma(u)}$ is increasing in $u$, we obtain

$$\left(1 - \beta(\alpha,k)\xi\right) E\{Z_{\gamma(u)}\} \leq 3E\left\{ \sup_{s \leq \gamma(u)} |U_s^1 - U_s^2|^2 \right\} + \beta(\alpha,k) \int_0^u E\{Z_{\gamma(s)-}\} ds.$$

The result now follows by multiplying by $\frac{1}{1-\beta(\alpha,k)\xi}$ and applying Gronwall's inequality. $\quad \square$

**Lemma 3.11**  *Assume* $Y = Y^{T_1-}$, $\rho_n(s) := \rho_n^1(s)$, $\gamma(u) := \gamma^1(u)$, *and* $c = \|f\|_{L^\infty}$, *where* $T_1$, $\rho_n^1$, *and* $\gamma^1$ *are given by* (3.4), (3.6), *and* (3.5). *Then,*

$$E\left\{ \sup_{s \leq \gamma(u)} |X_s - X_s^n|^2 \right\} \leq (8 \vee 2\alpha)k^2 c\left\{ \xi \rho_n(u) - \int_0^u \rho_n(s)ds \right\}.$$

**Proof**  We write the equation for $X^n$ as:

$$X_t^n = X_0 + U_t^n + \int_0^t f(X_{s-}^n)dY_s,$$

where $U_t^n = \int_0^t f(X_{\eta_n(s-)}^n) - f(X_{s-}^n)dY_s$. In the corresponding equation for $X$ take $U_s^0 \equiv 0$. By Lemma (3.10) it then suffices to estimate

$$E\left\{ \sup_{s \leq \gamma(u)} |U_s^n|^2 \right\} = E\left\{ \sup_{s \leq \gamma(u)} |U_s^n - U_s^0|^2 \right\}.$$

Using that $f$ is Lipschitz with constant $k$ and also bounded by $c$, and by Doob and Jensen's inequalities, we have:

$$E\left\{ \sup_{s \leq \gamma(u)} |U_s^n|^2 \right\} \leq (8k^2 \vee 2\alpha k^2)E\left\{ \int_0^{\gamma(u)} |X_{\eta_n(s-)}^n - X_{s-}^n|^2 dA_s \right\}$$

$$= (8 \vee 2\alpha)k^2 E\left\{ \int_0^{\gamma(u)} f(X_{\eta_n(s-)}^n)^2 (Y_{s-} - Y_{\eta_n(s-)})^2 dA_s \right\}$$

$$\leq (8 \vee 2\alpha)k^2 c E\left\{ \int_0^{\gamma(u)} (Y_{s-} - Y_{\eta_n(s-)})^2 dA_s \right\}$$

$$\leq h(\alpha,k,c)\left[ E\{(A_{\gamma(u)} - u)(Y_{\gamma(u-)} - Y_{\gamma(\eta_n(u-))})^2\} \right.$$

$$\left. + E\left\{ \int_0^u (Y_{\gamma(s-)} - Y_{\eta_n(\gamma(s-))})^2 ds \right\} \right],$$

where $h(\alpha, k, c) = (8 \vee 2\alpha)k^2 c$. Since $0 \leq A_{\gamma(u)} - u \leq \sup_t \Delta A_t \leq \sup_t A_t \leq \xi$, the preceding is:

$$\leq h(\alpha, k, c)\left[\xi E\{(Y_{\gamma(u-)} - Y_{\gamma(\eta_n(u-))})^2\} + \int_0^u E\{(Y_{\gamma(s-)} - Y_{\eta_n(\gamma(s-))})^2\}ds\right]$$

$$= h(\alpha, k, c)\left[\xi \rho_n(u) + \int_0^u \rho_n(s)ds\right]. \quad \square$$

**Lemma 3.12** *Assume* $Y = Y^{T_1}$. *Then*

$$\lim_{n \to \infty} P(\sup_{s \leq \gamma(u)} |X_s - X_s^n| > \delta) = 0,$$

*with rate of convergence* $\sqrt{\rho_n^1(u)}$.

**Proof** We have

$$P(\sup_{s \leq \gamma(u)} |X_s - X_s^n| > \delta)$$

$$\leq P\left(\sup_{s \leq \gamma(u)} |X_s^{T-} - (X_s^n)^{T-}| > \delta\right) + P(|X_R - X_R^n|^2 > \delta). \tag{3.13}$$

where $R = \gamma(u) \wedge T$. The first term on the right side of (3.13) tends to 0 at rate $\sqrt{\rho_n^{-1}(u)}$ by Lemma (3.11) and Chebyshev's inequality. For the second term on the right side of (3.13), note that

$$X_T - X_T^n = X_{T-} - X_{T-}^n + \big(f(X_{T-}) - f(X_{\eta_n(T-)}^n)\big)\Delta Y_T,$$

and using the Lipschitz property of $f$:

$$|X_T - X_T^n| \leq |X_{T-} - X_{T-}^n| + k|X_{T-} - X_{\eta_n(T-)}^n||\Delta Y_T|$$

$$\leq \sup_{s \leq \gamma(u)} |X_s^{T-} - (X_s^n)^{T-}|(1 + k|\Delta Y_T|),$$

and since $1 + k|\Delta Y_T| < \infty$ a.s., $X_T - X_T^n$ tends to 0 in probability at least as fast as $\sup_{s \leq \gamma(u)} |X_s^{T-} - (X_s^n)^{T-}|$. $\quad \square$

**Proof of Theorem 3.8** Lemmas (3.10), (3.11), and (3.12) establish that for any sequence $m_n$ increasing to $\infty$, if we assume $Y = Y^{T_1}$, then

$$\lim_{n \to \infty} \frac{1}{m_n \sqrt{\rho_n^1(u)}} \sup_{t \leq \gamma^1(u)} |X_t^n - X_t| = 0, \tag{3.14}$$

with convergence in probability, for each $u > 0$. By definition of $T_1$ and $\gamma^1$, if we take $u \geq \xi$, then $\gamma^1(u) \geq T_1$ a.s. Thus we can rewrite (3.14) as:

$$\lim_{n \to \infty} \frac{1}{m_n \sqrt{\bar{\rho}_n^1(u)}} \sup_{t \leq 1} |X_{t \wedge T_1}^n - X_{t \wedge T_1}| = 0 \tag{3.15}$$

with convergence in probability. We now let $Y^2 = Y^{T_2-} - Y^{T_1}$, and consider the equations

$$\begin{aligned}
X_t &= X_{t \wedge T_1} + \int_0^t f(X_{s-}) dY_s^2, \\
X_t^n &= X_{t \wedge T_1}^n + \int_0^t f(X_{\eta_n(s-)}^n) dY_s^2,
\end{aligned} \tag{3.16}$$

which are the same as equations (3.2) but stopped at $T_2-$. We rewrite these equations as

$$\begin{aligned}
X_t &= U_t^0 + \int_0^t f(X_{s-}) dY_s^2, \\
X_t^n &= U_t^n + \int_0^t f(X_{s-}^n) dY_s^2.
\end{aligned}$$

Since $Y$ is assumed to have square integrable jumps, we can immediately apply Lemma (3.10), and a slight modification of Lemmas (3.11) and (3.12) give the inductive step. This is a finite induction, only needing to be performed at most $[\alpha/\xi]+1$ times. Theorem (3.8) follows. $\square$

**Acknowledgements**   Research of A. Kohatsu is supported in part by an ROA grant from NSF, awarded to Grant #9103454. He wishes to thank the Mathematics Department of Purdue University for its hospitality during his stay under an ROA grant from NSF. Research of Philip Protter is supported in part by NSF Grant #9103454. He wishes to thank Tom Kurtz and Denis Talay for discussions during his work on this paper, as well as Professors H. Kunita and H. -H. Kuo for their invitation to the U.S.-Japan Bilateral Seminar.

## References

[1] Kloeden, P. E. and Platen, E.: *Numerical Solution of Stochastic Differential Equations*. Springer-Verlag, New York, 1992

[2] Kurtz, T. and Protter, P.: Wong-Zakai corrections, random evolutions, and simulation schemes for SDEs; in: *Stochastic Analysis*, Mayer-Wolf, Merzbach, and Schwartz (eds.), Academic Press, Boston (1991) 331–345

[3] Mitrinovic, D. S., Pečarić, J. E., and Fink, A. M.: *Inequalities Involving Functions and Their Integrals and Derivatives.* Kluwer, Dordrecht, 1991

[4] Protter, P.: *Stochastic Integration and Differential Equations: A New Approach.* Springer-Verlag, New York, 1990

[5] Sharpe, M.: *General Theory of Markov Processes.* Academic Press, Boston, 1988

[6] Talay, D.: Simulation of stochastic differential systems;in:*Effective Stochastic Analysis*, P. Krée and W. Wedig (eds.), Springer-Verlag, New York, 1994 (to appear)

Arturo Kohatsu-Higa
CINVESTAV-IPN
Departamento de Matematicas
Avenida Instituto Politecnico Nacional 2508
Colonia san Pedro Zacatenco
Mexico D. F., MEXICO
*E-mail*: higa@math.cinvestav.mx

Philip Protter
Mathematics and Statistics Departments
Purdue University
West Lafayette, IN 47907-1395, U.S.A.
*E-mail*: protter@math.purdue.edu

I. KUBO

# A direct setting of white noise calculus

## 1. Introduction and Preliminaries

In [7], the author has given a new setting of Hida distributions, which is based on a weighted number operator. The idea came from the joint work [8] with H. -H. Kuo on finite dimensional Hida distributions. Here we are going to give a more direct construction along this idea, and we will prepare some estimations of Hida testing functionals. After that we will introduce an infinite dimensional Brownian motion and discuss the action of its 'semi-group'.

Let $E_0$ be a separable real Hilbert space with inner product $(\cdot, \cdot)_0$ and norm $\|\cdot\|_0 \equiv \sqrt{(\cdot, \cdot)_0}$. Let $A$ be a positive self-adjoint operator on $E_0$. Suppose that $A$ has complete orthonormal eigenvectors $\{\zeta_k\}_{k=1}^{\infty}$ such that

$$A\zeta_k = \lambda_k \zeta_k, \quad k = 1, 2, \ldots, \tag{1.1}$$

$$1 < \rho^{-1} \equiv \lambda_1 \leq \lambda_2 \leq \cdots, \quad 2\rho^{2\kappa} < 1 \quad \text{and} \quad \sum_{k=1}^{\infty} \lambda_k^{-2\kappa} < \infty \tag{1.2}$$

with some $\kappa > 0$. The choice of $\kappa$ is a little bit tricky and makes some formulae simpler. Since $A$ is positive, we can define inner products and norms by

$$(\xi, \eta)_p \equiv (A^p \xi, A^p \eta)_0 \quad \text{and} \quad \|\xi\|_p \equiv \|A^p \xi\|_0 \quad \text{for } p \in \mathbb{R}. \tag{1.3}$$

For $p \in \mathbb{R}$, let $E_p$ be the completion of the linear span of $\{\zeta_k\}_{k=1}^{\infty}$ with respect to the norm $\|\cdot\|_p$. Then naturally $E_{-p}$ is the dual space of $E_p$. For $p > q$, $E_p$ is naturally densely imbedded in $E_q$. Moreover, if $p - q \geq \kappa$, then the injection $\iota_{q,p}$ from $E_p$ into $E_q$ is of Hilbert-Schmidt type with the Hilbert-Schmidt norm

$$\|\iota_{q,p}\|_{H.S.} = \left( \sum_{k=1}^{\infty} \lambda_k^{-2(p-q)} \right)^{1/2} < \infty \tag{1.4}$$

by Ineq. (1.2) and Eq. (1.3). Ineq. (1.2) implies that the following constants are finite for $p \geq \kappa$ ;

$$c_p \equiv \prod_{k=1}^{\infty} (1 - \lambda_k^{-2p})^{-1/2} \quad \text{and} \quad c_p(t) \equiv \prod_{k=1}^{\infty} (1 - t\lambda_k^{-2p})^{-1/2} \quad \text{for } t < \rho^{-2(p-\kappa)}. \tag{1.5}$$

Set $\mathcal{E} \equiv \cap E_p$ and $\mathcal{E}^* \equiv \cup E_p$. Topologize $\mathcal{E}$ by the projective limit topology of $E_p$ and $\mathcal{E}^*$ by the inductive limit topology of $E_p$. Then we have a Gel'fand triplet $\mathcal{E} \subset E_0 \subset \mathcal{E}^*$. Throughout this article, canonical bilinear forms are denoted simply by $\langle \cdot, \cdot \rangle$. For example, $\langle \xi, \eta \rangle = (\xi, \eta)_0$ holds for any $\xi, \eta \in E_p$ with $p \geq 0$.

By Bochner-Minlos theorem, there exists a Gaussian measure $\mu$ on $\mathcal{E}^*$ such that

$$\int_{\mathcal{E}^*} \exp\left[\sqrt{-1}\langle x, \xi \rangle\right] d\mu(x) = \exp\left[-\frac{1}{2}\|\xi\|_0^2\right].$$

The measure $\mu$ is called the *measure of Gaussian white noise*. Denote by $(L^2)$ the complex Hilbert space $L^2(\mathcal{E}^*, \mu)$ of all complex valued square integrable functions with respect to the measure $\mu$. Our first purpose is to construct a Gel'fand triplet $(\mathcal{E}) \subset (L^2) \subset (\mathcal{E}^*)$ and discuss its calculi.

Our fundamental tools are formulae of Hermite polynomials and those with parameter which are defined by the following generating functions :

$$\sum_{n=0}^{\infty} \frac{t^n}{n!} H_n(u) = \exp\left[tu - \frac{1}{2}t^2\right] \quad \text{and} \quad \sum_{n=0}^{\infty} \frac{t^n}{n!} H_n(u; \gamma) = \exp\left[tu - \frac{\gamma}{2}t^2\right].$$

Then we see the following relationships:

$$H_n(u; \beta^2) = \beta^n H_n\left(\frac{u}{\beta}\right) \quad \text{and} \quad H_n(u; -\beta^2) = \beta^n H_n\left(\frac{u}{\beta}; -1\right),$$

$$H_n(u; \gamma) = \sum_{2k \leq n} (-\gamma)^k \frac{n! 2^{-k}}{(n - 2k)! k!} u^{n-2k}, \tag{1.6}$$

$$\frac{d}{du} H_n(u; \gamma) = n H_{n-1}(u; \gamma), \tag{1.7}$$

$$(\gamma \frac{d^2}{du^2} - u \frac{d}{du}) H_n(u; \gamma) = -n H_n(u; \gamma), \tag{1.8}$$

$$H_n(u + v; \gamma) = \sum_{m=0}^{n} \frac{n!}{(n - m)! m!} v^m H_{n-m}(u; \gamma). \tag{1.9}$$

Each monomial $u^n$ is expanded as

$$u^n = H_n(u; 0) = \sum_{2k \leq n} \frac{n! 2^{-k}}{(n - 2k)! k!} \gamma^k H_{n-2k}(u; \gamma). \tag{1.10}$$

$$\sum_{n=0}^{\infty} \frac{t^n}{n!} H_n(u; \gamma)^2 = (1 - \gamma^2 t^2)^{-1/2} \exp\left[\frac{t}{1 + \gamma t} u^2\right] \quad \text{for } |\gamma t| < 1. \tag{1.11}$$

The orthogonality of $\{H_n(u;\gamma)\}$ is well known ;

$$\frac{1}{\sqrt{2\pi\gamma}}\int_{\mathbb{R}} H_n(u;\gamma)H_m(u;\gamma)\exp[-\frac{1}{2\gamma}u^2]du = \delta_{n,m}\gamma^n n! \quad \text{for } \gamma > 0. \qquad (1.12)$$

We will use the following notations of multi-indices : Put

$$\mathbb{N}_0^\infty \equiv \{\vec{n} = (n_1, n_2, \ldots, n_d, 0, \ldots); n_k \text{ non-negative integer}, 1 \le k \le d < \infty\}.$$

For $\vec{n} = (n_1, \ldots, n_k, \ldots) \in \mathbb{N}_0^\infty$ and $\vec{u} = (u_1, \ldots, u_k, \ldots) \in \mathbb{C}^\infty$, put

$$|\vec{n}| \equiv \sum_{k=1}^\infty n_k, \quad \vec{n}! \equiv \prod_{k=1}^\infty n_k!, \quad \vec{0} \equiv (0, 0, \ldots),$$

$$\vec{u}^{\vec{n}} \equiv \prod_{k=1}^\infty u_k^{n_k}, \quad |\vec{u}^{\vec{n}}| \equiv \prod_{k=1}^\infty |u_k|^{n_k}, \quad \vec{u}\cdot\vec{v} \equiv \sum_{k=1}^\infty u_k v_k.$$

We denote $\vec{m} \le \vec{n}$ if $0 \le m_k \le n_k$ for any $k = 1, 2, \ldots$. The following two formulae
are used often

$$\sum_{m=0}^\infty \frac{(n+m)!}{n!m!}t^m = (1-t)^{-n-1}, \qquad (1.13)$$

$$\sum_{n=0}^\infty \sum_{m=0}^\infty \frac{(n+m)!}{n!(m!)^2}s^n t^m = (1-s)^{-1}\exp\left[(1-\log(1-s))t\right]. \qquad (1.14)$$

## 2. A Construction of Hida Distributions

We know several kinds of constructions of Hida distributions. Here we introduce a
direct method based on a weighted number operator, which is defined bellow.

Let $\mathcal{P}(\mathcal{E}^*)$ be the set of all polynomials in $\{\langle x, \xi\rangle; \xi \in \mathcal{E}\}$. For $y \in \mathcal{E}^*$ define a
differential operator $D_y$ on $\mathcal{P}(\mathcal{E}^*)$ by Gateaux derivative

$$D_y\varphi(x) \equiv \lim_{h\to 0}\frac{1}{h}\big(\varphi(x+hy) - \varphi(x)\big).$$

It is obvious that

$$D_y\varphi(x) = \sum_{k=1}^\infty \langle y, \zeta_k\rangle D_{\zeta_k}\varphi(x) \in \mathcal{P}(\mathcal{E}^*) \quad \text{for any } x \in \mathcal{E}^*. \qquad (2.1)$$

We now introduce a *weighted number operator* $N_A$ acting on $\mathcal{P}(\mathcal{E}^*)$ by

$$N_A\varphi(x) \equiv \sum_{k=1}^\infty \log \lambda_k \big(-D_{\zeta_k}^2 + \langle x, \zeta_k\rangle D_{\zeta_k}\big)\varphi(x), \qquad (2.2)$$

154

where $\{\zeta_k\}$ is the complete orthonormal basis of $E_0$ with $A\zeta_k = \lambda_k\zeta_k$ as in Eq. (1.1). By Eq. (1.8), we have the following two propositions.

**Proposition 2.1** *Let*

$$\mathfrak{H}_{\vec{n}}(x) \equiv \prod_{k=1}^{\infty} H_{n_k}(\langle x, \zeta_k \rangle)$$

*for $x \in \mathcal{E}^*$. Then $\mathfrak{H}_{\vec{n}}(x)$ is an eigenfunction of $N_A$ such that*

$$N_A \mathfrak{H}_{\vec{n}}(x) = \log(\vec{\lambda}^{\vec{n}})\mathfrak{H}_{\vec{n}}(x) \quad and \quad \exp[pN_A]\mathfrak{H}_{\vec{n}}(x) = \vec{\lambda}^{p\vec{n}}\mathfrak{H}_{\vec{n}}(x) \quad for\ p \in \mathbb{R}.$$

By Eq. (1.12) and the above proposition, we have

**Proposition 2.2** *The operator $N_A$ on $\mathcal{P}(\mathcal{E}^*) \subset (L^2)$ is symmetric and can be extended to a self-adjoint operator on $(L^2)$ with the complete orthonormal eigenvectors $\{\vec{n}!^{-1/2}\mathfrak{H}_{\vec{n}}(x); \vec{n} \in \mathbb{N}_0^{\infty}\}$ and eigenvalues $\{\log(\vec{\lambda}^{\vec{n}})\}$. (The extension will be denoted by the same symbol $N_A$.)*

Define inner products and norms on $\mathcal{P}(\mathcal{E}^*)$ for $p \in \mathbb{R}$ by

$$(\varphi, \psi)_{(\mathcal{E})_p} \equiv (\exp[pN_A]\varphi, \exp[pN_A]\psi)_{(L^2)} \quad and \quad \|\varphi\|_{(\mathcal{E})_p} \equiv \|\exp[pN_A]\varphi\|_{(L^2)}.$$

Let $(\mathcal{E})_p$ be the completion of $\mathcal{P}(\mathcal{E}^*)$ with respect to the norm $\|\cdot\|_{(\mathcal{E})_p}$. Then we see the inclusions and the dualities :

$$(\mathcal{E})_p \subset (\mathcal{E})_q \quad for \quad p > q \quad and \quad (\mathcal{E})_{-p} = (\mathcal{E})_p^* \ (= \text{the dual of } (\mathcal{E})_p).$$

Let us topologize $(\mathcal{E}) \equiv \cap_p(\mathcal{E})_p$ by the projective limit topology and $(\mathcal{E})^* \equiv \cup_p(\mathcal{E})_p$ by the inductive limit topology.

**Theorem 2.3** *The triplet $(\mathcal{E}) \subset (L^2) \subset (\mathcal{E})^*$ is a Gel'fand triplet and each $\varphi \in (\mathcal{E})_p$ is expanded as*

$$\varphi(x) = \sum_{\vec{n}} a_{\vec{n}}\mathfrak{H}_{\vec{n}}(x) \quad with \quad \|\varphi\|_{(\mathcal{E})_p}^2 = \sum_{\vec{n}} \vec{n}!\vec{\lambda}^{2p\vec{n}}|a_{\vec{n}}|^2. \tag{2.3}$$

*Moreover, if $\varphi \in (\mathcal{E})_p$ with $p \geq \kappa$, then the following function $\widetilde{\varphi}$ is a unique continuous version of $\varphi$ on $E_{-p}$ (i.e., it is continuous on $E_{-p}$ and $\mu(\{x; \varphi(x) = \widetilde{\varphi}(x)\}) = 1$):*

$$\widetilde{\varphi}(x) = \begin{cases} \sum_{\vec{n}} a_{\vec{n}}\mathfrak{H}_{\vec{n}}(x) & \textit{if the series converges absolutely,} \\ 0 & \textit{otherwise,} \end{cases} \tag{2.4}$$

*and the following holds*

$$|\widetilde{\varphi}(x)| \le c_p \|\varphi\|_{(\mathcal{E})_p} \exp\left[\|x\|^2_{-p}\right] \quad \textit{for } x \in E_{-p}.$$

**Proof** The first part is obvious from Proposition 2.2. By Eqs. (1.6) and (1.11), we see the following for $|t| < 1$

$$\sum_{n=0}^{\infty} \left|\frac{t^n}{n!} H_n(u)^2\right| \le \sum_{n=0}^{\infty} \frac{|t|^n}{n!} H_n(|u|; -1)^2 \le (1 - |t|^2)^{-1/2} \exp\left[\frac{|t|}{1 - |t|}|u|^2\right] \qquad (2.5)$$

(even for complex t, u). Therefore, for $p \ge \kappa$ and $\widetilde{\varphi} = \sum_{\vec{n}} a_{\vec{n}} \mathfrak{H}_{\vec{n}}(x) \in (\mathcal{E})_p$, we have

$$|\widetilde{\varphi}(x)| \le \sum_{\vec{n}} |a_{\vec{n}} \mathfrak{H}_{\vec{n}}(x)| \le \left(\sum_{\vec{n}} \vec{n}! \vec{\lambda}^{2p\vec{n}} |a_{\vec{n}}|^2 \sum_{\vec{n}} \frac{\vec{\lambda}^{-2p\vec{n}}}{\vec{n}!} |\mathfrak{H}_{\vec{n}}(x)|^2\right)^{1/2}$$

$$\le \|\varphi\|_{(\mathcal{E})_p} \prod_{k=1}^{\infty} (1 - \lambda_k^{-4p})^{-1/4} \exp\left[\frac{1}{2}\frac{\lambda_k^{-2p}}{1 - \lambda_k^{-2p}}\langle x, \zeta_k\rangle^2\right] \qquad (2.6)$$

$$\le c_{2p}^{1/2} \|\varphi\|_{(\mathcal{E})_p} \exp\left[\frac{1}{2}\frac{1}{1 - \rho^{2p}}\|x\|^2_{-p}\right] < \infty \qquad (2.7)$$

for $x \in E_{-p}$ by Ineq. (1.2) and Eqs. (1.3), (1.5), and (2.3). Since each $\mathfrak{H}_{\vec{n}}(x)$ is continuous on $\mathcal{E}^*$ and the series converges uniformly on any bounded subset of $E_{-p}$ ($p \ge \kappa$) by Ineq. (2.6), $\widetilde{\varphi}(x)$ is continuous on $E_{-p}$. Since the measure $\mu$ is supported by $E_{-\kappa}$ ($\subset E_{-p}$), i.e. $\mu(E_{-\kappa}) = 1$ and since the series in (2.3) converges in $(L^2)$, $\widetilde{\varphi}(x) = \varphi(x)$ $\mu$-a.e. Since any non-empty open set in $E_{-p}$ is of positive measure, $\widetilde{\varphi}(x)$ is a unique continuous version of $\varphi(x)$ on $E_{-p}$. Ineq. (2.7) implies the last assertion because $c_{2p} < c_p^2$ and $1 - \rho^{2\kappa} > 1/2$. $\quad\square$

**Remark 1** The above theorem shows that the triplet coincides with the usual triplet for the space of *Hida test functionals* and the space of *Hida distributions*. Usually in Hida calculus, the basic Hilbert space $E_0$ and the operator $A$ are so chosen as $E_0 \equiv L^2(\mathbb{R})$ and $A \equiv -d^2/du^2 + u^2 + 1$. Then $\mathcal{E} = \mathcal{S}(\mathbb{R})$ (= Schwartz' rapidly decreasing functions) and $\mathcal{E}^* = \mathcal{S}'(\mathbb{R})$ (= temper distributions) hold.

**Remark 2** For $\varphi \in (\mathcal{E})_p$ with $p \ge \kappa$, we will assume that $\varphi(x)$ itself means the continuous version $\widetilde{\varphi}(x)$ on $E_{-p}$ defined by Eq. (2.4) for the rest of the paper.

Denoting $\langle y, \vec{\zeta}\rangle^{\vec{m}} \equiv \prod_{k=1}^{\infty} \langle y, \zeta_k\rangle^{m_k}$ and $D_{\vec{\zeta}}^{\vec{m}} \equiv \prod_{k=1}^{\infty} D_{\zeta_k}^{m_k}$, we have that

$$D_y^{\ell} \mathfrak{H}_{\vec{n}}(x) = \sum_{\vec{m} \le \vec{n}, |\vec{m}|=\ell} \frac{\ell! \vec{n}!}{\vec{m}!(\vec{n} - \vec{m})!} \langle y, \vec{\zeta}\rangle^{\vec{m}} \mathfrak{H}_{\vec{n}-\vec{m}}(x) \qquad (2.8)$$

by Eq. (1.7), and hence that

$$\|D_y^\ell \varphi\|_{(\mathcal{E})_p} \leq \|\varphi\|_{(\mathcal{E})_{p+r}} \Big( \sum_{\vec{n}} \sum_{\vec{m} \leq \vec{n}, |\vec{m}| = \ell} \frac{\ell!^2 \vec{n}!}{\vec{m}!^2 (\vec{n} - \vec{m})!} \vec{\lambda}^{-2r\vec{n}} \vec{\lambda}^{-2p\vec{m}} |\langle y, \vec{\zeta} \rangle^{2\vec{m}}| \Big)^{1/2}$$

$$\leq \|\varphi\|_{(\mathcal{E})_{p+r}} \Big( \sum_{|\vec{m}|=\ell} \sum_{\vec{n}} \frac{\ell!^2 (\vec{n} + \vec{m})!}{\vec{m}!^2 \vec{n}!} \vec{\lambda}^{-2r\vec{n}} \vec{\lambda}^{-2(p+r)\vec{m}} |\langle y, \vec{\zeta} \rangle|^2 \Big)^{1/2}$$

$$\leq c_r \|\varphi\|_{(\mathcal{E})_{p+r}} \Big( \ell! \sum_{k=1}^{\infty} (1 - \lambda_k^{-2r}) \lambda_k^{-2(p+r)} |\langle y, \zeta_k \rangle|^2 \Big)^{\ell/2}$$

$$\leq c_r \|\varphi\|_{(\mathcal{E})_{p+r}} \sqrt{\ell!} (1 - \rho^{2r})^{-\ell/2} \|y\|_{-p-r}^{\ell} \tag{2.9}$$

by Ineq. (1.2), Eqs. (1.5) and (1.13) and the multinomial theorem. Thus we have

**Proposition 2.4** *If $y \in E_{-p-r}$, then $D_y^\ell$ can be extended to a map from $(\mathcal{E})_{p+r}$ to $(\mathcal{E})_p$. Moreover, for any $y \in \mathcal{E}^*$, $D_y^\ell$ is a continuous operator on $(\mathcal{E})$ and for any $\eta \in \mathcal{E}$, $D_\eta^\ell$ is a continuous operator on $(\mathcal{E})^*$. In particular, the weighted number operator $N_A$ is expressed in the form (2.2) as a continuous operator acting on $(\mathcal{E})^*$.*

This suggests that $D_y$ can be given by the Fréchet derivative on $(\mathcal{E})_p$. To check it, we need to observe the shifts $\{\mathfrak{s}_y; \varphi(x) \mapsto \varphi(x + y)\}$. By Eq. (1.9),

$$\mathfrak{s}_y \mathfrak{H}_{\vec{n}}(x) \equiv \mathfrak{H}_{\vec{n}}(x + y) = \sum_{\vec{m} \leq \vec{n}} \frac{\vec{n}!}{(\vec{n} - \vec{m})! \vec{m}!} \langle y, \vec{\zeta} \rangle^{\vec{m}} \mathfrak{H}_{\vec{n}-\vec{m}}(x) \tag{2.10}$$

holds for any $x, y \in \mathcal{E}^*$ pointwise and also in $(\mathcal{E})$. Therefore, for $r \geq \kappa$, $\varphi \in (\mathcal{E})_{p+r}$, we have

$$\|\mathfrak{s}_y \varphi\|_{(\mathcal{E})_p} \leq \sum_{\vec{n}} |a_{\vec{n}}| \|\mathfrak{s}_y \mathfrak{H}_{\vec{n}}\|_{(\mathcal{E})_p}$$

$$\leq \Big( \sum_{\vec{n}} \vec{n}! \vec{\lambda}^{2(p+r)\vec{n}} |a_{\vec{n}}|^2 \Big)^{1/2} \Big( \sum_{\vec{n}} \sum_{\vec{m} \leq \vec{n}} \frac{\vec{n}!}{(\vec{n} - \vec{m})! (\vec{m}!)^2} \vec{\lambda}^{-2r\vec{n} - 2p\vec{m}} |\langle y, \vec{\zeta} \rangle^{2\vec{m}}| \Big)^{1/2}$$

$$\leq \|\varphi\|_{(\mathcal{E})_{p+r}} \Big( \sum_{\vec{n}} \sum_{\vec{m}} \frac{(\vec{n} + \vec{m})!}{\vec{n}! (\vec{m}!)^2} \vec{\lambda}^{-2r\vec{n} - 2(p+r)\vec{m}} |\langle y, \vec{\zeta} \rangle^{2\vec{m}}| \Big)^{1/2}$$

$$= \|\varphi\|_{(\mathcal{E})_{p+r}} c_r \exp \Big[ \frac{1}{2} \sum_{k=1}^{\infty} (1 - \log(1 - \lambda_k^{-2r})) \lambda_k^{-2(p+r)} |\langle y, \zeta_k \rangle|^2 \Big]$$

$$\leq c_r \|\varphi\|_{(\mathcal{E})_{p+r}} \exp \Big[ \frac{1}{2} (1 - \log(1 - \rho^{2r})) \|y\|_{-p-r}^2 \Big] \tag{2.11}$$

by Ineq. (1.2) and Eqs. (1.3), (1.5), (1.14), and (2.3). Thus we have

**Proposition 2.5** (1) *If $r \geq \kappa$, then the shifts $\{\mathfrak{s}_y; y \in E_{-p-r}\}$ act as continuous maps from $(\mathcal{E})_{p+r}$ to $(\mathcal{E})_p$ continuously in $y \in E_{-p-r}$.*

(2) *If $p \geq 0$ and $r \geq \kappa$, then for $\varphi \in (\mathcal{E})_{p+r}$*

$$\mathfrak{s}_y\varphi(x) = \varphi(x+y) \quad \mu\text{-a.s. } x \quad \text{for any } y \in E_{-p-r}.$$

(3) *If $p, \; r \geq \kappa$, then for $\varphi \in (\mathcal{E})_{p+r}$*

$$\mathfrak{s}_y\varphi(x) = \varphi(x+y) \quad \text{for any } x \in E_{-p}, \; y \in E_{-p-r}.$$

**Proof** The first assertion in (1) is obvious by Ineq. (2.11). From Eq. (2.10), it follows that

$$\|\mathfrak{s}_y\varphi - \varphi\|_{(\mathcal{E})_p} \leq \|\varphi\|_{(\mathcal{E})_{p+r}} \left( \sum_{\vec{n}} \sum_{\vec{m} \neq \vec{0}} \frac{(\vec{n}+\vec{m})!}{\vec{n}!(\vec{m}!)^2} \vec{\lambda}^{-2r\vec{n}-2(p+r)\vec{m}} |\langle y, \vec{\zeta}\rangle^{2\vec{m}}| \right)^{1/2}$$

$$\leq c_r\|\varphi\|_{(\mathcal{E})_{p+r}} \left( \exp\left[(1 - \log(1-\rho^{2r}))\|y\|^2_{-p-r}\right] - 1 \right)^{1/2} \tag{2.12}$$

by using Ineq. (1.2), Eqs. (1.5) and (1.14) again. If $p \geq 0$, then both series

$$\varphi(x) = \sum_{\vec{n}} a_{\vec{n}} \mathfrak{H}_{\vec{n}}(x) \quad \text{and} \quad \mathfrak{s}_y\varphi(x) = \sum_{\vec{n}} a_{\vec{n}} \mathfrak{s}_y \mathfrak{H}_{\vec{n}}(x) = \sum_{\vec{n}} a_{\vec{n}} \mathfrak{s}_y \mathfrak{H}_{\vec{n}}(x+y)$$

converge in $(\mathcal{E})_0 = (L^2)$. By choosing a suitable subsequence we have (2). (3) is true by the same reason together with Theorem 2.3 because of Remark 2. $\quad\square$

Since we have the relationship $\mathfrak{s}_y \mathfrak{H}_{\vec{n}}(x) = \sum_{\ell=0}^{\infty} \frac{1}{\ell!} D_y^{\ell} \mathfrak{H}_{\vec{n}}(x)$ by Eqs. (2.8) and (2.10), we see that for $r \geq \kappa$

$$\left\| \sum_{\ell=k}^{\infty} \frac{1}{\ell!} D_y^{\ell} \mathfrak{H}_{\vec{n}}(x) \right\|^2_{(\mathcal{E})_p} = \frac{\vec{n}!}{\ell!} \sum_{\ell=k}^{\infty} \sum_{\vec{m} \leq \vec{n}, |\vec{m}|=\ell} \frac{\ell! \vec{n}!}{\vec{m}!^2 (\vec{n}-\vec{m})!} \langle y, \vec{\zeta}\rangle^{2\vec{m}} \vec{\lambda}^{2p(\vec{n}-\vec{m})}$$

and

$$\left\| \mathfrak{s}_{ty}\varphi - \sum_{\ell=0}^{k-1} \frac{t^{\ell}}{\ell!} D_y^{\ell}\varphi \right\|_{(\mathcal{E})_p} \leq c_r\|\varphi\|_{(\mathcal{E})_{p+r}} \sum_{\ell=k}^{\infty} \frac{|t|^{\ell}}{\sqrt{\ell!}} (1-\rho^{2r})^{-\ell/2} \|y\|^{\ell}_{-p-r}$$

$$\leq c_r \left(\frac{1-\rho^{2r}}{1-2\rho^{2r}}\right)^{1/2} \|\varphi\|_{(\mathcal{E})_{p+r}} \frac{|t|^k \|y\|^k_{-p}}{\sqrt{k!}} \exp\left[\frac{|t|^2}{2} \|y\|^2_{-p}\right]. \tag{2.13}$$

**Proposition 2.6** *For $\varphi \in (\mathcal{E})_{p+r}$ and $y \in E_{-p-r}$ with $r \geq \kappa$, $D_y\varphi$ is the Fréchet derivative of $\varphi$ in $(\mathcal{E})_p$ and*

$$\mathfrak{s}_{ty}\varphi = \sum_{\ell=0}^{\infty} \frac{t^{\ell}}{\ell!} D_y^{\ell}\varphi \quad in \ (\mathcal{E})_p \ for \ any \ t \in \mathbb{R}.$$

*Furthermore, if $p \geq \kappa$, then*

$$\varphi(x + ty) = \sum_{\ell=0}^{\infty} \frac{t^{\ell}}{\ell!} D_y^{\ell}\varphi(x) \quad for \ any \ x \in E_{-p}, \ t \in \mathbb{R}.$$

## 3. Brownian Motion on $\mathcal{E}^*$

In this section, we treat an infinite dimensional Brownian motion moving in $\mathcal{E}^*$ and the action of its 'semi-group' on $(\mathcal{E})$. Since each element $\varphi$ of $(\mathcal{E})$ is not bounded as seen in Theorem 2.3, properties of the action are different from usual semi-groups on bounded continuous functions.

Let $\{\vec{b}(t,\omega) = (b_1(t,\omega), \ldots, b_k(t,\omega), \ldots)\}$ be an infinite dimensional Brownian motion defined on a standard probability space $(\Omega, \mathcal{B}, P)$; that is, $\{b_k\}_{k=1}^{\infty}$ are independent standard Brownian motions. Let $\mathcal{B}_t$ be the $\sigma$-field generated by $\{\vec{b}(s,\omega); s \leq t\}$. Define a process $\{\beta(t); 0 \leq t < \infty\}$ in $\mathcal{E}^*$ by

$$\beta(t,\omega) \equiv \sum_{k=1}^{\infty} b_k(t,\omega)\zeta_k. \tag{3.1}$$

Then $\beta(t,\omega)$ lives in $\mathcal{E}_{-\kappa}$ almost surely (see [3]). Actually, we see that for any $p \geq \kappa$ and any $T > 0$, $\sup_{0 \leq t \leq T} \|\beta(t,\omega)\|_{-p} < \infty$ $P$-a.s. by Ineq. (1.2), Eq. (1.3) and Doob's inequality. Moreover, $\beta(t,\omega)$ is continuous $P$-a.s. by Totoki's theorem [13]. By Proposition 2.5 (3), for $p \geq \kappa$, $r \geq \kappa$ and $\varphi \in (\mathcal{E})_{p+r}$,

$$\varphi(x + \beta(t)) = \mathfrak{s}_{\beta(t)}\varphi(x) \quad for \ any \ x \in E_{-p} \ \ P\text{-a.s.}$$

holds and $\varphi(x + \beta(t,\omega)) \in (\mathcal{E})_p$ $P$-a.s. By the equality

$$\frac{1}{\sqrt{2\pi t}} \int_{\mathbb{R}} \exp[\frac{a}{2}(u+v)^2] \exp[-\frac{1}{2t}u^2]du = (1 - at)^{-1/2} \exp[\frac{a}{2(1-at)}v^2] \tag{3.2}$$

for $t > 0$, $at < 1$, and by Ineqs. (2.11) and (1.2), and Eq. (1.5), we have

$$\mathbf{E}\big[\|\varphi(x + \beta(t))\|_{(\mathcal{E})_p}\big]$$

$$\leq c_r \|\varphi\|_{(\mathcal{E})_{p+r}} \mathbf{E}\Big[\exp\Big[\frac{1}{2}\sum_{k=1}^{\infty}(1 - \log(1 - \lambda_k^{2r}))\lambda_k^{2(p+r)}b_k(t)^2\Big]\Big]$$

$$\leq c_r \|\varphi\|_{(\mathcal{E})_{p+r}} \prod_{k=1}^{\infty}\Big(1 - t(1 - \log(1 - \lambda_k^{2r}))\lambda_k^{-2(p+r)}\Big)^{-1/2}$$

$$\leq c_r c_{p+r}(t(1 - \log(1 - \rho^{2r})))\|\varphi\|_{(\mathcal{E})_{p+r}}, \tag{3.3}$$

if $r \geq \kappa$, $r + p \geq \kappa$ and $t(1 - \log(1 - \rho^{2r}))\rho^{2(r+p)} < 1$. Therefore,

$$\mathbf{BE}\big[\mathfrak{s}_{\beta(t)}\varphi(x)\big] \equiv \int_{\Omega} \mathfrak{s}_{\beta(t)}\varphi(x)dP$$

defines an element of $(\mathcal{E})_p$ in the sense of Bochner integral. We denote it simply by

$$T_t\varphi(x) = \mathbf{BE}\big[\varphi(x + \beta(t))\big] \equiv \mathbf{BE}\big[\mathfrak{s}_{\beta(t)}\varphi(x)\big]. \tag{3.4}$$

Then $T_t : (\mathcal{E})_{p+r} \mapsto (\mathcal{E})_p$ gives a continuous map. By (2.6) and (3.2), we have that

$$\mathbf{E}\big[|\varphi(x + \beta(t))|\big] \leq \|\varphi\|_{(\mathcal{E})_p} c_p(t) \exp\Big[\frac{1}{2}\frac{1}{1 - (t+1)\rho^{2p}}\|x\|_{-p}^2\Big] \tag{3.5}$$

for $p \geq \kappa$ and $0 \leq t < \rho^{-2(p-\kappa)}$, and that

$$\mathbf{E}\big[|\varphi(x + \beta(t))|^2\big] \leq \|\varphi\|_{(\mathcal{E})_p} c_p c_p(2t) \exp\Big[\frac{1}{2}\frac{2}{1 - (2t+1)\rho^{2p}}\|x\|_{-p}^2\Big] \tag{3.6}$$

for $p \geq \kappa$ and $0 \leq 2t < \rho^{-2(p-\kappa)}$, because of $c_{2p} \leq c_p^2$.

**Proposition 3.1** (1) *If $r \geq \kappa$ and $0 \leq 2t < \rho^{-2(p+r)}$, then $T_t$ acts continuously from $(\mathcal{E})_{p+r}$ to $(\mathcal{E})_p$ with $\|T_t\varphi\|_p \leq c_r c_{p+r}(2t)\|\varphi\|_{(\mathcal{E})_{p+r}}$.*
(2) *If, in addition, $p \geq \kappa$, then $T_t\varphi(x) = \mathbf{E}\big[\varphi(x + \beta(t))\big]$ for any $x \in E_{-p}$.*

**Proof** Since $1 - \log(1 - \rho^{2r}) \leq 1 - \log(1 - 1/2) \leq 2$ holds by $r \geq \kappa$ and by Ineq. (1.2), Ineq. (3.3) implies the first assertion. If $p \geq \max\{\kappa, -\log 2t/2\log\rho\}$, we see that $\varphi(x + \beta(t))$ belongs to $L^2(\mu \times P)$ and that $\varphi(x + \beta(t))$ belongs to $L^1(\Omega, P) \bigcap L^2(\Omega, P)$ for any $x \in E_{-p}$ by Ineqs. (3.5) and (3.6). Therefore, its Bochner integral in $(L^2)$ coincides with the usual Fubini integral $\mathbf{E}\big[\varphi(x + \beta(t))\big]$ of $\varphi(x + \beta(t, \omega))$ as an element in $L^2(P \times \mu)$. Since both are continuous on $E_{-p}$, we have the assertion. $\square$

As mentioned in Remark 2, we always mean by $\mathfrak{s}_y\varphi$ and $T_t\varphi$ the continuous representatives in their equivalence classes, if they are in $(\mathcal{E})_p$ with $p \geq \kappa$.

**Theorem 3.2** $\{T_t; t \geq 0\}$ *is a semi-group acting continuously on* $(\mathcal{E})$. *For* $\varphi \in (\mathcal{E})$, $\varphi(x + \beta(t))$ *is well defined for each* $x \in \mathcal{E}^*$ *P-a.s. and* $T_t\varphi(x) = \mathbf{E}[\varphi(x + \beta(t))]$. *Moreover, it can be extended to a continuous semi-group on* $(\mathcal{E})^*$.

**Proof** Let us fix a $t > 0$. Since for any $p \geq \kappa - \log 2t/2 \log \rho$, $\varphi \in (\mathcal{E})_{p+\kappa}$, $T_t\varphi \in (\mathcal{E})_p$ by Theorem 3.2. Hence $T_t$ acts on $(\mathcal{E})$. By Ineq. (2.12) and Schwarz inequality,

$$\|T_t\varphi - \varphi\|_{(\mathcal{E})_p} \leq \mathbf{E}\big[\|\mathfrak{s}_{\beta(t)}\varphi - \varphi\|_{(\mathcal{E})_p}\big]$$

$$\leq c_\kappa\|\varphi\|_{(\mathcal{E})_{p+\kappa}} \big(\mathbf{E}\big[\exp\big[(1 - \log(1 - \rho^{2\kappa}))\|\beta(t)\|^2_{-p-\kappa}\big] - 1\big]\big)^{1/2}$$

$$\leq c_r(c_{p+\kappa}(4t)^2 - 1)^{1/2}\|\varphi\|_{(\mathcal{E})_{p+\kappa}}.$$

Since $c_{p+\kappa}(4t) \to 1$ as $t \downarrow 0$ by Eq. (1.5), we have the continuity of $T_t$ in $t$.

For any $p \geq \kappa + \max\{0, -\log 2(t+s)/2\log\rho\}$, $\varphi$ belongs to $(\mathcal{E})_{p+2\kappa}$. By Proposition 3.1, $T_t\varphi \in (\mathcal{E})_{p+\kappa}$ and so $T_t\varphi(x) = \mathbf{E}[\varphi(x + \beta(t))]$ for any $x \in E_{-\kappa}$. Since $\beta(t+s) - \beta(t) \in E_{-\kappa}$ $P$-a.s. and $\varphi(x + \beta(t+s))$ is integrable for any fixed $x \in E_{-\kappa}$ by Ineq. (3.5) and since $\beta(t+s) - \beta(s)$ is independent of $\mathcal{B}_s$, we have

$$T_t\varphi(x + \beta(s)) = \mathbf{E}\big[\varphi(x + \beta(t) - \beta(s) + \beta(s))\big|\mathcal{B}_s\big] \quad P\text{-a.s.} \quad \text{for any } x \in E_{-\kappa}.$$

By Proposition 3.1 (2), for any $x \in E_{-\kappa}$, we have

$$T_s(T_t\varphi)(x) = \mathbf{E}\big[T_t\varphi(x + \beta(s))\big] = \mathbf{E}\big[\mathbf{E}\big[\varphi(x + \beta(t + s))\big|\mathcal{B}_s\big]\big] = T_{t+s}\varphi(x).$$

Since $T_s(T_t\varphi)$, $T_{t+s}\varphi(x) \in (\mathcal{E})_p$, we have $T_s(T_t\varphi) = T_{t+s}\varphi$ in $(\mathcal{E}_p)$. Since $T_t$, $T_s$ and $T_{t+s}$ act on $(\mathcal{E})^*$ continuously by Ineq. (3.3) and since the semi-group property holds on a dense subset, $\{T_t; t \geq 0\}$ is a semi-group on $(\mathcal{E})^*$. $\square$

**Proposition 3.3** *The generator of the semi-group of* $\{T_t; t \geq 0\}$ *on* $(\mathcal{E})$ *is the Gross Laplacian ; that is,*

$$\lim_{t \downarrow 0} \frac{T_t\varphi(x) - \varphi(x)}{t} = \frac{1}{2}\Delta_G\varphi(x) \equiv \frac{1}{2}\sum_{k=1}^\infty D^2_{\zeta_k}\varphi(x).$$

**Proof** By Eq. (2.1),

$$\mathbf{E}\big[D^\ell_{\beta(t)}\varphi(x)\big] = \sum_{|\vec{n}|=\ell} \mathbf{E}\big[\langle\beta(t), \vec{\zeta}\rangle^{\vec{n}}\big] D^{\vec{n}}_{\vec{\zeta}}\varphi(x).$$

Hence the mean is 0, if $\ell$ is odd, and $\mathbf{E}\big[D^2_{\beta(t)}\varphi(x)\big] = \Delta_G\varphi(x)$. On the other hand, by Ineq. (2.13) and Eq. (1.4), we have

$$\mathbf{E}\big\| \mathfrak{s}_{\beta(t)}\varphi - D_{\beta(t)}\varphi - \frac{1}{2}D^2_{\beta(t)}\varphi - \frac{1}{6}D^3_{\beta(t)}\varphi \big\|_{(\mathcal{E})_p}$$

$$\leq const.\|\varphi\|_{(\mathcal{E})_p}\,\mathbf{E}\Big[\|\beta(t)\|^4_{-p}\exp\big[\frac{1}{2}\|\beta(t)\|^2_{-p}\big]\Big]$$

$$\leq const.c_p(t)\|\iota_{0,p}\|^4_{H.S.}\|\varphi\|_{(\mathcal{E})_p}\Big(\frac{t}{1-t\rho^{2p}}\Big)^{3/2}.$$

Therefore,

$$0 = \lim_{t\downarrow 0}\frac{1}{t}\mathbf{E}\big[\mathfrak{s}_{\beta(t)}\varphi(x) - D_{\beta(t)}\varphi(x) - \frac{1}{2}D^2_{\beta(t)}\varphi(x) - \frac{1}{6}D^3_{\beta(t)}\varphi(x)\big]$$

$$= \lim_{t\downarrow 0}\frac{1}{t}\big(T_t\varphi(x) - \varphi(x) - t\Delta_G\varphi(x)\big). \qquad \square$$

**Remark 3**  In [9], the one-parameter group $\{\exp[\frac{t}{2}\Delta_V]; t \in \mathbb{R}\}$ generated by the Volterra Laplacian $\Delta_V$ was discussed. The Volterra Laplacian, in view of Remark 2, coincides with the Gross Laplacian on $(\mathcal{E})$. Therefore, $\{\exp[\frac{t}{2}\Delta_V]; t \geq 0\}$ coincides with $\{T_t; t \geq 0\} = \{\exp[\frac{t}{2}\Delta_G]; t \geq 0\}$ on $(\mathcal{E})$. A representation of $\{\exp[\frac{t}{2}\Delta_G]; t < 0\}$ by using Brownian motion will be discussed in the next section.

## 4. Complexifications

It is well known that each $\varphi(x) \in (\mathcal{E})$ is a real entire functional on $\mathcal{E}^*$, as seen directly from (2.13). Furthermore, $\varphi(x)$ can be extended to an entire function $\varphi(z)$ in $z \in \mathcal{E}^*_c \equiv \mathcal{E}^* + \sqrt{-1}\mathcal{E}^*$. The following estimate can be derived in several ways (e.g., by using Ineq. (2.5)) :

$$|\varphi(z)| \leq \|\varphi\|_{(\mathcal{E})_p}\, c_{2p}^{1/2}\exp\big[\frac{1}{2}\frac{1}{1-\rho^{2p}}\|z\|^2_{-p}\big] \qquad (4.1)$$

for $p \geq \kappa$. This estimate gives a criterion for a function $\varphi$ to belong to $(\mathcal{E})$. To state the criterion, we prepare some notations.

Put $E_{c,p} \equiv E_p + \sqrt{-1}E_p, \mathcal{E}_c \equiv \mathcal{E}^* + \sqrt{-1}\mathcal{E}$, and $\mathcal{E}^*_c \equiv \mathcal{E}^* + \sqrt{-1}\mathcal{E}^*$. Inner products are given as usual :

$$(\zeta, \zeta')_p \equiv (\xi, \xi')_p + (\eta, \eta')_p + \sqrt{-1}\big(-(\xi, \eta')_p + (\eta, \xi')_p\big)$$

for $\zeta = \xi + \sqrt{-1}\eta$ and $\zeta' = \xi' + \sqrt{-1}\eta'$ with $\xi, \xi', \eta$ and $\eta' \in E_p$. The duality $\langle \cdot, \cdot \rangle$ is defined as a bilinear form (not sesquilinear) :

$$\langle z, \zeta \rangle \equiv \langle x, \xi \rangle - \langle y, \eta \rangle + \sqrt{-1}(\langle y, \xi \rangle + \langle x, \eta \rangle)$$

for $\zeta = \xi + \sqrt{-1}\eta$ and $z = x + \sqrt{-1}y$ with $\xi, \eta \in E_p$ and $x, y \in E_{-p}$.

For a continuous functional $\varphi$ on $E_{c,-p}$, we define weighted supremum norms by

$$\|\varphi\|_p \equiv \sup_{z \in E_{c,-p}} |\varphi(z)| \exp\left[-\frac{1}{2}\|z\|^2_{-p}\right]. \tag{4.2}$$

Here we say that $\varphi(z)$ is an *entire functional on* $E_{c,-p}$, if $\varphi(z)$ is continuous in $z \in E_{c,-p}$ and $\varphi(\sum_{k=1}^{d} w_k \zeta_k)$ is entire in $\vec{w} \in \mathbb{C}^d$ for any $d$. Of course, the second condition is implied by the condition that $\varphi(z + wx)$ is entire in $w \in \mathbb{C}$ for any $x \in E_{-p}$ and $z \in E_{c,-p}$. Following [5, 11], we denote by $\mathcal{A}_p$ the set of all entire functionals $\psi(z)$ on $E_{c,-p}$ such that $\|\psi\|_p < \infty$. A given entire functional $\psi(z)$ on $E_{c,-p}$ is formally expanded as

$$\psi(z) = \sum_{\vec{n}} b_{\vec{n}} \langle z, \vec{\zeta} \rangle^{\vec{n}}.$$

Let $\mathcal{F}_p$ be the set of all entire functionals $\psi(z)$ whose coefficients satisfy

$$\|\psi\|^2_{\mathcal{F}_p} \equiv \sum_{\vec{n}} \vec{n}! \vec{\lambda}^{2p\vec{n}} |b_{\vec{n}}|^2 < \infty.$$

Then $\mathcal{F}_p$ is a Hilbert space with norm $\|\psi\|_{\mathcal{F}_p}$ and can be identified with a Bargmann space. The following is given by several authors in various ways.

**Proposition 4.1** *For any $p \in \mathbb{R}$, the inclusions*

$$\mathcal{A}_{p+\kappa} \subset \mathcal{F}_p \subset \mathcal{A}_p$$

*hold with the inequalities*

$$\|\psi\|_p \leq \|\psi\|_{\mathcal{F}_p} \leq c_\kappa^2 \|\psi\|_{p+\kappa}. \tag{4.3}$$

**Proof** For $\psi \in \mathcal{F}_p$, we have

$$|\psi(z)| \leq \left(\sum_{\vec{n}} \vec{n}! |b_{\vec{n}}|^2 \vec{\lambda}^{2p\vec{n}}\right)^{1/2} \left(\sum_{\vec{n}} \frac{1}{\vec{n}!} |\langle z, \vec{\zeta} \rangle|^{2\vec{n}} \vec{\lambda}^{-2p\vec{n}}\right)^{1/2} \leq \|\psi\|_{\mathcal{F}_p} \exp[\frac{1}{2}\|z\|^2_{-p}]$$

for $z \in E_{-p}$. On the other hand, by the theory of entire functionals of exponential type of order 2, we have

$$|b_{\vec{n}}| \leq \|\psi\|_p \frac{\vec{\lambda}^{\vec{n}}}{\sqrt{\vec{n}!}} \prod_{k=1}^{\infty} \sqrt{n_k + 1}. \tag{4.4}$$

Hence we get the inequalities in (4.3). If $q > p$ and $\psi \in \mathcal{F}_q$ (or $\mathcal{A}_q$), then $\psi(z)$ is defined on $E_{c,-p}$ and is entire. Hence the inequalities imply the inclusions. $\square$

A given $\psi \in \mathcal{F}_p$ (or $\in \mathcal{A}_p$) is characterized only by the values of $\psi(\xi)$, $\xi \in \mathcal{E}$. Therefore, we sometimes think that these spaces are of functionals on $\mathcal{E}$. On the other hand, Bargmann space is usually defined as an $L^2$-space with respect to a complex Gaussian measure on $\mathcal{E}^*$ (see [1, 5, 11, 14]). Unfortunately, the measure of $\mathcal{E}_c$ is null. Therefore, we have to be careful for the observations as $L^2$-space.

**Proposition 4.2** (1) *A continuous functional $\psi(x)$ in $x \in \mathcal{E}^*$ belongs to $(\mathcal{E})$ if and only if it can be extended to an entire functional $\psi(z)$ in $z \in \mathcal{E}_c^*$ such that $\|\psi\|_p < \infty$ for any $p$.*
(2) *For $p \geq \kappa$, a continuous functional $\psi(x)$ in $x \in E_{-p}$ belongs to $(\mathcal{E})_p$ if it can be extended to an entire functional in $z \in E_{c,-p}$ such that $\|\psi\|_{p+2\kappa} < \infty$. Furthermore,*

$$\|\psi\|_{(\mathcal{E})_p} \leq C_{\kappa,p+\kappa}\|\psi\|_{(\mathcal{F})_p} \leq c_\kappa^2 C_{\kappa,p+\kappa}\|\psi\|_{p+2\kappa} \quad \text{and} \quad \|\psi\|_p \leq c_{p+\kappa}\|\psi\|_{(\mathcal{E})_{p+\kappa}},$$

*where $C_{r,p} \equiv c_r \exp[\frac{1}{2} \sum_{k=1}^{\infty} \left(1 - \log(1 - \lambda_k^{-2r})\right)\lambda_k^{-2p}] < \infty$ for $p, r \geq \kappa$.*

**Proof** By Proposition 4.1, if $\|\psi\|_{p+2\kappa} < \infty$, then $\psi \in \mathcal{F}_{p+\kappa}$ and $\psi$ is expanded as

$$\psi(z) = \sum_{\vec{n}} b_{\vec{n}} \langle z, \vec{\zeta} \rangle^{\vec{n}} \quad \text{with} \quad \|\psi\|_{\mathcal{F}_{p+\kappa}}^2 \equiv \sum_{\vec{n}} \vec{n}! \vec{\lambda}^{2(p+2\kappa)\vec{n}} |b_{\vec{n}}|^2 < \infty.$$

By the relationship (1.10) and the inequality $(2m)! \leq 2^{2m} m!^2$, we have

$$\|\langle z, \vec{\zeta} \rangle^{\vec{n}}\|_{(\mathcal{E})_p}^2 = \sum_{2\vec{m} \leq \vec{n}} \frac{\vec{n}!^2 2^{-2\vec{m}}}{(\vec{n} - 2\vec{m})! \vec{m}!^2} \lambda^{2p(\vec{n} - 2\vec{m})} \leq \sum_{\vec{m} \leq \vec{n}} \frac{\vec{n}!^2}{(\vec{n} - \vec{m})! \vec{m}!^2} \vec{\lambda}^{2p(\vec{n} - \vec{m})}.$$

Therefore, by Eq. (1.14)

$$\|\psi\|_{(\mathcal{E})_p} \leq \sum_{\vec{n}} |b_{\vec{n}}| \sqrt{\vec{n}!} \left( \sum_{\vec{m} \leq \vec{n}} \frac{\vec{n}!}{(\vec{n} - \vec{m})! \vec{m}!^2} \vec{\lambda}^{2p(\vec{n} - \vec{m})} \right)^{1/2}$$

$$\leq \left( \sum_{\vec{n}} \vec{n}! |b_{\vec{n}}|^2 \vec{\lambda}^{2(p+\kappa)\vec{n}} \right)^{1/2} \left( \sum_{\vec{n}} \sum_{\vec{m}} \frac{(\vec{n} + \vec{m})!}{\vec{n}! \vec{m}!^2} \vec{\lambda}^{-2\kappa\vec{n}} \vec{\lambda}^{-2(p+\kappa)\vec{m}} \right)^{1/2}$$

$$= \|\psi\|_{\mathcal{F}_{p+\kappa}} c_\kappa \exp\Big[\frac{1}{2}\sum_{k=1}^\infty (1 - \log(1 - \lambda_k^{-2\kappa}))\lambda_k^{-2(p+\kappa)}\Big] = C_{\kappa,p}\|\psi\|_{\mathcal{F}_{p+\kappa}}.$$

By using Ineqs. (4.1) and (1.2) and $\|z\|^2_{-p-\kappa} \leq \rho^{2\kappa}\|z\|^2_{-p}$, we derive

$$|\psi(z)| \leq \|\psi\|_{(\mathcal{E})_{p+\kappa}} c_{p+\kappa} \exp[\frac{1}{2}\|z\|^2_{-p}]$$

for any $z \in E_{-p}$. Thus we have the conclusion. $\qquad \square$

The following discussions are closely related to the direct and the inverse Gauss transform in [5, 14] and to the operator $\mathcal{F}_{\alpha,\beta}$ in [11]. For $\varphi \in (\mathcal{E})$, $w \in \mathbb{C}$ and $t \geq 0$, we can define $\varphi(z + w\beta(t))$ and its restriction to $\mathcal{E}^*$ belongs to $(\mathcal{E})$ again. Put

$$T_{w,t}\varphi(x) = \mathbf{E}\big[\varphi(x + w\beta(t))\big] \tag{4.5}$$

Then by an analogous discussion to Section 2, we see that $\{T_{w,t}\}$ acts continuously on $(\mathcal{E})$ and has semi-group property in t. By the equality

$$\frac{1}{\sqrt{2\pi t}} \int_{\mathbb{R}} H_n(u + \omega v; \gamma) \exp[-\frac{v^2}{2t}]dv = H_n(u; \gamma - \omega^2 t) \quad \text{for } t > 0,$$

we have

$$T_{\sqrt{-1},s}\Big(T_t \prod_{k=1}^\infty H_{n_k}(\langle x, \zeta_k\rangle; \gamma)\Big) = \prod_{k=1}^\infty H_{n_k}(\langle x, \zeta_k\rangle; \gamma - t + s)$$

which assures that $T_{\sqrt{-1},t}$ is the inverse of $T_t$. Therefore, we have

**Theorem 4.3** *Let $T_{-t} \equiv T_{\sqrt{-1},t}$ for $t > 0$. Then $\{T_t; t \in \mathbb{R}\}$ is a one-parameter group acting on $(\mathcal{E})$. Furthermore,*

$$T_t = \exp\Big[\frac{t}{2}\Delta_G\Big] \quad \text{for any } t \in \mathbb{R}.$$

**Remark 4** The $S$-transform was defined for $\xi \in \mathcal{E}$ by

$$S\Phi(\xi) \equiv \langle\Phi, \exp\big[\langle x, \xi\rangle - \frac{1}{2}\|\xi\|_0^2\big]\rangle$$

in [9]. It was shown that $S\Psi(\xi)$ can be extended to an entire functional on $\mathcal{E}_c$ (see [4, 11, 12] etc.). Characterizations of $S(\mathcal{E})^*$ have been given in ([11, 12] etc.). The relationship $\exp[\frac{1}{2}\Delta_V]\varphi(\xi) = S\varphi(\xi)$ for $\varphi \in (\mathcal{E})$ in [9] is now clear by

$$S\varphi(\xi) = \int_{\mathcal{E}^*} \varphi(x + \xi)d\mu(x) = \mathbf{E}\big[\varphi(\xi + \beta(1))\big] = T_1\varphi(\xi) = \exp[\frac{1}{2}\Delta_G]\varphi(\xi).$$

# References


[1] Bargmann, V.: On a Hilbert space of analytic functions and an associated integral transform I, II; *Comm. Pure Appl. Math.* **14** (1961) 187–214, *ibid.* **20** (1967) 1–101

[2] Hida, T.: *Analysis of Brownian Functionals.* Carleton Mathematical Lecture Notes, no. **13**, 1975

[3] Itô, K.: *Foundations of stochastic differential equations in infinite dimensional spaces.* SIAM Reg. Conf. Series **47**, 1984

[4] Ito, Y., Kubo, I., and Takenaka, S.: Calculus on Gaussian white noise and Kuo's Fourier transformation; in: *White Noise Analysis–Mathematics and Applications*, World Scientific (1990) 180–207

[5] Kondratiev, Yu. G.: Nuclear spaces of entire functions in problems of infinite-dimensional analysis; *Soviet Math. Dokl.* **22** (1980) 588–592

[6] Krée, P.: Solutions faibles d'équations aux dérivées fonctionnelles I,II; *Lecture Notes in Math.* **410** (1974) 142–181, Springer-Verlag, *ibid.* **474** (1975) 16–47

[7] Kubo, I.: The structure of Hida distributions; in: *Mathematical Approach to Fluctuations* **1**, *Proc. IIAS Workshop*, World Scientific (1994) 49–114

[8] Kubo, I. and Kuo, H. -H.: Finite dimensional Hida distributions; *J. Func. Anal.* (1994) to appear

[9] Kubo, I. and Takenaka, S.: Calculus on Gaussian white noise I, II, III, IV; *Proc. Japan Acad.* **56A** (1980) 376–380, *ibid.* **56A** (1980) 376–380, *ibid.* **56A** (1980) 411–416, *ibid.* **57A** (1981) 433–437, *ibid.* **58A** (1982) 186–189

[10] Kuo, H. -H., Potthoff, J., and Streit, L.: A characterization of white noise test functionals; *Nagoya Math. J.* **121** (1991) 185–194

[11] Lee, Y. -J.: Analytic version of test functionals, Fourier transform and a characterization of measures in white noise calculus; *J. Func. Anal.* **100** (1991) 459–380

[12] Potthoff, J. and Streit, L.: A characterization of Hida distributions; *J. Func. Anal.* **101** (1991) 212–229

[13] Totoki, H.: A method of construction of measures on function spaces and its applications to stochastic processes; *Mem. Facul. Sci., Kyushu Univ. Ser.A Math.* **15** (1962) 177–190

[14] Yokoi, Y.: Simple setting for white noise calculus using Bargmann space and Gauss transform; *Hiroshima Math. J.* (1993) to appear



Izumi Kubo
Department of Mathematics
Faculty of Science
Hiroshima University
Higashi-Hiroshima 724, JAPAN
*E-mail*: kubo@math.sci.hiroshima-u.ac.jp

H. KUNITA

# Stable Lévy processes on nilpotent Lie groups

## 1. Introduction

One dimensional stable Lévy processes are well studied objects: Their laws are completely characterized through Lévy-Khinchine's formula. Also, the stable property of multi-dimensional Lévy processes called "operator-stable" processes have been studied with some details. See [7, 10]. In this paper we are concerned with the stable property of Lévy processes on simply connected, nilpotent Lie groups. Stable laws on the Heisenberg group or more generally on a homogeneous group were studied by [3, 5] and others. Our stable process is motivated by these works. However, our definition is more general than theirs, thereby including all strictly operator-stable laws in case where the underlying group is a Euclidean space.

In Section 2, we will define a stable Lévy process on a Lie group $G$. In order to do so, the underlying Lie group should be equipped with a *dilation* $\{\gamma_r\}_{r>0}$, which is by definition a one parameter group of automorphisms of the Lie group such that for any $\sigma \in G$, $\gamma_r(\sigma)$ converges to $e$ (the unit element of $G$) as $r$ tends to 0. Then a Lévy process $\varphi_t, t > 0$ on $G$ is called stable with respect to the dilation, if the law of $\varphi_{rt}$ coincides with the law of $\gamma_r(\varphi_t)$ for any $t, r > 0$. We remark that not every Lie group admits a dilation. Indeed, if a Lie group admits a dilation, it is necessarily simply connected and nilpotent, although not every such Lie group admits a dilation. Thus our stable Lévy process can only be defined on a certain simply connected, nilpotent Lie group.

In Section 3, we consider the case where the underlying group $G$ is a Euclidean space. Then $\{\gamma_r\}_{r>0}$ is a dilation if and only if it is represented by $\gamma_r = \exp(\log r)Q, r > 0$, where $Q$ is a matrix with eigenvalues of positive real parts. Hence our definition of the stable process will coincide with the strict operator-stable process in [10]. In Theorem 3.1, we will characterize the stable law through Lévy-Khinchine's formula.

Section 4 is the main part of this paper. In Theorem 4.1, we show that every Lévy process on a nilpotent Lie group $G$ is obtained by integrating a certain Lévy process on the associated nilpotent Lie algebra $\mathcal{G}$. Further, in Theorem 4.2 we show that every stable process on a simply connected, nilpotent Lie group $G$ with respect to a dilation $\{\gamma_r\}_{r>0}$ is obtained by integrating a certain strictly operator-stable process on the Lie algebra $\mathcal{G}$ (regarding it as a vector space). A remarkable fact is that not all strictly operator-stable processes on the Lie algebra generate stable processes on the Lie group. Indeed, the strictly operator-stable process should also be stable with respect to the dilation $\{d\gamma_r\}_{r>0}$, where $d\gamma_r$ is the differential of the map $\gamma_r$, which defines an automorphism of $\mathcal{G}$.

## 2. Dilations and the Associated Stable Lévy Processes on Lie Groups

We begin by defining a dilation of a Lie group. Let $G$ be a connected Lie group of dimension $d$. Let $\{\gamma_r\}_{r>0}$ be a one parameter group of automorphisms of $G$, i.e.,

   (i) For each $r > 0$, $\gamma_r$ is a one to one map of $G$ onto itself and satisfies $\gamma_r(\tau\sigma) = \gamma_r(\tau)\gamma_r(\sigma)$ for any $\tau, \sigma \in G$.

  (ii) $\gamma_r\gamma_s = \gamma_{rs}$ holds for any $r, s > 0$.

 (iii) $\gamma_r$ is continuous in $r \in (0, \infty)$.

It is called a *dilation* if it satisfies, in addition,

  (iv) $\gamma_r(\sigma) \to e$ as $r \to 0$ for any $\sigma$, where $e$ is the unit element of $G$.

Let $\mathcal{G}$ be the left invariant Lie algebra of $G$. Let $d\gamma_r$ be the differential of the automorphism $\gamma_r$. Then $d\gamma_r$ defines an automorphism of $\mathcal{G}$ i.e., $d\gamma_r$ is a one to one linear map of $\mathcal{G}$ onto itself and satisfies $d\gamma_r[X, Y] = [d\gamma_r X, d\gamma_r Y]$ for any $X, Y \in \mathcal{G}$, where $[\ ,\ ]$ is the Lie bracket. Therefore, $\{d\gamma_r\}_{r>0}$ is a one parameter group of automorphisms of $\mathcal{G}$. Given a basis $\{X_1, ..., X_d\}$ of the Lie algebra $\mathcal{G}$, the linear map $d\gamma_r$ is represented by $d\gamma_r = \exp(\log r)Q$ where $Q$ is a $d \times d$-matrix whose eigenvalues have positive real parts. Further, it satisfies $Q[X, Y] = [QX, Y] + [X, QY]$ for all $X, Y \in \mathcal{G}$. The matrix $Q$ is called the *exponent* of the dilation $\{\gamma_r\}_{r>0}$. In later discussions of the stable property, eigenvalues of the exponent $Q$, in particular, their real parts, play an important role. In case where $Q$ is semisimple and its eigenvalues are all real numbers, $\{\gamma_r\}_{r>0}$ is called a *strict dilation with exponents* $(\alpha_1, ..., \alpha_d)$.

If $G$ is a Euclidean space (commutative and simply connected Lie group), then for any $d \times d$-matrix $Q$ having eigenvalues with positive real parts, $\gamma_r \equiv \exp(\log r)Q, r > 0$ defines a dilation of $G$. In particular, for any positive numbers $\alpha_i, i = 1, \ldots, d$,

$$\gamma_r(x) \equiv (r^{\alpha_1}x_1, \ldots, r^{\alpha_d}x_d), \quad x = (x_1, \ldots, x_d) \in G$$

is a strict dilation with exponents $(\alpha_1, \ldots, \alpha_d)$. However, if $G$ is not commutative, we can not choose the exponent of the dilation freely since $d\gamma_r$ should satisfies additional conditions: $d\gamma_r[X,Y] = [d\gamma_r X, d\gamma_r Y]$ for $X, Y \in \mathcal{G}$. Further, not every Lie group admits a dilation. Indeed, it is known that if a dilation exists in a Lie group $G$, the Lie group is necessarily simply connected and nilpotent, though not every simply connected, nilpotent Lie group admits a dilation. See [2].

Conversely, suppose that we are given a one parameter group of linear maps $\{\gamma_r'\}_{r>0}$ of the Lie algebra $\mathcal{G}$. It is represented by $\gamma_r' \equiv \exp(\log r)Q$ with a $d \times d$-matrix $Q$, when a basis $\{X_1, ..., X_d\}$ of $\mathcal{G}$ is given. If the matrix $Q$ satisfies $Q[X,Y] = [QX,Y]+[X,QY]$ for all $X, Y \in \mathcal{G}$, $\gamma_r'$ is an automorphism of the Lie algebra $\mathcal{G}$. If all eigenvalues of $Q$ have positive real parts, $\{\gamma_r'\}_{r>0}$ is called a *dilation with exponent* $Q$ of the Lie algebra. A Lie algebra admitting a dilation is necessarily nilpotent.

**Theorem 2.1** *Let $G$ be a simply connected, nilpotent Lie group with the Lie algebra $\mathcal{G}$. For any dilation $\{\gamma_r'\}_{r>0}$ of $\mathcal{G}$, there exists a unique dilation $\{\gamma_r\}_{r>0}$ of $G$ such that $d\gamma_r = \gamma_r'$ holds for all $r > 0$.*

**Proof** It is known that the map $\exp : \mathcal{G} \to G$ is a diffeomorphism (See [4, Chapter XII]), so that every $\sigma \in G$ is represented uniquely by $\sigma = \exp X$. Define a map $\gamma_r : G \to G$ by $\gamma_r(\exp X) = \exp \gamma_r' X$. Noting that $\gamma_r'$ is an automorphism of $\mathcal{G}$, it is easy to see that $\gamma_r$ is an automorphism of $G$ and satisfies $d\gamma_r = \gamma_r'$. Therefore, $\{\gamma_r\}_{r>0}$ is a dilation of $G$. Next, let $\{\tilde{\gamma}_r\}_{r>0}$ be another dilation of $G$ such that $d\tilde{\gamma}_r = \gamma_r'$ holds for all $r > 0$. Then $d\gamma_r = d\tilde{\gamma}_r$ holds for all $r > 0$. Since $\gamma_r(e) = \tilde{\gamma}_r(e)$ holds, we have $\gamma_r(\sigma) = \tilde{\gamma}_r(\sigma)$ for all $\sigma \in G$ for all $r > 0$. Therefore, the uniqueness of the dilation follows.

It follows from the above theorem that the exponent determines the dilation uniquely.

In the sequel we give examples of nilpotent Lie algebras admitting dilations. A Lie algebra $\mathcal{G}$ is called *graded* if it is endowed with a vector space decomposition $\mathcal{G} = \oplus_1^m V_k$ such that $[V_i, V_j] \subset V_{i+j}$. A graded Lie algebra is nilpotent. If $\mathcal{G}$ is graded, there exists a natural strict dilation. Indeed, for any given positive number $\beta$, we define $\gamma_r' = \exp(\log r)Q$, where $QX = \beta k X$ for $X \in V_k$. Then $\{\gamma_r'\}_{r>0}$ is a strict dilation of $\mathcal{G}$. Let $n_k$ be the dimension of the space $V_k$. Then $n_1 + \cdots + n_m = d$. The exponents of $\{\gamma_r'\}$ are $\alpha_1 = \cdots = \alpha_{n_1} = \beta$, $\alpha_{n_1+1} = \cdots = \alpha_{n_1+n_2} = 2\beta, ...,$ $\alpha_{n_1+\cdots+n_{m-1}} = \cdots = \alpha_d = m\beta$.

A Lie algebra $\mathcal{G}$ is called *stratified* if $\mathcal{G}$ is graded and $V_1$ generates $\mathcal{G}$ as an algebra. In this case $\mathcal{G}$ is nilpotent of step $m$. We have further $\mathcal{G} = \oplus_1^m V_k$ and $[V_j, V_l] = V_{j+l}$. Therefore, if $\{\gamma_r'\}_{r>0}$ is a dilation of $\mathcal{G}$ such that $\gamma_r'(X) = r^\beta X$ holds for $X \in V_1$, then $\gamma_r'(X) = r^{j\beta} X$ holds for $X \in V_j$. Hence its exponents are $\beta, 2\beta, ..., m\beta$ as before.

Here is an example of a stratified Lie group. Let $G$ be the Lie group consisting of all real $d \times d$-matrices $C = (c_{ij})$ such that $c_{ij} = 0$ if $j < i$ and $c_{11} = \cdots = c_{dd}$ (upper triangular matrices with the same diagonal elements). It is a simply connected nilpotent Lie group. Let $\mathcal{G}$ be its Lie algebra. Then $\mathcal{G}$ is the set of all $d \times d$-matrices $X = (x_{ij})$ such that $x_{ij} = 0$ if $j \leq i$ (upper triangular matrices whose diagonals are 0). As a basis of $\mathcal{G}$, we will choose $X_{ij}, 1 \leq i < j \leq d$ where $X_{ij}$ is a $d \times d$-matrix such that its $(ij)$-element is 1 and all other elements are 0. These basis satisfies $[X_{ij}, X_{jk}] = X_{ik}$ and $[X_{ij}, X_{j',k}] = 0$ if $j \neq j'$. Given positive numbers $\alpha_1, ..., \alpha_{d-1}$, define $\gamma_r' : \mathcal{G} \to \mathcal{G}$ by $\gamma_r'(X_{ij}) = r^{(\alpha_i + \cdots + \alpha_{j-1})} X_{ij}$. Then it satisfies $[\gamma_r'(X_{ij}), \gamma_r'(X_{j',k})] = \gamma_r'[X_{ij}, X_{j',k}]$ for all $i < j, j' < k$. Therefore, $\{\gamma_r'\}_{r>0}$ is a strict dilation of $\mathcal{G}$ with exponents $(\alpha_1, ..., \alpha_{d-1}, ..., \alpha_1 + \cdots + \alpha_{d-1})$. Conversely, any strict dilation with respect to the basis $\{X_{ij}\}$ is of the above form.

Let $\varphi_t = \varphi_t(\omega), t \geq 0$ be a cadlag process with values in $G$ defined on a probability space $(\Omega, \mathcal{F}, P)$. It is called a *Lévy process* if it satisfies the following three properties:

(1) It has independent increments, i.e., for any $0 = t_0 < \cdots < t_n$, the random variables $\varphi_{t_i}^{-1} \varphi_{t_{i+1}}, i = 0, \ldots, n - 1$ are independent.

(2) It is continuous in probability and $\varphi_0 = e$ a.s. $P$, where $e$ is the unit element of the Lie group $G$.

(3) It is time homogeneous, i.e., the law of $\varphi_s^{-1} \varphi_t$ equals that of $\varphi_{s+h}^{-1} \varphi_{t+h}$ for any $t > s$ and $h > 0$.

Further, if $\varphi_t(\omega), t \geq 0$ is continuous with respect to $t$ a.s., it is called a *Brownian motion*.

Let $X_t$ be a cadlag process with values in $\mathcal{G}$. Then with a basis $\{X_1, ..., X_d\}$ of $\mathcal{G}$, it is represented by $X_t = \sum_j X_j Z_t^j$. The process $X_t$ is called a *Lévy process* (or a *Brownian motion*) if $(Z_t^1, ..., Z_t^d)$ is an $\mathbf{R}^d$-valued Lévy process (or a Brownian motion).

**Definition** Let $\varphi_t, t \geq 0$, be a Lévy process (or a Brownian motion) on a simply connected, nilpotent Lie group $G$ equipped with a dilation $\{\gamma_r\}_{r>0}$. It is said to be *stable with respect to the dilation* $\{\gamma_r\}_{r>0}$ if the law of $\gamma_r(\varphi_1)$ coincides with the law of $\varphi_r$ for any $r > 0$.

**Definition** Let $X_t, t \geq 0$, be a Lévy process (or a Brownian motion) on the Lie algebra $\mathcal{G}$ equipped with a dilation $\{\gamma_r'\}_{r>0}$. It is said to be *stable with respect to the dilation* $\{\gamma_r'\}_{r>0}$ if the law of $\gamma_r'(X_1)$ coincides with the law of $X_r$ for any $r > 0$.

## 3. Case of Euclidean Space. Strictly Operator-stable Processes

Let $\{\gamma_r\}_{r>0}$ be a dilation on a Euclidean space. Then it is represented by $\gamma_r = \exp(\log r)Q$, where $Q$ is a matrix with eigenvalues of positive real parts. Conversely, for an arbitrary given matrix $Q$ with eigenvalues of positive real parts, set $\gamma_r' = \exp(\log r)Q$. Then $\{\gamma_r'\}_{r>0}$ is a dilation of the Euclidean space. Therefore, a Lévy process $X_t$ on a Euclidean space is stable process with respect to a certain dilation if and only if it is a strictly operator-stable process in the sense of [10].

We shall characterize a strictly operator-stable process through the Lévy-Khinchine formula, assuming that its exponent $Q$ is semisimple. Let

$$\alpha_1, \ldots, \alpha_{j_1}, \alpha_{j_1+1} + i\beta_{j_1+1}, \ldots, \alpha_{j_1+2j_2} + i\beta_{j_1+2j_2}, \quad (j_1 + 2j_2 = d)$$

be eigenvalues of $Q$, where $\alpha_j, \beta_j$ are real numbers such that

$$\alpha_{j_1+2k-1} = \alpha_{j_1+2k}, \quad \beta_{j_1+2k-1} = -\beta_{j_1+2k}, \quad k = 1, ..., j_2.$$

There exists a real nonsingular matrix $C$ such that $C^{-1}\gamma_r C$ is decomposed as $\delta_r \theta_r$, where $\delta_r$ is a diagonal matrix with diagonal elements $r^{\alpha_1}, ..., r^{\alpha_{j_1+2j_2}}$ and

$$\theta_r = \begin{pmatrix} B_1 & 0 & \cdots & 0 \\ 0 & B_2 & \cdots & 0 \\ \vdots & \vdots & \ddots & \vdots \\ 0 & \cdots & \cdots & B_{j_1+j_2}, \end{pmatrix} \tag{3.1}$$

where $B_1 = \cdots = B_{j_1} = 1$ and

$$B_{j_1+k} = \begin{pmatrix} \cos(\beta_{j_1+2k} \log r), & \sin(\beta_{j_1+2k} \log r) \\ -\sin(\beta_{j_1+2k} \log r), & \cos(\beta_{j_1+2k} \log r) \end{pmatrix} \tag{3.2}$$

for $k = 1, ..., j_2$. We classify the index set $\{1, ..., d\}$ according to the exponents of the dilation.

$$J_0 = \{j; 0 < \alpha_j < 1/2\}, \quad I_{1/2} = \{j; \alpha_j = 1/2\}, \quad J_{1/2} = \{j; 1/2 < \alpha_j < 1\},$$
$$I_1 = \{j; \alpha_j = 1\}, \quad J_1 = \{j; \alpha_j > 1\}, \quad J = J_{1/2} \cup J_1 \cup I_1. \tag{3.3}$$

Then it holds $J_0 \cup I_{1/2} \cup J = \{1, ..., d\}$. Further, set

$$\mathbf{R}_J = \{x = (x_1, ..., x_d) \in \mathbf{R}^d; x_j = 0 \text{ if } j \notin J\}, \quad S = \{x \in \mathbf{R}_J; |x| = 1\}. \qquad (3.4)$$

Then $\{\theta_r\}_{r>0}$ map $S$ into itself.

Note that if $\{X_t\}_{t\geq 0}$ is stable with respect to the dilation $\{\gamma_r\}_{r>0}$, then $\tilde{X}_t \equiv CX_t$ is stable with respect to the dilation $\{\delta_r\theta_r\}_{r>0}$. So we shall only consider a stable process with respect to the dilation $\{\delta_r\theta_r\}_{r>0}$.

**Theorem 3.1** (c.f. [10]) *Let $\{\delta_r\theta_r\}_{r>0}$ be a semisimple dilation on $\mathbf{R}^d$.*

*(i) If $\{X_t\}_{t\geq 0}$ is a stable Lévy process on $\mathbf{R}^d$ with respect to the dilation $\{\delta_r\theta_r\}_{r>0}$, then its characteristic function $\psi_{X_t}(z) \equiv E[e^{i\langle z, X_t\rangle}]$ is represented by*

$$\exp\left[-\frac{1}{2}\langle z, Az\rangle + \int_{\mathbf{R}_J}\left(e^{i\langle z,x\rangle} - 1 - i\langle z, x_{J_{1/2}} + x_{I_1}\chi_{(0,1)}(|x_{I_1}|)\rangle\right)\nu(dx) + i\langle b, z\rangle\right]t, \qquad (3.5)$$

*where $x_I = (x_1', ..., x_d')$ is defined from $x = (x_1, ..., x_d)$ by the relation $x_i' = x_i$ if $i \in I$ and $x_i' = 0$ if $i \notin I$. Further, $A, \nu, b$ satisfy the following properties:*

*(a) $A = (a_{ij})$ is a symmetric nonnegative definite matrix such that $a_{ij} = 0$ holds if $i \notin I_{1/2}$ or $j \notin I_{1/2}$. Further, $A$ is $\{\theta_r\}_{r>0}$-invariant, i.e., $\theta_r A \theta_r^T = A$ holds for all $r > 0$.*

*(b) There exists a bounded $\{\theta_r\}_{r>0}$-invariant measure $\lambda$ on $S$ satisfying*

$$\int_S \xi_{I_1}\lambda(d\xi) = 0. \qquad (3.6)$$

*Further, the measure $\nu$ is represented by*

$$\nu(B) = \int_S \lambda(d\xi)\int_0^\infty \chi_B(\gamma_r\xi)r^{-2}dr. \qquad (3.7)$$

*(c) $b$ is a d-vector such that $b_j = 0$ if $j \notin I_1$.*

*(ii) Conversely, suppose that we are given a $d \times d$-matrix $A$ satisfying (a) and a bounded $\{\theta_r\}_{r>0}$-invariant measure $\lambda$ on $S$ satisfying (3.6). Define the measure $\nu$ by (3.7). Suppose further that we are given a d-vector $b$ satisfying (c). Then there exists a unique stable Lévy process with respect to the dilation $\{\delta_r\theta_r\}_{r>0}$ whose characteristic function is given by (3.5).*

**Proof** Assume that $\{X_t\}_{t\geq 0}$ is stable with respect to the dilation $\{\delta_r\theta_r\}_{r>0}$. The characteristic function $\psi_{X_t}(z)$ is represented by

$$\exp\left[-\frac{1}{2}\langle z, Az\rangle + \int_{\mathbf{R}^d}\left(e^{i\langle z,x\rangle} - 1 - \frac{i\langle z, x\rangle}{1 + |x|^2}\right)\nu(dx) + i\langle z, b\rangle\right]t, \qquad (3.8)$$

172

by Lévy-Khinchine's formula, where $A$ is a nonnegative definite symmetric matrix, $\nu$ is a Lévy measure on $\mathbf{R}^d - \{0\}$ satisfying $\int |x|^2/(1+|x|^2)\nu(dx) < \infty$ and $b$ is a d-vector. Then the characteristic function of $\gamma_r(X_t)$ is equal to $\psi_{X_t}(\gamma_r^T z)$ and is represented by

$$\exp\left[-\frac{1}{2}\langle z, \gamma_r A\gamma_r^T z\rangle + \int_{\mathbf{R}^d}\left(e^{i\langle z,x\rangle} - 1 - \frac{i\langle z,x\rangle}{1+|\gamma_r^{-1}x|^2}\right)\gamma_r^*\nu(dx) + i\langle z, \gamma_r b\rangle\right]t, \quad (3.9)$$

where $\gamma_r^T$ is the transpose of the matrix $\gamma_r$ and $\gamma_r^*\nu$ is the pullback of the measure $\nu$ by the map $\gamma_r$, i.e., it is a measure on $\mathbf{R}^d - \{0\}$ such that $\gamma_r^*(E) = \nu(\gamma_r^{-1}E)$ holds for all Borel sets $E$. Compare this with the characteristic function of $X_{rt}$. Then we have $rA = \gamma_r A\gamma_r^T$ and $r\nu = \gamma_r^*\nu$. The first equality implies (a) of the theorem.

We shall prove the latter half of (b). Let $C$ be a Borel set in $S$ and set $\tilde{C} = \bigcup_{r\in[1,\infty)}\gamma_r(C)$. Then $\tilde{C}$ is a Borel set in $\mathbf{R}_J$. We define a measure $\lambda$ on $(S, \mathcal{B}(S))$ by $\lambda(C) = \nu(\tilde{C})$. It is $\{\theta_r\}_{r>0}$-invariant. We show that $\nu$ is represented by (3.7). Define a measure $\tilde{\nu}$ on $\mathbf{R}_J$ by the right hand side of (3.7). For $C \in \mathcal{B}(S)$ and $a > 0$ we set $\tilde{C}_a = \bigcup_{r\in[a,\infty)}\gamma_r(C)$. Then we have $\tilde{\nu}(\tilde{C}_a) = \lambda(C)a^{-1} = \nu(\tilde{C})a^{-1} = (\gamma_a^{-1})^*\nu(\tilde{C}) = \nu(\tilde{C}_a)$. Since the family of sets $\{\tilde{C}_a; C \in \mathcal{B}(S), a > 0\}$ generates the Borel sets of $\mathbf{R}_J$, we have $\nu = \tilde{\nu}$.

We shall next prove the former half of (b) and (c). Let us observe that if $\alpha_j > 1$,

$$\int_{|x_j|\leq 1}|x_j|\nu(dx) = \int_S \lambda(d\xi)\int_0^1 |\delta_r\xi_j|r^{-2}dr = \lambda(S)\int_0^1 r^{\alpha_j-2}dr < \infty.$$

Similarly if $\alpha_j < 1$, we have

$$\int_{|x_j|\geq 1}|x_j|\nu(dx) = \lambda(S)\int_1^\infty r^{\alpha_j-2}dr < \infty.$$

These two inequalities enable us to prove the existence and finiteness of the integral in the expression (3.5). Then we can rewrite the Lévy-Khinchine formula (3.8) as (3.5). Therefore, the characteristic function of $\gamma_r(X_t)$ is represented by

$$\exp\left[-\frac{1}{2}\langle z, \gamma_r A\gamma_r^T z\rangle\right.$$
$$\left.+\int_{\mathbf{R}_J}\left(e^{i\langle z,x\rangle} - 1 - i\langle z, x_{J_{1/2}} + x_{I_1}\chi_{(0,1)}(|\gamma_r^{-1}x_{I_1}|)\rangle\right)\gamma_r^*\nu(dx) + i\langle z, \gamma_r b\rangle\right]t.$$

Compare this with the characteristic function of $X_{rt}$. Note that $0 < |\gamma_r^{-1} x_{I_1}| < 1$ is equivalent to $0 < |x_{I_1}| < r$. Then we obtain

$$\gamma_r b + r \int_{1 \le |x_{I_1}| < r} x_{I_1} \nu(dx) = rb, \quad \forall r > 0.$$

Since

$$\int_{1 < |x_{I_1}| < r} x_{I_1} \nu(dx) = \int_S \xi_{I_1} \lambda(d\xi) \int_1^r r^{-1} dr = \log r \int_S \xi_{I_1} \lambda(d\xi)$$

holds, we have $\gamma_r b = rb$ and $\int_S \xi_{I_1} \lambda(d\xi) = 0$. The first equality holds if and only if $b_j = 0$ for $j \notin I_1$. Therefore, (c) and (3.6) are established. We have thus proved the first assertion (i).

We shall next prove (ii). Let $(A, \nu, b)$ be the triple given in (ii). Consider the function (3.5) associated with these $(A, \nu, b)$. We denote it by $\psi_t(z)$. It is a characteristic function of an infinitely divisible law. Then there exists a Lévy process $\{X_t\}_{t \ge 0}$ such that its characteristic function is equal to $\psi_t(z)$. A direct computation yields that $\psi_t(\gamma_r^T z) = \psi_{rt}(z)$, which proves that the law of $\gamma_r(X_t)$ coincides with the law of $X_{rt}$. Therefore, $\{X_t\}$ is stable with respect to $\{\gamma_r\}_{r>0}$. The proof is complete.

Theorem 3.1 shows that the property of the strictly stable process $X_t$ depends crucially on the real parts $\alpha_1, ..., \alpha_d$ of the eigenvalues of the exponent of the dilation. For example, if $\alpha_j < 1/2$ holds for all $i$, there is no nontrivial operator-stable process except $X_t \equiv 0$. If $\alpha_j = 1/2$ holds for all $j$, $X_t$ is a Brownian motion and if $\alpha_j > 1/2$ holds for all $j$, it is a jump type Lévy process. In general, the strictly operator-stable process is a certain combination of these three cases. Indeed, we have the following.

**Corollary 3.2** *Let $\{X_t\}_{t \ge 0}$ be a stable process with respect to the dilation $\{\delta_r \theta_r\}$. Then we have the following :*

$$(X_t^j)_{j \in J_0} \equiv 0. \tag{3.10}$$

$$(X_t^j)_{j \in I_{\frac{1}{2}}} = \text{Brownian motion with mean 0 and covariance } A_{I_{\frac{1}{2}}} t. \tag{3.11}$$

$$(X_t^j)_{j \in J} = \text{jump type stable process with dilation } \{\delta_r^J \theta_r^J\}_{r>0}. \tag{3.12}$$

*Here $A_{I_{1/2}} = (a_{jk})_{j,k \in I_{1/2}}$, $\delta_r^J = E_J^T \delta_r E_J$, and $\theta_r^J = E_J^T \theta_r E_J$, where $E_J$ is the projection to $\mathbf{R}_J$. Further, $(X_t^j)_{j \in I_{1/2}}$ and $(X_t^j)_{j \in J}$ are independent.*

**Proof** Let $\nu$ be the Lévy measure of $X_t$. We shall prove

$$\int x_j^2 \nu(dx) = 0 \quad \text{if} \quad j \in J_0 \cup I_{1/2}. \tag{3.13}$$

For any $c > 0$, we have

$$\int_{|x^j|<c} x_j^2 \nu(dx) = r^{-2\alpha_j} \int_{|x^j|<r^{\alpha_j}c} x_j^2 \delta_r^* \nu(dx) = r^{1-2\alpha_j} \int_{|x^j|<r^{\alpha_j}c} x_j^2 \nu(dx)$$

for any $r > 0$. Let $r$ tend to 0. Then we get $\int_{|x^j|<c} x_j^2 \nu(dx) = 0$ if $2\alpha_j \le 1$. Since $c$ is arbitrary, we obtain (3.13).

Now (3.13) shows that the Lévy measure $\nu$ is supported by $\mathbf{R}_J$. Consequently, the characteristic function of the Lévy process $(X_t^j)_{j \in J_0}$ is identically 1, proving that $(X_t^i)_{i \in J_0} \equiv 0$. Next, the characteristic function of $(X_t^j)_{j \in I_{1/2}}$ is $\exp(-\langle z, Az\rangle/2)$. Therefore, it is a Brownian motion with mean 0 and covariance $At$. Finally $(X_t^j)_{j \in J}$ is a Lévy process whose characteristic function is of the form (3.5) with $A \equiv 0$. Hence it is a pure jump process. The processes $(X_t^j)_{j \in I_{1/2}}$ and $(X_t^j)_{j \in J}$ are independent. The proof is complete.

The exponent of the dilation is not equal to the exponent of a stable process on Euclidean space. We give here a few examples.

**Examples** (1) A stable process on $\mathbf{R}^d$ with exponent $\alpha$ $(0 < \alpha \le 2)$ is stable with respect to the strict dilation $\gamma_r = r^{\frac{1}{\alpha}}I, r > 0$ where $I$ is the identity matrix.

(2) Let $X_t^1, ..., X_t^d$ be independent one dimensional stable processes with exponents $\alpha_1, ..., \alpha_d$, respectively, where $0 < \alpha_i \le 2$. Then the vector process $X_t = (X_t^1, ..., X_t^d)$ is stable with respect to the strict dilation with exponents $\alpha_1^{-1}, ..., \alpha_d^{-1}$. Its Lévy measure $\nu$ is supported by each axis of $\mathbf{R}^d$, i.e.,

$$\text{Support}(\nu) \subset \cup_{i=1}^d \{(0, ..., 0, x_i, 0, ..., 0); x_i \in \mathbf{R}^1\},$$

where $x_i$ appears in the $i$-th component of the vector. Then the measure $\lambda$ of Theorem 2.1 is discrete and is supported by $d$-points $(0, ..., 0, 1, 0, ..., 0)$ (1 appears at the $i$-th component of the vector.)

## 4. Characterization of Stable Lévy Processes by Stochastic Differential Equations

We shall consider a Lévy process on a nilpotent Lie group. Let $G$ be a nilpotent Lie group of dimension $d$ and let $\mathcal{G}$ be its Lie algebra. We will fix a basis $\{X_1, ..., X_d\}$ of $\mathcal{G}$. The exponential map $\xi : \mathbf{R}^d \to G$ is defined by $\xi(x) = \xi(x_1, ..., x_d) = \exp(\sum_j x_j X_j)$. We will first show that every Lévy process on a nilpotent Lie group is obtained by integrating a certain Lévy process on the Lie algebra.

**Theorem 4.1**  *Let $\varphi_t$ be a Lévy process on a nilpotent Lie group $G$. Then there exists an $\mathbf{R}^d$-valued Lévy process $Z_t = (Z_t^1, \ldots, Z_t^d)$ such that $\varphi_t$ is represented as a solution of the following stochastic differential equation governed by a $G$-valued Lévy process $X_t = \sum_j Z_t^j X_j$ :*

$$f(\varphi_t) = f(e) + \sum_{j=1}^d \int_0^t X_j f(\varphi_{s-}) dZ_s^j + \frac{1}{2} \sum_{i,j=1}^d a_{ij} \int_0^t X_i X_j f(\varphi_{s-}) ds$$

$$+ \sum_{s \le t} \{ f(\varphi_{s-}\xi(\triangle Z_s)) - f(\varphi_{s-}) - \sum_j X_j f(\varphi_{s-}) \triangle Z_s^j \} \tag{4.1}$$

*for any $f \in C_\infty(G)$. Here $A = (a_{ij})$ is the covariance of $t^{-1} B_t$, where $B_t$ is the Brownian motion part of the Lévy process $Z_t$ and $\triangle Z_s = Z_s - Z_{s-}$.*

*In particular, if $G$ is simply connected, then the Lévy process $Z_t$ is uniquely determined by $\varphi_t$ and it satisfies*

$$\sigma(Z_s; s \le t) = \sigma(\varphi_s; s \le t) \quad \forall t. \tag{4.2}$$

*Conversely, for any given $\mathbf{R}^d$-valued Lévy process $Z_t$, the stochastic differential equation (4.1) has a unique global solution $\varphi_t$ which is a Lévy process on the Lie group $G$.*

**Proof**  [1] has shown that for any given Lévy process $\varphi_t$ on a Lie group, there exists a unique independent pair of an $\mathbf{R}^d$- valued Brownian motion $B_t = (B_t^1, \ldots, B_t^d)$ with mean 0 and covariance $(a_{jk})t$ and a Poisson random measure $N(dsd\sigma)$ on $[0, \infty) \times G$ with intensity $ds\nu(d\sigma)$ such that

$$\sigma(B_s, N((0,s] \times E); s \le t, \ E \in \mathcal{B}(G - \{e\})) = \sigma(\varphi_s; s \le t) \tag{4.3}$$

and $\varphi_t$ satisfies

$$f(\varphi_t) = f(e) + \sum_j \int_0^t X_j f(\varphi_{s-}) dB_s^j$$

$$+ \frac{1}{2} \sum_{i,j} a_{ij} \int_0^t X_i X_j f(\varphi_{s-}) ds + \sum_j a_j \int X_j f(\varphi_{s-}) ds$$

$$+ \int_0^t \int_{U_\delta} (f(\varphi_{s-}\sigma) - f(\varphi_{s-}))(N(dsd\sigma) - dsd\nu(\sigma))$$

$$+ \int_0^t \int_{U_\delta^c} (f(\varphi_{s-}\sigma) - f(\varphi_{s-}))N(dsd\sigma)$$

$$+ \int_0^t \int_{U_\delta} \left( f(\varphi_{s-}\sigma) - f(\varphi_{s-}) - \sum_j x_j(\sigma) X_j f(\varphi_{s-}) \right) \nu(d\sigma) ds, \tag{4.4}$$

where $U_\delta = \{\sigma : |\xi^{-1}(\sigma)| < \delta\}$.

Now suppose that the underlying Lie group is simply connected and nilpotent. Then every $\sigma \in G$ is represented uniquely by $\sigma = \exp \sum_{j=1}^d x_j X_j$. Therefore, for the Poisson random measure $N(dsd\sigma)$, there exists a Poisson point process $p_s = (p_s^1, \ldots, p_s^d)$ on $\mathbf{R}^d$ such that $\varphi_{s-}\varphi_s = \exp \sum_j p_s^j X_j$. Define

$$J((s,t] \times E) = \sharp\{u \in (s,t]; p_u \in E\}.$$

It is a Poisson random measure on $\mathbf{R}^d \times [0,\infty)$. Let $m$ be the intensity measure of $J$, i.e., $m(F) = E[J((0,1] \times F)]$. Then $\nu$ is the pullback of $m$ by $\xi$, i.e., $\nu = \xi^* m$. Further,

$$N((s,t] \times E) = J((s,t] \times \xi^{-1}(E)). \tag{4.5}$$

Consequently, (4.4) is written as

$$f(\varphi_t) = f(e) + \sum_j \int_0^t X_j f(\varphi_{s-})dB_s^j$$

$$+ \frac{1}{2}\sum_{i,j} a_{ij} \int_0^t X_i X_j f(\varphi_{s-})ds + \sum_j a_j \int_0^t X_j f(\varphi_{s-})ds$$

$$+ \int_0^t \int_{|x|<\delta_0} (f(\varphi_{s-}\xi(x)) - f(\varphi_{s-}))(J(dsdx) - dsm(dx))$$

$$+ \int_0^t \int_{|x|\geq\delta_0} (f(\varphi_{s-}\xi(x)) - f(\varphi_{s-}))J(dsdx)$$

$$+ \int_0^t \int_{|x|<\delta_0} \Big(f(\varphi_{s-}\xi(x)) - f(\varphi_{s-}) - \sum_j x_j X_j f(\varphi_{s-})\Big)m(dx)ds. \tag{4.6}$$

Set

$$Z_t = B_t + at + \int_0^t \int_{0<|x|<\delta_0} x\Big(J(dsdx) - dsm(dx)\Big)$$

$$+ \int_0^t \int_{|x|\geq\delta_0} xJ(dsdx). \tag{4.7}$$

Then (4.6) is written as (4.1).

For the proof of the uniqueness of the process $Z_t$, we proceeds as follows. Let $\tilde{Z}_t$ be any Lévy process on $\mathbf{R}^d$ satisfying (4.1) (replacing $Z_t$ by $\tilde{Z}_t$). It has the Lévy-Itô decomposition of the form (4.7). We denote the corresponding Brownian motion and Poisson random measure by $\tilde{B}_t$ and $\tilde{J}(dsdx)$, respectively. Define the Poisson random measure $\tilde{N}(dsd\sigma)$ on the Lie group $G$ through the formula (4.5) using $\tilde{J}(dsdx)$. Then $\varphi_t$ satisfies (4.4) replacing $B_t$ and $N(dsd\sigma)$ by $\tilde{B}_t$ and $\tilde{N}(dsd\sigma)$, respectively. Then we have $B_t = \tilde{B}_t$ and $N(dsd\sigma) = \tilde{N}(dsd\sigma)$ by the uniqueness theorem in [1]. This proves the uniqueness of the process $Z_t$. Now since $\sigma(\varphi_s; s \leq t) = \sigma(B_s, N(dsd\sigma); s \leq t)$ holds by [1], we have the relation (4.2) in case where the underlying Lie group $G$ is simply connected.

We next consider the case where a Lévy process $\varphi_t$ is defined on a general nilpotent Lie group $G$. Let $G'$ be the universal covering group of $G$ and let $p : G' \to G$ be the covering map. It is a real analytic map. The Lie algebra $\mathcal{G}'$ of $G'$ is isomorphic to the Lie algebra $\mathcal{G}$ of $G$. We shall identify them. We shall prove that there exists a Lévy process $\varphi'_t$ on $G'$ such that $p(\varphi'_t) = \varphi_t$ holds for any $t$ a.s. Let $B_t = (B_t^1, ..., B_t^d)$ be a Brownian motion and $N(dsd\sigma)$ be a Poisson random measure on $G$ satisfying (4.3) and (4.4). Let $N'(dsd\sigma)$ be a Poisson random measure on $G'$ such that $N'(ds, p^{-1}(E)) = N(ds, E)$ holds for any Borel set $E$ of $G$. Consider the stochastic differential equation (4.4) on $G'$, where $N(dsd\sigma)$ is replaced by $N'(dsd\sigma)$. Let $\varphi'_t$ be its solution. Define a $G$-valued process by $\tilde{\varphi}_t = p(\varphi'_t)$. Note that $\varphi'_t$ satisfies (4.4) for the function $\tilde{f} \equiv f \circ p$ on $G'$, where $f$ is a $C^\infty$-function on $G$. Then $\tilde{\varphi}_t$ satisfies (4.4) for $f$. Therefore, we have $\tilde{\varphi}_t = \varphi_t$ by the uniqueness of the solution of equation (4.4). Now for $\varphi'_t$, there exists an $\mathbf{R}^d$-valued Lévy process $Z_t$ satisfying (4.2), since $G'$ is simply connected. Then $\tilde{\varphi}_t \equiv p(\varphi_t)$ also satisfies (4.2).

The third assertion of the theorem is proved in [1]. The proof is complete.

Theorem 4.1 shows that there is a one to one correspondence between a Lévy process $\varphi_t$ on a simply connected, nilpotent Lie group $G$ and a Lévy process $X_t$ on the Lie algebra $\mathcal{G}$ through the stochastic differential equation (4.2). However, the equation looks complicated at the first sight, since the relation between two processes $\varphi_t$ and $X_t$ is not direct. We shall first consider the case where $\varphi_t$ is a Brownian motion on $G$. Then the associated process $X_t$ is also a Brownian motion on $\mathcal{G}$ and equation (4.2) is written simply as

$$f(\varphi_t) = f(e) + \sum_{j=1}^{d} \int_0^t X_j f(\varphi_{s-}) dZ_s^j + \frac{1}{2} \sum_{i,j=1}^{d} a_{ij} \int_0^t X_i X_j f(\varphi_{s-}) ds$$

$$= f(e) + \sum_{j=1}^{d} \int_0^t X_j f(\varphi_{s-}) \circ dZ_s^j,$$

(4.8)

where the last integral is the Stratonovich integral. It is well-known that the solution $\varphi_t$ of (4.8) can be obtained by the following approximating method. Define

$$\varphi_t^n = \exp(X_{1/n})\exp(X_{2/n} - X_{1/n})\cdots\exp(X_{[nt]/n} - X_{([nt]-1)/n}), \qquad (4.9)$$

where $X_t = \sum_j Z_t^j X_j$. Then the sequence $\{\varphi_t^n\}$ converges to $\varphi_t$ uniformly in t on any finite interval as $n \to \infty$ in the sense of the convergence in probability. See [9]. The similar approximation is valid for the Lévy process with jumps.

**Theorem 4.2** ([8]) *Let $X_t$ be a $\mathcal{G}$-valued Lévy process of Theorem 4.1. Define a sequence of G-valued processes by (4.9). Then the sequence $\{\varphi_t^n\}$ converges to $\varphi_t$ in probability with respect to the Skorohod topology.*

The proof is omitted since it is found in [8].

By the above theorem, we may denote the correspondence between $\varphi_t$ and $X_t$ symbolically as

$$\varphi_t = \bigcap_{s \leq t} \exp(dX_s), \qquad (4.10)$$

similarly as in [9], although $X_t$ has jumps. The $X_t$ is sometimes called the *driving process* of $\varphi_t$.

The following theorem shows that there is a one to one correspondence between a stable Lévy process on a Lie group with respect to a dilation $\{\gamma_r\}_{r>0}$ and a stable Lévy process on the Lie algebra with respect to the dilation $\{d\gamma_r\}_{r>0}$ through the formula (4.10).

**Theorem 4.3** *Let $G$ be a simply connected, nilpotent Lie group equipped with a dilation $\{\gamma_r\}_{r>0}$. Let $\varphi_t$ be a Lévy process on $G$ and let $X_t$ be its driving process. Then the Lévy process $\varphi_t$ is stable with respect to the dilation $\{\gamma_r\}_{r>0}$ if and only if the driving Lévy process $X_t$ is stable with respect to the dilation $\{d\gamma_r\}_{r>0}$.*

**Proof** For a fixed $r > 0$, set $\tilde{\varphi}_t = \gamma_r(\varphi_t)$. Then since $\varphi_t$ satisfies (4.1), $\tilde{\varphi}_t$ satisfies

$$f(\tilde{\varphi}_t) = f \circ \gamma_r(\varphi_t)$$

$$= f(e) + \sum_j \int_0^t X_j(f \circ \gamma_r)(\varphi_{s-})dZ_s^j + \frac{1}{2}\sum_{i,j=1}^m a_{ij} \int_0^t X_i X_j(f \circ \gamma_r)(\varphi_{s-})ds$$

$$+ \sum_{s \leq t}\left\{ f \circ \gamma_r(\varphi_{s-}\xi(\triangle Z_s)) - f \circ \gamma_r(\varphi_{s-}) - \sum_j X_j(f \circ \gamma_r)(\varphi_{s-})\triangle Z_s^j \right\}.$$

We have

$$X_j(f \circ \gamma_r)(\varphi_s) = d\gamma_r X_j f(\tilde{\varphi}_s)$$

and

$$\gamma_r(\varphi_{s-}\xi(\triangle Z_s)) = \tilde{\varphi}_{s-}\xi(d\gamma_r\triangle Z_s).$$

Therefore, setting $\tilde{Z}_t = d\gamma_r Z_t$ and $\tilde{A} = d\gamma_r A(d\gamma_r)^T$, we have

$$f(\tilde{\varphi}_t) = f(e) + \sum_j \int_0^t X_j f(\tilde{\varphi}_{s-})d\tilde{Z}_s^j + \frac{1}{2}\sum_{i,j=1}^m \tilde{a}_{ij}\int_0^t X_i X_j f(\tilde{\varphi}_{s-})ds$$
$$+ \sum_{s\leq t}\left\{ f(\tilde{\varphi}_{s-}\xi(\triangle\tilde{Z}_s)) - f(\tilde{\varphi}_{s-}) - \sum_j X_j f(\tilde{\varphi}_{s-})\triangle\tilde{Z}_s^j \right\}.$$

This shows that $\tilde{\varphi}_t$ is driven by $\sum_j \tilde{Z}_t^j X_j$.

Now if $\{X_t\}_{t\geq 0}$ is a stable Lévy process on $\mathcal{G}$ with respect to the dilation $\{d\gamma_r\}_{r>0}$, then for any $r > 0$ the law of the Lévy process $\tilde{X}_t \equiv d\gamma_r X_t, t \geq 0$ coincides with the law of the Lévy process $\bar{X}_t \equiv X_{rt}, t \geq 0$. This implies that the law of the process $\tilde{\varphi}_t \equiv \gamma_r(\varphi_t), t \geq 0$ coincides with the law of the process $\bar{\varphi}_t \equiv \varphi_{rt}, t \geq 0$ for any $r > 0$. Therefore, $\{\varphi_t\}_{t\geq 0}$ is stable with respect to the dilation $\{\gamma_r\}_{r>0}$.

Next suppose that $\varphi_t$ is stable with respect to the dilation $\{\gamma_r\}_{r>0}$. Then the law of $\{\gamma_r(\varphi_t)\}_{t\geq 0}$ coincides with the law of $\{\varphi_{rt}\}_{t\geq 0}$ for any $r > 0$. Then the law of the driving process $\tilde{X}_t \equiv d\gamma_r X_t, t \geq 0$ of $\{\gamma_r(\varphi_t)\}_{t\geq 0}$ coincides with the law of the driving process $\bar{X}_t \equiv X_{rt}, t \geq 0$ of $\{\varphi_{rt}\}_{t\geq 0}$. See [1]. This proves that $\{X_t\}_{t\geq 0}$ is stable with respect to the dilation $\{d\gamma_r\}_{r>0}$. The proof is complete.

The above theorem tells us a method of constructing all possible stable Lévy processes on a simply connected, nilpotent Lie group $G$ associated with a given dilation $\{\gamma_r\}_{r>0}$. In particular, we find that if at least one of the real parts of eigenvalues of its exponents is greater than or equal to $1/2$, there exists a non trivial stable Lévy process on $G$ associated with the dilation $\{\gamma_r\}_{r>0}$. Here a Lévy process $\varphi_t$ on $G$ is called *trivial* if $\varphi_t = e$ holds for all $t$ a.s. However, if all real parts of eigenvalues of its exponents are less than $1/2$, there is no nontrivial stable process on $G$ associated with the dilation.

The theorem indicates that not all Brownian motions on a simply connected, nilpotent Lie group are stable with respect to a suitable dilation, i.e. for a certain Brownian motion $\varphi_t$ there is no dilation for which $\varphi_t$ is stable. We will give a few examples.

**Examples** (1) We shall consider a stable Brownian motion and a stable Lévy process on the Heisenberg group. For $G \equiv \mathbf{R}^3$, we define the group law by

$$(x_1, x_2, x_3)(y_1, y_2, y_3) = (x_1 + y_1, x_2 + y_2, x_3 + y_3 - y_2 x_1). \qquad (4.11)$$

Then it is a Lie group called a Heisenberg group. Define

$$X_1 = \frac{\partial}{\partial x_1} + 2x_2 \frac{\partial}{\partial x_3}, \quad X_2 = \frac{\partial}{\partial x_2} - 2x_1 \frac{\partial}{\partial x_3}, \quad X_3 = \frac{\partial}{\partial x_3}. \qquad (4.12)$$

Then it is a basis of the Lie algebra of $G$. Since $[X_1, X_2] = X_3, [X_1, X_3] = [X_2, X_3] = 0$ hold, it is a nilpotent Lie algebra of step 2.

Let $Z_t = (Z_t^1, Z_t^2, Z_t^3)$ be a Levy process on $\mathbf{R}^3$ and let $\varphi_t$ be the Lévy process on the Heisenberg group driven by $X_t \equiv X_1 Z_t^1 + X_2 Z_t^2 + X_3 Z_t^3$. Then $\varphi_t$ can be represented by

$$\varphi_t = \begin{pmatrix} Z_t^1 \\ Z_t^2 \\ Z_t^3 + 2(Z_t^1 Z_t^2 - 2Z_t^{1,2}) \end{pmatrix},$$

where $Z_t^{12} = \int_0^t Z_s^1 \circ dZ_s^2$. See [11]. It should be noted that if $(Z_t^1, Z_t^2)$ is stable with respect to the strict dilation with exponent $(\alpha_1, \alpha_2)$, then $Z_t^1 Z_t^2 - 2Z_t^{12}$ is a one dimensional stable process with exponent $\alpha_1 + \alpha_2$. Therefore, $\varphi_t$ is a stable Brownian motion if and only if $(Z_t^1, Z_t^2)$ is a 2-dimensional brownian motion and $Z_t^3 = 0$. Further, $\varphi_t$ is a stable Lévy process if and only if $Z_t$ is a stable process in $\mathbf{R}^3$ with respect to a strict dilation with exponent $(\alpha_1, \alpha_2, \alpha_1 + \alpha_2)$, where $\alpha_1, \alpha_2 \geq 1/2$.

(2) Let $G$ be the Lie group of $d \times d$ upper triangular matrices $C = (c_{ij})$ with the same diagonal elements and let $\{X_{ij}\}$ be the basis of the Lie algebra $\mathcal{G}$ of $G$ introduced in Section 2. Let $Z_t = (Z_t^{ij}), 1 \leq i < j \leq d$ be a Lévy process on $\mathbf{R}^{d(d-1)/2}$. Let $\varphi_t$ be the Lévy process on $G$ driven by $X_t = \sum X_{ij} Z_t^{ij}$. Then if $X_t$ is stable with respect to a strict dilation with exponent $(\alpha_1, ..., \alpha_{d-1}, ..., \alpha_1 + \cdots + \alpha_{d-1})$, $\varphi_t$ is also stable with respect to the same dilation by Theorem 4.3. A simple example of the exponent is $(\frac{1}{2}, ..., \frac{1}{2}, 1, ..., 1, 2, ..., 2, ..., d-2)$. In such a case a Brownian motion $X_t$ is stable with respect to this dilation if and only if $Z_t^{ij} = 0$ for any $j - i \geq 2$.

## References


[1] Applebaum, D. and Kunita, H.: Lévy flows on manifolds and Lévy processes on Lie groups; *Kyoto J. Math.* **33** (1993) 1103–1123

[2] Folland, G. B. and Stein, E. M.: *Hardy Spaces on Homogeneous Groups.* Math. Notes, Princeton Univ. Press, 1982

[3] Glowacki, P.: Stable semi-groups of measures on the Heisenberg group; *Studia Math.* **79** (1984) 105–138

[4] Hochschild, G.: *The Structure of Lie Groups.* Holden-Day Inc., San Francisco, 1965

[5] Hulanicki, A.: A class of convolution semi-groups of measures on a Lie group; *Lecture Notes in Math.* **828** (1980) 82–101

[6] Hunt, G. A.: Semigroups of measure on Lie groups; *Trans. Amer. Math. Soc.* **81-2** (1956) 264–293

[7] Jurek, Z. J. and Mason, J. D.: *Operator-limit Distributions in Probability Theory.* John Wiley and Sons, New York, 1993

[8] Kunita, H.: Some problems concerning Lévy processes on Lie groups; in: *Proc. of AMS Summer Institute, Stochastic Analysis*, to appear (1994)

[9] McKean, H. P.: *Stochastic Integrals.* Academic Press, New York and London, 1969

[10] Sato, K.: *Lectures on Multivariate Infinitely Divisible Distributions and Operator-stable Processes.* Carleton Univ., 1985

[11] Yamato, Y.: Stochastic differential equations and nilpotent Lie algebras; *Z. W.* **47** (1979) 213–229

Hiroshi Kunita
Department of Applied Science
Faculty of Engineering 36
Kyushu University
Fukuoka 812, JAPAN
*E-mail*: f77532a@kyu-cc.cc.kyushu-u.ac.jp

Y. -J. LEE

# Transformations of white noise functionals and applications

## 1. Introduction

Let $\mathcal{S}$ be the Schwartz space of rapidly decreasing functions on $\mathbb{R}^1$ with dual space $\mathcal{S}'$, known as the tempered distributions. The Gel'fand triple $\mathcal{S} \subset L^2(\mathbb{R}^1) \subset \mathcal{S}'$ then serves as the underlying infinite dimensional space in the investigation of white noise analysis. It is well-known that the space $\mathcal{S}'$, as the dual space of the nuclear space $\mathcal{S}$, carries a Gaussian measure $\mu$ with characteristic function $C(\xi) = \exp(-\frac{1}{2}|\xi|^2)$, where $\xi \in \mathcal{S}$ and $|\cdot| = \sqrt{\langle \cdot, \cdot \rangle}$ denotes the norm of $L^2(\mathbb{R}^1)$. Let $\mathcal{B}$ denote the Borel field of $\mathcal{S}'$, then $(\mathcal{S}', \mathcal{B}, \mu)$ is a probability space, which is also called white noise space. The Brownian motion $B(t)$ is then defined by

$$B(t, x) = \begin{cases} \langle x, 1_{(0,t]} \rangle, & t \geq 0 \\ -\langle x, 1_{(t,0]} \rangle, & t < 0, \end{cases}$$

for almost all $x \in \mathcal{S}'$. Consequently, the White noise $\dot{B}(t)$ is given formally by the expression $\dot{B}(t, x) = x(t)$. For rigorous definition of $\dot{B}(t)$, we refer the reader to [5, 6, 8, 11, 14, 15, 16].

The well-known Wiener-Ito theorem states that, for $f \in L^2[\mathcal{S}', \mu]$, $f$ can be written uniquely as the orthogonal direct sum $\sum_{n=0}^{\infty} \oplus I_n(f_n)$ of multiple Wiener integrals $I_n(f_n)$ with square integrable symmetric kernels $f_n \in L^2(\mathbb{R}^n)$. In notation, we write $f \sim (f_n)$. Moreover, we have

$$\|f\|_{L^2[\mathcal{S}',\mu]}^2 = \sum_{n=0}^{\infty} n! |f_n|^2, \tag{1}$$

where $|\cdot|$ denotes the $L^2(\mathbb{R}^n)$-norm.

The Wiener-Ito theorem may be reformulated in the following way. For a continuous $n$-linear symmetric operator $T$ on $\mathcal{S}'$, define

$$: Tx^n := \int_{\mathcal{S}'} T(x + iy)^n \mu(dy), \tag{2}$$

where $Tx^n := T(x, \ldots, x)$ ($n$-copies) and $T$ extends linearly to the complexification $\mathcal{C}\mathcal{S}'$ of $\mathcal{S}'$.

Then the mapping $T \to \frac{1}{\sqrt{n!}}(: Tx^n :)$ extends to an isometric isomorphism from the class of all symmetric $n$-linear Hilbert-Schmidt operators on $L^2 = L^2(\mathbb{R}^1)$ onto the homogeneous chaos of order $n$ on $\mathcal{S}'$. For any $f \in L^2[\mathcal{S}', \mu]$, $\mu f(h) = \int_{\mathcal{S}'} f(x+h)\mu(dx)$ is infinitely Fréchet differentiable at the origin in the direction of $L^2$ [4] and its $n$-th derivative $D^n \mu f(0)$ at zero is a Hilbert-Schmidt type $n$-linear operator on $L^2$ so that $: D^n \mu f(0) x^n :$ exists for almost all $x \in \mathcal{S}'$ with respect to $\mu$. Further, $f$ can be written uniquely as the orthogonal direct sum

$$f(x) = \sum_{n=0}^{\infty} \frac{1}{n!} : D^n \mu f(0) x^n : \tag{3}$$

with

$$\|f\|_{L^2[\mathcal{S}',\mu]} = \left\{ \sum_{n=0}^{\infty} \frac{1}{n!} \|D^n \mu f(0)\|_{HS}^2 \right\}^{\frac{1}{2}}, \tag{4}$$

where $\| \cdot \|_{HS}$ denotes the Hilbert-Schmidt norm for multilinear operators on $L^2$. (see [10, 12]).

It is easy to see that $f_n$ is nothing but the kernel function of the operator $\frac{1}{n!} D^n \mu f(0)$ and

$$\frac{1}{n!} : D^n \mu f(0) x^n := I_n(f_n)(x) \tag{5}$$

for almost all $x \in \mathcal{S}'$ with respect to $\mu$.

Another formulation of Wiener-Ito decomposition which is frequently used in white noise analysis can be found in [1, 5, 6, 9, 11, 25].

In this paper we concern ourselves with the transformation $Tf$ of $f$ induced from a sequence of operators $\{T_n\}$ acting on the kernel functions $\{f_n\}$ and we ask what would be the function $Tf$ such that $Tf \sim (T_n f_n)$? For example, if $A$ is an operator on $L^2(\mathbb{R}^1)$ and if $T_n = A^{\otimes n}$, then the kernels $(A^{\otimes n} f_n)$ correspond to the second quantization $\Gamma(A)f$. Apply the reformulation form (3) of Wiener-Ito decomposition, it is shown that, for any test white noise functional $f \in (\mathcal{S})$ and for $A \in \mathcal{L}[\mathcal{S}, \mathcal{S}]$, $\Gamma(A)f$ enjoys the following integral representation:

$$\Gamma(A)f(x) = \int_{\mathcal{S}'} \int_{\mathcal{S}'} \tilde{f}(A^* x + i A^* y + z)\mu(dz)\mu(dy), \tag{6}$$

where $\tilde{f}$ denotes the analytic version of $f$ and $A^* : \mathcal{S}' \to \mathcal{S}'$ is the adjoint of $A$.

We find that the transform $f \to f_{(\lambda)}$ (where $f_{(\lambda)} \sim (\lambda^n f_n)$), the Fourier-Wiener transform, the Ornstein-Uhlenbeck semigroup and the conditional expectation $E[f|(x, e_j); 1 \le j \le n]$, $e_j \in \mathcal{S}$, can be represented by (6)

Applying the representation (6), we show that the two test white noise functional spaces $(S)$ and $\mathcal{A}_\infty$ are topologically equivalent. Similarly we also establish a topological equivalence between two new spaces of white noise functionals $\mathcal{M}$ and $\mathcal{E}$, introduced respectively by Meyer and Yan [22] and Lee [14, 19].

The results of this paper can be extended to more general Gel'fand triples $E \subset H \subset E^*$. For details, we refer the reader to [20].

**Notations:**

(1) $CX$: The complexification of $X$, i.e., $CX = X + iX$. If $X$ is a Banach space, the norm $\|\cdot\|_{CX}$ is defined by $\|Z\|_{CX} = \left(\|x\|_X^2 + \|y\|_X^2\right)^{\frac{1}{2}}$ for $z = x + iy$, $x, y \in X$. For notational convenience, we shall use $\|\cdot\|_X$ also for the norm of $CX$.

(2) $\mathcal{HS}^n[H]$: Hilbert-Schmidt $n$-linear operators on a Hilbert space $H$ with norm $\|\cdot\|_{\mathcal{HS}^n[H]}$.

## 2. Preliminaries

**The spaces $(S)$ and $(S^*)$:**

Let $A$ denote the operator $A = -\Delta + (1 + t^2)$ on $L^2(\mathbb{R}^1)$ and $\Gamma(A) = \sum_{n=0}^\infty \oplus A^{\otimes n}$, the second quantization operator of $A$. For each $p \geq 0$, we set $S_p := \mathcal{D}(A^p)$, the domain of $A^p$ and $(S_p) := \mathcal{D}(\Gamma(A^p))$. Denote by $S_{-p}$ (resp. $(S_{-p})$) the dual of $S_p$ (resp. $(S_p)$). Put $(S) = \bigcap_p (S_p)$. Endow $(S)$ with the projective limit topology, $(S)$ becomes a topological convex space. $(S)$ serves as the test functionals in white noise analysis. The dual space of $(S)$ is denoted by $(S)^*$. Members of $(S)^*$ are called generalized white noise functionals (GWNF, for abbrev.). The following relationships hold:

$$S = \bigcap_p S_p, \quad S' = \bigcup_p S_{-p}, \quad (S)^* = \bigcup_p (S_{-p}),$$

$$S \subset S_p \subset L^2(\mathbb{R}^1) \subset S_{-p} \subset S',$$

$$(S) \subset (S_p) \subset (L^2) \subset (S_{-p}) \subset (S)^*,$$

where $(L^2) = L^2[S', \mu]$. Let $S_p(\mathbb{R}^n)$ denote the space spanned by $S_p^{\otimes n}$ and let $\widehat{S_p(\mathbb{R}^n)}$ be the symmetric member in $S_p(\mathbb{R}^n)$. For $f \in (L^2)$, let $f \sim (f_n)$. Then for $p \geq 0$, $(S_p)$ is a Hilbert space with norm given by

$$\|f\|_{2,p} = \left\{ \sum_{n=0}^\infty n! |f_n|_p^2 \right\}^{\frac{1}{2}}, \tag{7a}$$

where $|\cdot|_p$ denotes the $\mathcal{S}_p(\mathbb{R}^n)$ norm and $f_n \in \widehat{\mathcal{S}(\mathbb{R}^n)}$.

The space $(\mathcal{S}_p)$ may be reformulated in terms of our language as follows.

Recall that if $\varphi \sim (\varphi_n)$, then $I_n(\varphi_n)(x) = \frac{1}{n!} : D^n \mu \varphi(0) x^n :$ for almost all $x \in \mathcal{S}'$ with respect to $\mu$ and that $|\varphi_n|_p = \frac{1}{n!} \| D^n \mu \varphi(0) \|_{\mathcal{HS}^n[\mathcal{S}_{-p}]}$ (see [13]). We have

$$\|\varphi\|_{2,p} = \left\{ \sum_{n=0}^{\infty} \frac{1}{n!} \| D^n \mu \varphi(0) \|^2_{\mathcal{HS}^n[\mathcal{S}_{-p}]} \right\}^{\frac{1}{2}}. \tag{7b}$$

The following lemma is basic.

**2.1 Lemma** [16]  *If $f \in (\mathcal{S}_p)$, then the series (3) converges to an analytic function $\tilde{f}_p$ on $\mathcal{CS}_{-p}$ absolutely and uniformly on bounded subsets of $\mathcal{CS}_{-p}$. If $f \in (\mathcal{S})$ and, for every $z \in \mathcal{CS}'$, we let*

$$\tilde{f}(z) = \sum_{n=0}^{\infty} \frac{1}{n!} \int_{\mathcal{S}'} D^n \mu f(0)(z + iy)^n \mu(dy), \tag{8}$$

*then the series converges to $\tilde{f}$ absolutely and uniformly on bounded subsets of $\mathcal{CS}'$ and $\tilde{f} = \tilde{f}_p$ on $\mathcal{CS}_{-p}$. ($\tilde{f}$ is called the analytic version of $f$).*

**The space $\mathcal{A}_\infty$:**

For $p \in \mathbb{R}^1$, let $\mathcal{A}_p$ denote the space of analytic functions $f$ which are defined on $\mathcal{CS}_{-p}$ and satisfy the following property:

(A) There exists a constant $c$, depending only on $f$, such that

$$|f(z)| \leq c \exp\left(\frac{1}{2}|z|^2_{-p}\right)$$

for all $z \in \mathcal{CS}_{-p}$.

Define a norm $\|\cdot\|_{\mathcal{A}_p}$ on $\mathcal{A}_p$ by

$$\|f\|_{\mathcal{A}_p} = \sup_{z \in \mathcal{CS}_{-p}} \left[ |f(z)| \exp(-\tfrac{1}{2}|z|^2_{-p}) \right].$$

Then $(\mathcal{A}_p, \|\cdot\|_{\mathcal{A}_p})$ forms a Banach space and $\mathcal{A}_p \subset \mathcal{A}_q$ for $p \geq q$. Put $\mathcal{A}_\infty = \bigcap_{p>0} \mathcal{A}_p$ and endow $\mathcal{A}_\infty$ with the projective limit topology induced by the family $\{\mathcal{A}_p, p > 0\}$. Then $\mathcal{A}_p$ becomes a locally convex space as well as a topological algebra (see [16]).

**2.2 Lemma** [16]  *If $\varphi \in (\mathcal{S})$, then the analytic version $\tilde{\varphi} \in \mathcal{A}_\infty$. $\mathcal{A}_\infty \subset (\mathcal{S})$ and the embedding from $\mathcal{A}_\infty$ into $(\mathcal{S})$ is continuous. Moreover, the mapping $\varphi \to \tilde{\varphi}$ from*

$(S)$ into $\mathcal{A}_\infty$ is also continuous. Thus if we identify $\varphi$ with $\tilde{\varphi}$, then the two families $\{\|\cdot\|_{2,p}, p > 0\}$ and $\{\|\cdot\|_{\mathcal{A}_p}, p > 0\}$ are equivalent in $(S)$ (or $\mathcal{A}_\infty$).

## 3. Integral Representation of the Second Quantization

Let $K$ be a linear operator on $L^2(\mathbb{R}^1)$ such that $K[S] \subset S$ and $K$ is continuous on $S$. Recall that the second quantization of $K$ is given by $\Gamma(K) = \sum_{n=0}^\infty \oplus K^{\otimes n}$. If $\varphi \in (S)$ with $\varphi \sim (\varphi_n)$, then $\Gamma(K)\varphi \sim (K^{\otimes n}\varphi_n)$. Observe that, for a function $g \in \mathcal{S}(\mathbb{R}^n)$, $g$ and $K^{\otimes n}g$ define continuous $n$-linear functionals on $S \times \cdots \times S$ ($n$-fold) such that

$$(K^{\otimes n}g)(x_1, x_2, \ldots, x_n) = g(K^*x_1, K^*x_2, \ldots, K^*x_n) = (g, \bigotimes_{k=1}^n K^*x_k). \qquad (9)$$

If $G$ is the $n$-linear operator with kernel function $g$, then according to (9) we must have $I_n(K^{\otimes n}g)(x) = \frac{1}{n!} : G(K^*x)^n :$. It follows that if $\varphi \in (S)$, we then have

$$\Gamma(K)\varphi(x) = \sum_{n=0}^\infty \frac{1}{n!} : D^n \mu\tilde{\varphi}(0)(K^*x)^n :$$

$$= \sum_{n=0}^\infty \frac{1}{n!} \int_{S'} D^n \mu\tilde{\varphi}(0)(K^*x + iK^*y)^n \mu(dy)$$

$$= \int_{S'} \sum_{n=0}^\infty \frac{1}{n!} D^n \mu\tilde{\varphi}(0)(K^*x + iK^*y)^n \mu(dy)$$

$$= \int_{S'} \mu\tilde{\varphi}(K^*x + iK^*y)\mu(dy)$$

$$= \int_{S'} \int_{S'} \tilde{\varphi}(K^*x + iK^*y + z)\mu(dz)\mu(dy).$$

We have proved the following identity for $\varphi \in (S)$

$$\Gamma(K)\varphi(x) = \int_{S'} \int_{S'} \tilde{\varphi}(K^*x + iK^*y + z)\mu(dz)\mu(dy). \qquad (10)$$

**Example 1** Let $K = \lambda I$ ($\lambda \in \mathbb{C}$) and apply (10), we have

$$\Gamma(\lambda I)\varphi(x) = \int_{S'} \int_{S'} \tilde{\varphi}(\lambda x + i\lambda y + z)\mu(dz)\mu(dy)$$

$$= \int_{S'} \tilde{\varphi}(\lambda x + \sqrt{1 - \lambda^2}y)\mu(dy) \qquad \text{(by [16, Lemma 4.3])}.$$

On the other hand, for $\varphi \sim (\varphi_n)$, let $\varphi_{(\lambda)}$ be the function such that $\varphi_{(\lambda)} \sim (\lambda^n \varphi_n)$. Then it follows from Lemma 2.1 and the formula (8) that we have

$$
\begin{aligned}
\varphi_{(\lambda)}(z) &= \sum_{n=0}^{\infty} \frac{\lambda^n}{n!} \int_{S'} D^n \mu \tilde{\varphi}(0)(z+iy)^n \mu(dy) \\
&= \int_{S'} \mu \tilde{\varphi}(\lambda z + iy) \mu(dy) \\
&= \int_{S'} \tilde{\varphi}(\lambda z + \sqrt{1-\lambda^2} y) \mu(dy) \quad \text{(by [16, Lemma 4.3])}. \qquad (11a)
\end{aligned}
$$

Thus we have shown that

$$
\varphi_{(\lambda)}(z) = \Gamma(\lambda I)\varphi(z) = \int_{S'} \tilde{\varphi}(\lambda z + \sqrt{1-\lambda^2} y) \mu(dy).
$$

In particular, when $\lambda = i$, $\Gamma(iI)\varphi$ becomes the Fourier-Wiener transform.

**3.1 Remark** Besides $\varphi_{(\lambda)}$, Yan [27] also introduced the following transforms as tools for studying semigroup and renormalizations.

(i) For $\varphi \in (S)$ and $\lambda \in \mathbb{C}$, define

$$
\varphi^{(\lambda)}(z) = \tilde{\varphi}(\lambda z).
$$

(ii) Let $\lambda \in \mathbb{C} \setminus \{0\}$. For $\varphi \in (S)$, define

$$
\mathcal{R}_\lambda \varphi = \left(\varphi_{(\lambda)}\right)^{(\frac{1}{\lambda})}, \quad \mathcal{R}_\lambda^{-1} \varphi = \left(\varphi^{(\lambda)}\right)_{(\frac{1}{\lambda})}.
$$

By applying Lemma 2.1 and formulas (8) and (11a), we immediately have following formulas:

$$
\left(\varphi^{(\lambda)}\right)_{(\beta)}(y) = \int_{S'} \tilde{\varphi}(\lambda\sqrt{1-\beta^2}x + \lambda\beta y)\mu(dx). \qquad (11b)
$$

$$
\left(\varphi_{(\lambda)}\right)^{(\beta)}(y) = \int_{S'} \tilde{\varphi}(\sqrt{1-\lambda^2}x + \lambda\beta y)\mu(dx). \qquad (11c)
$$

$$
\mathcal{R}_\lambda \varphi(y) = \int_{S'} \tilde{\varphi}(\sqrt{1-\lambda^2}x + y)\mu(dx). \qquad (11d)
$$

$$
\mathcal{R}_\lambda^{-1} \varphi(y) = \int_{S'} \tilde{\varphi}(\sqrt{\lambda^2-1}x + y)\mu(dx). \qquad (11e)
$$

$$
\int_{S'} \tilde{\varphi}(\alpha x + \beta y)\mu(dx) = \left(\varphi^{(\sqrt{\alpha^2+\beta^2})}\right)_{(\beta/\sqrt{\alpha^2+\beta^2})} = \left(\varphi_{(\sqrt{1-\alpha^2})}\right)^{(\beta/\sqrt{1-\alpha^2})}. \qquad (11f)
$$

For their proofs, we refer the reader to [18]. /// 

**Example 2** When $A$ is the Fourier-transform on $\mathbb{R}^1$, then $A^*x = \widehat{x}$, the Fourier-transform of $x \in \mathcal{S}'$. Hence we have

$$\Gamma(A)\varphi(x) = \int_{\mathcal{S}'} \int_{\mathcal{S}'} \tilde{\varphi}(\widehat{x} + i\widehat{y} + z)\mu(dy)\mu(dz).$$

In other words, if $\varphi \sim (\varphi_n)$, then $\Gamma(A)\varphi \sim (\widehat{\varphi}_n)$.

**Example 3** Let $A = e^{-tI}$ $(t > 0)$. Then

$$\begin{aligned}
\Gamma(e^{-tI})\varphi(x) &= \int_{\mathcal{S}'} \int_{\mathcal{S}'} \tilde{\varphi}(e^{-t}x + ie^{-t}y + z)\mu(dy)\mu(dz) \\
&= \int_{\mathcal{S}'} \tilde{\varphi}(e^{-t}x + \sqrt{1 - e^{-2t}}y)\mu(dy).
\end{aligned}$$

It is well-known that $\Gamma(e^{-tI})$ is nothing but the Ornstein-Uhlenbeck semigroup. It is easy to see that, as $t \to 0$,

$$\frac{1}{t}\big(\Gamma(e^{-tI}) - I\big)\varphi \to -\mathcal{N}\tilde{\varphi} \quad \text{in} \quad \mathcal{A}_\infty,$$

where $\mathcal{N}u(x) = -\text{trace}_{L^2} D^2u(x) + (x, Du(x))$ is the number of particle operator.

**Example 4** Let $\{e_i : i = 1, 2, \ldots, n\}$ be an O.N. set in $L^2(\mathbb{R}^1)$ consisting of elements of $\mathcal{S}$. Let $P_n(x) = \sum_{j=1}^{n}(x, e_j)e_j$. Then we have

$$\Gamma(P_n)\varphi = E[\tilde{\varphi}|\tilde{e}_j, j = 1, 2, \ldots, n], \tag{12a}$$

where $\tilde{e}_j(x) = (x, e_j)$.

In fact, for $\lambda_1, \lambda_2, \ldots, \lambda_n \in \mathbb{R}^1$, we have

$$\begin{aligned}
E\Big[E[\tilde{\varphi}|\tilde{e}_j; 1 \leq j &\leq n]\exp\big(-i\sum_{j=1}^{n}\lambda_j\tilde{e}_j\big)\Big] \\
&= \int_{\mathcal{S}'} \tilde{\varphi}(x)e^{-i\sum_{j=1}^{n}\lambda_j(x,e_j)}\mu(dx) \tag{12b} \\
&= e^{-\frac{1}{2}\sum_{j=1}^{n}\lambda_j^2} \int_{\mathcal{S}'} \tilde{\varphi}(x + i\sum_{j=1}^{n}\lambda_j e_j)\mu(dx).
\end{aligned}$$

On the other hand,

$$E\left[\Gamma(P_n)\varphi \exp\left(-i\sum_{j=1}^{n}\lambda_j\tilde{e}_j\right)\right]$$

$$= \int_{\mathcal{S}'}\int_{\mathcal{S}'}\int_{\mathcal{S}'}\tilde{\varphi}(P_nx+iP_ny+z)e^{-i\sum_{j=1}^{n}\lambda_j(x,e_j)}\mu(dy)\mu(dz)\mu(dx)$$

$$= \int_{\mathcal{S}'}\int_{\mathcal{S}'}\tilde{\varphi}(P_nx+(I-P_n)y)e^{-i\sum_{j=1}^{n}\lambda_j(x,e_j)}\mu(dy)\mu(dx) \tag{12c}$$

$$= e^{-\frac{1}{2}\sum_{j=1}^{n}\lambda_j^2}\int_{\mathcal{S}'}\int_{\mathcal{S}'}\tilde{\varphi}\left(P_nx+(I-P_n)y+i\sum_{j=1}^{n}\lambda_je_j\right)\mu(dy)\mu(dx)$$

$$= e^{-\frac{1}{2}\sum_{j=1}^{n}\lambda_j^2}\int_{\mathcal{S}'}\tilde{\varphi}\left(x+i\sum_{j=1}^{n}\lambda_je_j\right)\mu(dx).$$

Comparing the two identities (12b) and (12c), we obtain

$$\Gamma(P_n)\varphi(x) = E[\tilde{\varphi}|(x,e_j); 1\le j\le n]$$

$$= \int_{\mathcal{S}'}\int_{\mathcal{S}'}\tilde{\varphi}(P_nx+iP_ny+z)\mu(dy)\mu(dz) \tag{12d}$$

$$= \int_{\mathcal{S}'}\tilde{\varphi}(P_nx+(I-P_n)y)\mu(dy).$$

The identity (12a) was also observed by Kubo and Kuo [7]. By applying the integral representation (12d), one sees that the cylinder function $\Gamma(P_n)\varphi$ converges to $\tilde{\varphi}$ in $\mathcal{A}_\infty$ (hence in $(\mathcal{S})$). In the case that $\varphi$ is a GWNF, we define

$$\ll \Gamma(P_n)\varphi, \psi \gg = \ll \varphi, \Gamma(P_n)\psi \gg,$$

for $\psi \in (\mathcal{S})$, where $\ll \cdot, \cdot \gg$ denotes the $(\mathcal{S})^*$–$(\mathcal{S})$ pairing. Then it is easy to see that the finite dimensional Hida distribution $\Gamma(P_n)\varphi$ converges to $\varphi$ in $(\mathcal{S})^*$. This also proves [7, Theorem 3.11].

## 4. The Topological Equivalence of $(\mathcal{S})$ and $\mathcal{A}_\infty$

Let $f \in (\mathcal{S})$ and $\tilde{f}$ be its analytic version as given in Lemma 2.1. Let $p > \frac{1}{2}$. It has been proved in [16] that there exist some constants $\alpha_p$ and $\beta_p$ such that

$$\alpha_p\|\tilde{\varphi}\|_{\mathcal{A}_{p-1}} \le \|\varphi\|_{2,p} \le \beta_p\|\tilde{\varphi}\|_{\mathcal{A}_{p+3}}. \tag{13a}$$

The inequality (13a) obviously implies the topological equivalence between $(\mathcal{S})$ and $\mathcal{A}_\infty$. The first inequality can be improved easily. For example, it can be shown that, for $p > 1$, there exists some constant $\alpha_p > 0$ and $s > \frac{1}{2}$ such that $\alpha_p \|\varphi\|_{\mathcal{A}_{p-s}} \leq \|\varphi\|_{2,p}$. The second inequality of (13a) has been improved in [17] by using the same idea of the proof as given in [16]. In [17] it is shown that there exists some constant $\beta_p$ so that

$$\|\varphi\|_{2,p} \leq \beta_p \|\tilde{\varphi}\|_{\mathcal{A}_{p+2}}. \tag{13b}$$

The inequality (13b) can be further improved by applying the integral representation of second quantization $\Gamma(A^p)$ of $A^p$, where $A = -\Delta + (1 + t^2)$ as given in section 2.

**4.1 Lemma** [20]  *If $\varphi \in (\mathcal{S})$ and $\tilde{\varphi}$ be its analytic version in $\mathcal{A}_\infty$, then*

$$\Gamma(A^p)\varphi(x) = \int_{\mathcal{S}'} \tilde{\varphi}\big(A^p x + i\sqrt{A^{2p} - I}\, y\big)\mu(dy), \quad x \in \mathcal{S}'.$$

**Proof**  Let $\{e_n\}$ be a CONS of $L^2(\mathbb{R}^1)$ consisting of eigenvectors of $A$ and let $P_n x = \sum_{j=1}^n (x, e_j)e_j$. Then we have

$$\Gamma(A^p)\varphi(x) = \int_{\mathcal{S}'} \int_{\mathcal{S}'} \tilde{\varphi}(A^p x + i A^p y + z)\mu(dy)\mu(dz)$$

$$= \lim_{n \to \infty} \int_{\mathcal{S}'} \int_{\mathcal{S}'} \tilde{\varphi}(A^p x + i A^p P_n y + P_n z)\mu(dy)\mu(dz). \tag{14}$$

Since

$$\int_{\mathcal{S}'} \int_{\mathcal{S}'} \tilde{\varphi}(A^p x + i A^p P_n y + P_n z)\mu(dy)\mu(dz)$$

$$= \int_{\mathcal{S}'} \int_{\mathcal{S}'} \tilde{\varphi}\Big(A^p x + \sum_{j=1}^n \big(i(2j+2)^{2p} y + z, e_j\big)e_j\Big)\mu(dy)\mu(dz)$$

$$= \int_{\mathcal{S}'} \tilde{\varphi}\Big(A^p x + \sum_{j=1}^n i\sqrt{(2j+2)^{2p} - 1}\,(y, e_j)e_j\Big)\mu(dy)$$

$$= \int_{\mathcal{S}'} \tilde{\varphi}(A^p x + i\sqrt{A^{2p} - I}\, P_n y)\mu(dy),$$

it follows from (14) that we have

$$\Gamma(A^p)\varphi(x) = \lim_{n \to \infty} \int_{\mathcal{S}'} \tilde{\varphi}\big(A^p x + i\sqrt{A^{2p} - I}\, P_n y\big)\mu(dy)$$

$$= \int_{\mathcal{S}'} \tilde{\varphi}\big(A^p x + i\sqrt{A^{2p} - I}\, y\big)\mu(dy). \qquad ///$$

The main results of this section is given as follows.

**4.2 Proposition** *Let $p > \frac{1}{2}$ and $r > \frac{1}{2}$. Let $f$ and $\tilde{f}$ be functions as given in Lemma 4.1. Then we have*

$$\|\varphi\|_{2,p} \le C\|\tilde{\varphi}\|_{A_{p+r}}, \tag{15}$$

*where $C_r = \left\{ \int_{S'} \exp(|x|^2_{-r})\mu(dx) \right\}$.*

**Proof** Recall that $\|\varphi\|_{2,p} = \|\Gamma(A^p)\varphi\|_{L^2[S',\mu]}$ and apply Lemma 4.1 we obtain

$$\|\varphi\|^2_{2,p} \le \int_{S'} \int_{S'} \left|\tilde{\varphi}(A^p x + i\sqrt{A^{2p} - Iy})\right|^2 \mu(dy)\mu(dx)$$

$$\le \|\tilde{\varphi}\|^2_{A_{p+r}} \int_{S'} e^{|x|^2_{-r} + |y|^2_{-r}} \mu(dy)\mu(dx)$$

$$\le C_r^2 \|\tilde{\varphi}\|^2_{A_{p+r}}. \qquad\qquad ///$$

The inequality (15) clearly improves (13b).

**Locality:**

For $F \in (S)^*$, let $U_f$ denote the $U$–functional of $F$ defined as follows:

$$U_F(\xi) = \begin{cases} \mu * F(\xi), & \text{if } F \in L^2[S',\mu]; \\ e^{-\frac{1}{2}|\xi|^2} \ll F, e^{(\cdot,\xi)} \gg, & \text{if } F \in (S)^*, \end{cases}$$

where $\xi \in S$.

In [16], it has been shown that, for $p < 0$,

$$\|F\|^2_{2,p} = \sum_{n=0}^{\infty} \frac{1}{n!} \|D^n U_F(0)\|^2_{\mathcal{HS}^n[S_{-p}]}$$

$$= \lim_{n\to\infty} \int_{S'} \left| \int_{S'} U_F(A^p P_n x + iA^p P_n y)\mu(dy)\right|^2 \mu(dx),$$

where $P_n$'s are the projections as given in Lemma 4.1. By applying the similar arguments as given in the proof of Proposition 4.2, one can prove the following Proposition. For details, we refer the reader to [20].

**4.3 Proposition** *Let $p \in \mathbb{R}^1$ and $r > \frac{1}{2}$. Then we have*

$$\|F\|_{2,p} \le C_r\|U_F\|_{A_{p+r}}.$$

# 5. The Topological Equivalence of $\mathcal{E}$ and $\mathcal{M}$

Recently two new classes of test functionals, $\mathcal{M}$ and $\mathcal{E}$, were introduced respectively by Meyer and Yan [23] and [14, 19]. It can be shown that $\mathcal{E}$ is an analytic version of $\mathcal{M}$ and, as an application of the integral representation of second quantization, one can prove that $\mathcal{M}$ and $\mathcal{E}$ are topologically equivalent to each other. We first describe the construction of $\mathcal{M}$ and $\mathcal{E}$ as follows.

**The space $\mathcal{E}$:**

For $p > 0$, let $\mathcal{E}_p$ denote the class of analytic functions $f$ which are defined on $\mathcal{CS}_{-p}$ such that there exist constants $c$ and $c'$, depending only on $f$, such that

$$|f(z)| \leq c \exp(c'|z|_{-p}).$$

for $z \in \mathcal{CS}_{-p}$.

For $p > 0$, $m > 0$, and for $f \in \mathcal{E}_p$, define

$$|||f|||_{p,m} = \sup_{z \in \mathcal{CS}_{-p}} [|f(z)| \exp(-m|z|_p)].$$

Denote by $\mathcal{E}_{p,m}$ the class of functions $f \in \mathcal{E}_p$ such that $|||f|||_{p,m} < \infty$. Then $(\mathcal{E}_{p,m}, |||\cdot|||_{p,m})$ forms a Banach space and $\mathcal{E}_{p,m} \subset \mathcal{E}_{p,m'}$ for $m' > m$ so that $\mathcal{E}_p = \bigcup_{m>0} \mathcal{E}_{p,m}$. Endow $\mathcal{E}_p$ with the inductive limit topology, $\mathcal{E}_p$ becomes a locally convex space (see also [14]). Put $\mathcal{E} = \bigcap_{p>0} \mathcal{E}_p$ and endow $\mathcal{E}$ with the projective limit topology induced by $\mathcal{E}_p$, $\mathcal{E}$ becomes a locally convex space as well as a topological algebra (see [19]).

**The space $\mathcal{M}$:**

Let $f \in L^2[\mathcal{S}', \mu]$ with $f \sim (f_n)$. For $p, q > 0$, define $\|f\|_{\mathcal{M}_{p,q}}$ by

$$\|f\|_{\mathcal{M}_{p,q}} = \left\{ \sum_{n=0}^{\infty} (n!)^2 2^{-qn} |f_n|_p^2 \right\}^{\frac{1}{2}},$$

where $|f_n|_p$ denotes the $\mathcal{S}_p^{\otimes n}$–norm of $f_n$.

Denote by $\mathcal{M}_{p,q}$ the class of functions $f$ such that $\|f\|_{\mathcal{M}_{p,q}} < \infty$. Then $(\mathcal{M}_{p,q}, \|\cdot\|_{p,q})$ becomes a Hilbert space and $\mathcal{M}_{p,q} \subset \mathcal{M}_{p,q'}$ if $q' > q$. Put $\mathcal{M}_p = \bigcup_{q>0} \mathcal{M}_{p,q}$ and endow it with the inductive limit topology. Next, put $\mathcal{M} = \bigcap_{p>0} \mathcal{M}_p$ and then endow $\mathcal{M}$ with the projective limit topology induced by $\mathcal{M}_p$. $\mathcal{M}$ also becomes a locally convex space.

Since $\mathcal{M} \subset (\mathcal{S})$, every member $\varphi \in \mathcal{M}$ admits an analytic version $\tilde{\varphi} \in \mathcal{A}_\infty$. Write

$$\tilde{\varphi}(x) = \sum_{n=0}^{\infty} \frac{1}{n!} : D^n \mu(0) x^n : . \tag{16a}$$

Then we have

$$\|\varphi\|^2_{\mathcal{M}_{p,q}} = \sum_{n=0}^{\infty} 2^{-qn} \|D^n \mu\varphi(0)\|^2_{\mathcal{HS}^n[\mathcal{S}_{-p}]}. \tag{16b}$$

From (16a) and (16b), we immediately have the following.

**5.1 Theorem** [20]  *If $\varphi \in \mathcal{M}$, then $\tilde{\varphi} \in \mathcal{E}$ and, for $p > \frac{1}{2}$, there exists a constant $C_{p,q}$ such that*

$$\|\|\tilde{\varphi}\|\|_{p,(\sqrt{2})^q} \leq C_{p,q} \|\varphi\|_{\mathcal{M}_{p,q}}.$$

Conversely, as an application of the integral representation of second quantization, we also have the following theorem. We state it without proof. For details, we refer the reader to [20].

**5.2 Theorem**  *Let $r > \frac{1}{2}$ and $p \geq 0$. Let $\varphi \in \mathcal{E}_{p+r,m}$ and choose $q$ so large that $em < 2^q$. Then $\varphi \in \mathcal{M}_{p,q}$ and we have*

$$\|\varphi\|_{\mathcal{M}_{p,q}} \leq \beta_{p+r,m,q} \|\|\varphi\|\|_{p+r,m},$$

*where $\beta_{p+r,m,q}$ is a constant depending on $p + r$, $m$ and $q$.*

**5.3 Remark**  Theorem 5.2 also implies $\mathcal{E} \subset \mathcal{M}$ and Theorem 5.1 implies that the mapping $\varphi \to \tilde{\varphi}$ is a continuous embedding from $\mathcal{M}$ onto $\mathcal{E}$. Thus $\mathcal{E}$ and $\mathcal{M}$ are topologically equivalent. ///

### References


[1] Albeverio, S., Hida, T., Potthoff, J., Röchner, M., and Streit, L: Dirichlet form in terms of white noise analysis I – Construction and QFT examples; *Reviews in Math. Phys.*, Vol. 1, no. 2 & 3(1990), 219–312; II – Construction of infinite dimensional diffusions and QFT examples, ibid., 313–323.

[2] Gel'fand, I. M. and Vilenkin N. Ya.: *Generalized Functions*, Vol. 4 (English translation), 1964

[3] Gross, L.: Abstract Wiener spaces; *Proc. 5th Berkeley Sym. Math. Stat. Prob.* **2** (1965) 31–42

[4] Gross, L.: Potential theory on Hilbert space; *J. Func. Anal.* **1** (1967) 123–181

[5] Hida, T.: *Analysis of Brownian functionals.* Carleton Mathematical Lecture Notes, no. 13, 1975

[6] Hida, T., Kuo, H. -H., Potthoff, J., and Streit, L.: *White Noise : An Infinite Dimensional Calculus.* Kluwer Academic Publishers, Dordrecht/Boston/London, 1993

[7] Kubo, I. and Kuo, H. -H.: Finite dimensional Hida distributions; *Preprint* (1993)

[8] Kubo, I. and Takenaka, S.: Calculus on Gaussian white noise I: *Proc. Japan Acad.* **56A** (1980) 376–380; II: ibid. **56A** (1980) 411–416; III: ibid. **57A** (1981) 433–436; IV: ibid. **58A** (1982) 186–189

[9] Kubo, I. and Yokoi, Y.: A remark on the space of testing random variables in the white noise calculus; *Nagoya Math. J.* **115** (1989) 139–149

[10] Kuo, H. -H.: Integration in Banach spaces; in: *Notes in Banach spaces*, H. Elton Lacey (ed.) 1–38, Univ. of Texas, Austin, 1980

[11] Kuo, H. -H.: Brownian motion, diffusions and infinite dimensional calculus; in: *Stochastic Analysis and Related Topics*, H. Korezlioglu and A. S. Ustunel (eds.), Lecture Notes in Math. **1316** (1988) 130–169

[12] Lee, Y. -J.: Sharp inequalities and regularity of heat semigroup on infinite dimensional space; *J. Funct. Anal.* **71** (1987) 69–87

[13] Lee. Y. -J.: On the convergence of Wiener-Ito decomposition; *Bull. Inst. Math. Acad. Sinica* **17** (1989) 305–312

[14] Lee. Y. -J.: Generalized functions on infinite dimensional spaces and its application to white noise calculus; *J. Funct. Anal.* **82** (1989) 429–464

[15] Lee. Y. -J.: A reformulation of white noise calculus; in: *White Noise Analysis*, T. Hida et al. (eds.) 274–293, World Scientific (1990)

[16] Lee. Y. -J.: Analytic version of test functionals, Fourier transform and a characterization of measures in white noise calculus; *J. Func. Anal.* **100** (1991) 359–380.

[17] Lee. Y. -J.: A characterization of generalized functions on infinite dimensional spaces and Bargmann-Segal analytic functions; In *Gaussian Random Field*, The Third Nagoya Lévy Seminar, K. Itô and T. Hida (eds.) 272–284, World Scientific, 1991

[18] Lee. Y. -J.: Transformation and Wiener-Ito decomposition of white noise functionals; *Bull. Inst. Math. Academia Sinica* (Taiwan) **21** (1993) 279–291

[19] Lee. Y. -J.: Positive generalized functions; in: *Stochastic Processes*, a Festschrift in Honour of Gopinath Kallianpur 225–234, Springer-Verlag, 1993

[20] Lee, Y. -J.: Integral representation of second quantization and Ito application to white noise analysis; **Preprint** (1994)

[21] Meyer, P. A. and Yan, J. A.: A propos des distributions sur lespace de Wiener; *Séminaire de Probabilités XXI*, LN in Math. **1247** (1987) 8–26, Springer

[22] Meyer, P. A. and Yan, J. A.: Distributions sur l'espace de Wiener (suit); *Séminaire de Probab. XXIII*, LN in Math. **1372** (1989) 382–392, Springer

[23] Meyer, P. A. and Yan, J. A.: Les "fonctions caractéristiques" des distributions sur l'espace de Wiener; *Sém. Prob. XXV*, LN in Math. **1485** (1991) 61–78

[24] Nelson, E.: Probability theory and Euclidean field theory; in: *LN. in Physics* **25** (1974) 94–124, Springer-Verlay

[25] Potthoff, J. and Streit, L.: A characterization of Hida distributions; *J. Func. Anal.* **101** (1991) 212–229

[26] Simon B.: *The $P(\phi)_2$ Euclidean (Quantum) Field Theory.* Princeton Series in Physics, Princeton University Press, Princeton, New Jersey, 1974

[27] Yan, J. A.: Notes on Wiener semigroup and renormalization; in: *Seminaire de probabilitiés XXV* (1991)

[28] Yan, J. A.: Products and transformations of white noise functionals; *Preprint* (1993)

Yuh-Jia Lee
Department of Mathematics
National Cheng-Kung University
Tainan 70101, TAIWAN
*E-mail*: yjlee@mail.ncku.edu.tw

V. MANDREKAR & M. M. MEERSCHAERT

# Sample moments and symmetric statistics

## 1. Introduction

We study the asymptotic behavior of sample moments and symmetric statistics for i.i.d. random variables which belong to the domain of attraction of a stable law. The sample moments are analyzed using the theory of operator stable laws and generalized domains of attraction. Symmetric statistics are analyzed using the LePage representation. A generating function is defined which describes the joint limiting distribution of the elementary symmetric statistics of all orders.

Let $X$ be a real-valued random variable and $X_1, X_2, \ldots$ be i.i.d. as $X$. Then we say that $X$ belongs to the domain of attraction of $Y$ if there exists $a_n > 0$ and $b_n$ real so that

$$a_n^{-1}(X_1 + \cdots + X_n - b_n) \Rightarrow Y.$$

Necessarily, $Y$ is a stable random variable and has index $\alpha$. Our purpose in this work is two-fold. In section 2, we study the joint limit laws of the sample moments

$$\mu_n^k = \sum_{i=1}^{n} X_i^k \qquad k = 1, 2, \ldots, m.$$

In special cases, Tucker [19] and Seidl' [17] have given sufficient conditions for the existence of these limit laws. Our Theorem 1 gives necessary and sufficient condition in terms of the distribution of $X^j$. From this the results in the papers of Tucker and Seidl' can be derived.

The joint convergence above and Newton's method can be used to obtain the specific form of the limit laws for finite symmetric statistics of random variables in the domain of attraction of stable laws. This gives the specific form of the limit laws for certain triangular arrays constructed from a sequence of i.i.d. random variables.

For the general case of triangular arrays with i.i.d. rows, the problem of limit laws of symmetric statistics was considered by Avram and Taqqu [2]. However, no explicit form of the limit distribution was given. The existence of the limit can be shown rather simply (see Appendix).

The second problem considered here is the analogue of the Dynkin and Mandelbaum [5] work for the infinite order symmetric statistics without the assumption of second moment. These statistics are useful in applications (see Mandrekar and Patterson [11]). Because of the lack of a Fock space in the general problem, we can only consider limit theorems for the generating function. In fact, in the case of a finite second moment, the generating functions (related to the exponential) span the Fock space. It is hoped that the form of the limit may lead to defining exponentials of stable random variables and eventually to the analogue of the Fock space. In any case, from our result one can get the form of the limit law (in terms of series representation) for finite symmetric statistics.

## 2. Limit Laws of Sample Moments

Let $X, X_1, X_2, \ldots$ be i.i.d. real-valued random variables. Define sample moments

$$\mu_n^k = \sum_{i=1}^{n} X_i^k \qquad k = 1, 2, \ldots$$

As stated in the Introduction, sufficient conditions for the weak convergence of the laws of $M_n = (\mu_n^1, \ldots \mu_n^k)$ (suitably normalized) were considered in [19] in symmetric case, and [17] in case $X_1$ has regularly varying tails. In this section, we use the modern theory of operator stable laws in [13, 14, 15] to get necessary and sufficient conditions for $M_n$ to belong to the domain of attraction of an operator stable law. Let us start with some definitions.

Let $A$ be a $k \times k$ matrix and

$$\exp(A) = I + A + A^2/2! + \cdots$$

For a positive integer $n$ and a $k \times k$ matrix $B$, define

$$n^B = \exp(B \log n).$$

Let $U$ be an $\mathbf{R}^k$-valued random vector with full support i.e. the law of $U$ is not supported on a proper subspace. If $U_1, U_2, \ldots$ are i.i.d. as $U$ and satisfy for a (non-random) $\{u_n\} \subseteq \mathbf{R}^k$ and $B$, $k \times k$ matrix,

$$(U - u_n) =^d n^{-B}(U_1 + \cdots + U_n),$$

then we call $U$ operator stable with exponent $B$. A random $k$-vector $Z$ is said to belong to the *generalized domain of attraction* of a random vector $U$ if there exists a sequence of $k \times k$ matrices $\{A_n\}$ and vectors $\{v_n\} \subseteq \mathbf{R}^k$ such that

$$A_n(Z_1 + \cdots + Z_n) - v_n \Rightarrow U, \qquad (2.1)$$

where $Z_1, Z_2, \ldots$ are i.i.d. as $Z$.

**Theorem 1** *The random vector $Z = (X, X^2, \ldots, X^k)$ belongs to the generalized domain of attraction of some operator stable law on $\mathbf{R}^k$ if and only if each component $X^j$ belongs to the domain of attraction of some stable law on $\mathbf{R}^1$.*

**Proof** Suppose that $Z$ is in the generalized domain of attraction of some full random vector $U$, and that (2.1) holds. If $U$ is strictly normal then Klosowska [7] shows that $<Z, \theta>$ belongs to the domain of attraction of a normal law for every nonzero $\theta \in \mathbf{R}^k$. Then each component $X^j$ is attracted to a normal law. If $U$ has no normal component, and $B$ is an exponent of $U$, then according to Theorem 3.3 of [14] there exists an index function $\rho(\theta) \in (0, 2)$ defined for all nonzero $\theta \in \mathbf{R}^k$ such that $E<Z, \theta>$ is finite for all $0 < \rho < \rho(\theta)$ and infinite for all $\rho > \rho(\theta)$. This index function is related to the eigenvalues of $B$. If $\lambda$ is (the real part of) any eigenvalue of $B$ then there is some $\theta$ for which $\rho(\theta) = 1/\lambda$. Now let $e_1, \ldots, e_k$ denote the standard basis for $\mathbf{R}^k$ and define $\alpha = \rho(e_1)$. Then by definition of $Z$ we certainly have $\rho(e_i) = \alpha/i$ for all $i = 1, \ldots, k$. It follows that $B$ has $k$ distinct real eigenvalues $1/\alpha, 2/\alpha, \ldots, k/\alpha$. Then the spectral decomposition for operator stable laws [14, Theorem 2.2] yields that $U$ has $k$ distinct stable marginals with index $\alpha, \alpha/2, \ldots, \alpha/k$. Since $Z$ is also in the generalized domain of attraction of $TU$ for any invertible linear operator $T$ on $\mathbf{R}^k$, we may assume without loss of generality that these stable marginals coincide with projection onto the coordinate axes. Now apply the spectral decomposition for generalized domains of attraction (Theorem 4.2 of [14]) to see that the norming operators $A_n$ in (2.1) can be chosen to be diagonal. Then project onto the $j$th coordinate axis to see that $X^j$ is attracted to a nonnormal stable law with index $\alpha/j$. Finally suppose that $U$ has both a normal and a nonnormal component. Use the spectral decomposition to reduce to the case of a strictly nonnormal limit, and apply the previous argument.

Conversely, suppose that $X^j$ belongs to the domain of attraction of some stable law for each $j = 1, \ldots, k$. The proof that (2.1) holds for some choice of norming operators $A_n$ and centering vectors $\nu_n$ will be divided into three cases, based on the form of the operator stable limit $U$. In each case, we will obtain an explicit formula for the Lévy representation of $U$.

Case I: Strictly Nonnormal Limits. Suppose that $a_n^{-1}(X_1 + \cdots + X_n) - b_n \Rightarrow Y$ and that $Y$ is stable with index $\alpha < 2$. Then we have $nP[a_n^{-1}X \in dx] \to \phi(dx)$ where $\phi$ is the Lévy measure of the stable limit $Y$, defined by $\phi\{x : x < -r\} = ar^{-\alpha}$ and $\phi\{x : x > r\} = br^{-\alpha}$. Without loss of generality we may assume that $a + b = 1$. Define linear operators $A_n$ on $\mathbf{R}^k$ by their matrix representations $A_n = \mathrm{diag}(a_n^{-1}, a_n^{-2}, \ldots, a_n^{-k})$. Meerschaert (1986) shows that (2.1) holds for some choice of centering vectors $\nu_n$ and some full random vector $U$ having no normal component if and only if

$$nP[A_n Z \in dx] \to \Phi\{dx\}, \tag{2.2}$$

where $\Phi$ is the Lévy measure of $U$. Then $t^{-1}\Phi\{dx\} = \Phi\{t^B dx\}$ for all $t > 0$, where $B$ is an exponent of the operator stable random vector $U$. Let $\nu_+ = (1, 1, \ldots, 1)$ and $\nu_- = (-1, +1, \ldots, (-1)^k)$ and define $B = \mathrm{diag}(1/\alpha, 2/\alpha, \ldots, k/\alpha)$. Notice that $A_n Z$ is concentrated on the two orbits $\{t^B \nu_+ : t > 0\}$ and $\{t^B \nu_- : t > 0\}$. Then it is easy to verify that (2.2) holds, where the Lévy measure $\Phi$ is also concentrated on these two orbits, with

$$\Phi\{t^B \nu_- : t > r\} = ar^{-\alpha}, \qquad \Phi\{t^B \nu_+ : t > r\} = br^{-\alpha}. \tag{2.3}$$

Case II: Strictly Normal Limits. If $E|X|^{2k} < \infty$, then the central limit theorem applies. Suppose then that $E|X|^{2k} = \infty$ and $a_n^{-1}(X_1^k + \cdots + X_n^k) - b_n \Rightarrow Y$ where $Y$ is normal. Define

$$A_n = \mathrm{diag}(n^{-1/2}, \ldots, n^{-1/2}, a_n^{-1}).$$

If $EZ \neq 0$ then replace $Z$ by $Z - EZ$. Now [15] shows that (2.1) holds for some full normal random vector $U$ if and only if

$$nF(A_n x) \to Q(x) \tag{2.4}$$

as $n \to \infty$ where $Q$ is the quadratic form in the Lévy representation of the limit law $U$ and $F(x) = E<Z, x>^2 I\{|<Z, x>| < 1\}$. Write

$$nF(A_n x) = nE\left(\sum_{i=1}^{k-1} n^{-1/2} X^i x_i\right)^2 I\{|<A_n Z, x>| < 1\}$$

$$+ 2nE\left(\sum_{i=1}^{k-1} n^{-1/2} X^i x_i\right)(a_n^{-1} X^k x_k) I\{|<A_n Z, x>| < 1\}$$

$$+ nE(a_n^{-1} X^k x_k)^2 I\{|<A_n Z, x>| < 1\}.$$

By the one variable theory, we may assume that $(n/a_n^2)\mu(a_n) \to 1$ as $n \to \infty$ where $\mu(x) = E|X|^{2k} I\{|X|^k < x\}$, and also that $nP[a_n^{-1}|X| > r] \to 0$ for all $r > 0$. Use this to show that $nP[A_n Z \in dx] \to 0$. Then the indicator set in the third term can be replaced by $I\{|<A_n Z, x_k>| < 1\}$ without affecting the limit, and it follows that this term tends to $x_k^2$ as $n \to \infty$. The one variable theory also yields that $E|X|^\rho < \infty$ for all $\rho < 2k$. This together with the fact that $n/a_n^2 \to 0$ shows that the second term vanishes in the limit. Since the original indicator set tends to $\mathbf{R}^k$ as $n \to \infty$, the indicator set can be deleted from the first term without changing the limit. Let $m_{ij} = EX^{i+j}$. Then we have shown that (2.4) holds for all $x$, and so (2.1) also holds, where $U$ is normal with Lévy representation $[0, Q, 0]$ and

$$Q(x) = \sum_{1 \le i,j \le k-1} m_{ij} x_i x_j + x_k^2. \tag{2.5}$$

Case III: Mixed Limits. Suppose that $a_n^{-k}(X_1^k + \cdots + X_n^k) - b_n^{(k)} \Rightarrow Y^{(k)}$ where $Y^{(k)}$ is nonnormal stable with index $\alpha/k$ and $\alpha \ge 2$. Choose $j$ so that $2(j-1) \ge \alpha < 2j$. By essentially the same argument as in Case I, except that now $A_n = \mathrm{diag}(a_n^{-j}, \ldots, a_n^{-k})$ and $B = \mathrm{diag}(j/\alpha, \ldots, k/\alpha)$, we have that $(X^j, \ldots, X^k)$ is attracted to a nonnormal limit. Case II shows that $(X, X^2, \ldots, X^{j-1})$ is attracted to a normal limit. By [15], this is sufficient to establish (2.1) for some $U$ having both a normal and a nonnormal component.

**Corollary 1** *The random vector $Z_1 = (X, X^2, \ldots X^k)$ belongs to the generalized domain of attraction of a non-Gaussian operator stable law on $\mathbf{R}^k$ iff $X$ belongs to the domain of attraction of some stable law on $\mathbf{R}^1$.*

Tucker [19] showed that a sufficient condition for $Z = (X, \ldots X^k)$ to belong to the generalized domain of attraction of some full operator stable law is that $X$ has a symmetric distribution and $X^k$ belongs to the domain of attraction of some stable law. Another sufficient condition is due to Seidl' [17], who apparently was unaware of the work of Tucker. The Seidl' condition is that $T(t) = P[|X| > t]$ varies regularly and $P[X < t]/T(t) \to a$, $P[X < -t]/T(t) \to b$ as $t \to \infty$. Both conditions are contained in ours. The Seidl' condition implies that $X^j$ belongs to the domain of attraction of a stable law for any $j$. Also, for symmetric distributions (or if $k$ is odd), if $X^k$ belongs to some domain of attraction then $X^j$ also belongs to some domain of attraction for all $j = 1, \ldots, k-1$. Both Tucker and Seidl' also obtain a formula for the characteristic function of $U$. This formula can be recovered directly from equations (2.3) and (2.5) above, using the fact that an infinitely divisible law with Lévy representation $[a, Q, \Phi]$ has characteristic function $e^{f(t)}$ where

$$f(t) = i < a, t > -\frac{1}{2}Q(t) + \int_{x \neq 0} (e^{i<t,x>} - 1 - \frac{i < t, x >}{1+ < x, x >})\Phi(dx).$$

Operator stable laws on Banach spaces were studied by Krakowiak [8] and Siegel [18]. Since weak convergence in $\mathbf{R}^\infty$ is equivalent to the weak convergence of finite dimensional distributions, we immediately obtain from Theorem 1 that the random vector $Z = (X, X^2, X^3, \ldots)$ belongs to the generalized domain of attraction of some operator stable law on $\mathbf{R}^\infty$ if and only if $X^k$ belongs to the domain of attraction of a stable law for all $k$. If $E|X|^k < \infty$ for all $k$ then $U$ is a Gaussian random element of $\mathbf{R}^\infty$. If $X$ is attracted to a nonnormal stable law with index $\alpha$ then $U$ is a strictly nonnormal operator stable random element of $\mathbf{R}^\infty$ with exponent $B = \mathrm{diag}(1/\alpha, 2/\alpha, 3/\alpha, \ldots)$. Otherwise $U$ has a strictly nonnormal component of this nature along with a finite dimensional normal component.

The limit laws of elementary symmetric statistics $\sigma_n^k$ of finite order were considered by Avram and Taqqu [2] where

$$\sigma_n^k = \sum_{1 \leq i_1 < i_2 < \cdots < i_k \leq n} X_{i_1} \cdots X_{i_k}. \tag{2.6}$$

Using Newton's identity they show that

$$\sigma_n^k = P_k(\mu_n^1, \ldots, \mu_n^k),$$

where $P_k(y_1, \ldots, y_k) = (-1)^k \sum \prod (-y_i)^{m_i}/i^{m_i} \ m_i!$ where sum is over all nonnegative integers $m_1, \ldots, m_k$ such that $m_1 + 2m_2 + \cdots + km_k = k$ and the product is over $i = 1, 2, \ldots, k$. Suppose $\alpha < 1$ or $1 < \alpha < 2$ and $EX = 0$. Then by Theorem 1, we get that $P_k(a_n^{-1}\mu_n^1, \ldots, a_n^{-k}\mu_n^k) \Rightarrow P_k(Y^{(1)}, \ldots, Y^{(k)})$ and hence

$$\frac{\sigma_n^j}{\sum_{i=1}^n X_i^k} \Rightarrow \frac{P_j(Y^{(1)}, \ldots, Y^{(j)})}{Y^{(k)}} \qquad j, k \geq 1$$

and, in particular,

$$\frac{\sigma_n^1}{\left(\sum_{i=1}^k X_i^k\right)} \Rightarrow \frac{Y^{(1)}}{(Y^{(k)})^{1/k}}.$$

202

These results generalize the result of Logan, Mallows, Rice and Shepp [10]. The form of the limit given here can be shown to be equivalent to that of LePage, Woodroofe and Zinn [9]. In fact, for the limit laws in Theorem 1, we can get a representation of their type using results of Hahn, Hudson, and Veeh [6]. Given an operator stable law $\mu$ with exponent $B$ and Lévy representation $[a, 0, \Phi]$ we define the mixing measure

$$K(E) = \Phi\{t^B x : t > 1,\ x \in E\}$$

for all Borel subsets $E$ of the set

$$S = \{(x_1, \ldots, x_k) : x_1^2 + \cdots + x_k^2 = k\}.$$

It follows that for any Borel subset $F$ of $\mathbf{R}^k - \{0\}$ we have

$$\Phi(F) = \int_S \int_0^\infty I\{t^B x \in F\} t^{-2} dt K(dx).$$

In our case we have exponent $B = \mathrm{diag}(1/\alpha, 2/\alpha, \ldots k/\alpha)$. Let $\nu_+ = e_1 + \cdots + e_k$ and $\nu_- = -e_1 + e_2 + \cdots + (-1)^k e_k$ where $e_1, \ldots e_k$ are the standard basis vectors. By our earlier results, the Lévy measure $\Phi$ is concentrated on the two orbits $\{t^B \nu_+ : t > 0\}$ and $\{t^B \nu_- : t > 0\}$. In fact we have $\Phi\{t^B \nu_+ : t > r\} = ar^{-\alpha}$ and $\Phi\{t^B \nu_- : t > r\} = br^{-\alpha}$. Hence our mixing measure $K$ consists of a point mass $a$ at the point $\nu_+$ and a point mass $b$ at the point $\nu_-$, so $K$ is a probability measure. Now for $\alpha < 1$ the LePage representation for $\mu$ is

$$\sum_{n=1}^\infty \Gamma_n^{-B} U_n$$

where $\Gamma_n = E_1 + \cdots + E_n$, $\{E_n\}$ are i.i.d. negative exponential with mean one, and $\{U_n\}$ are i.i.d. according to $K$. Let $\{\delta_n\}$ be i.i.d. random variables with $P[\delta_n = 1] = a$ and $P[\delta_n = -1] = b$. Then we can take $U_n = (\delta_n, \ldots, \delta_n^k)$, and so $\mu$ is identically distributed with

$$\left( \sum_{n=1}^\infty \Gamma_n^{-1/\alpha} \delta_n, \ldots, \sum_{n=1}^\infty \Gamma_n^{-k/\alpha} \delta_n^k \right).$$

For $\alpha \geq 1$ there is an additional centering term

$$\sum_{n=1}^\infty (\Gamma_n^{-B} U_n - E\Gamma_n^{-B} I(\Gamma_n \geq 1) EU_n).$$

Notice that $\Gamma_n$ is the $n$th jump time of a Poisson process with rate one. Now let $N(t)$ be the Lévy process with Lévy representation $[0,0,K]$. In other words, $N(t)$ is a Poisson process with rate one, split according to probabilities $a$ and $b$, and laid out along the rays defined by the vectors $\nu_+, \nu_-$. Then in the case of no centering, the LePage representation can be rewritten in the form

$$\int_0^\infty t^{-B} N(dt).$$

Centerings are necessary in the cases $\alpha = 1$ and $\alpha > 1$, $a \neq b$ because in those cases, the integral does not converge a.s.

## 3. Asymptotic Distribution of Infinite Order Statistics

In Dynkin and Mandelbaum [5] the limit law of infinite order symmetric statistic was studied for the i.i.d. case. Using the techniques in Mandrekar and Rao [12], these results were extended to infinite exchangeable sequences by Mandrekar and Patterson [11], where some statistical applications are given.

In the first two papers, the infinite order symmetric statistic was defined by

$$Y_n(\underline{h}) = \sum_{k=1}^\infty n^{-k/2} \sigma_n^k(h_k), \qquad \sigma_n^k(h_k) = \sum_{1 \leq i_1 < \cdots < i_k \leq n} h_k(X_{i_1}, \ldots, X_{i_k}),$$

where $\underline{h} = (h_1, h_2, \ldots)$ is a sequence, with $h_k \in L^2(\mathcal{X}^k, \mu^k)$ and $\mu^k$, the product measure on the product space $\mathcal{X}^k$. In addition, $h_k$ is assumed to be symmetric with $\int h_k(x_1, x_2, \ldots, x_k) d\mu = 0$ for all $(x_1, x_2, \ldots x_{k-1})$ (canonical). One can treat $\underline{h}$ as element of a Fock space and the problem reduces to finding the limit law in case $h_k(x_1, \ldots x_k) = \varphi(x_1), \ldots, \varphi(x_k)$ with $\varphi \in L^2(\mathcal{X}, \mu)$ with $\int \varphi(x) d\mu(x) = 0$. In this case $Y_n(\underline{h})$ looks like the Doleans-Dade exponential of $\sum_{k=1}^n \varphi(x_k)/\sqrt{n}$. Using this point of view Avram [1] found the limit theorem for the case of a sequence of semimartingales. Because of the generality of the problem, it is not possible to identify the limit distribution. In addition, processes with independent increments are not semimartingales in general and even under the assumption that their sample paths are in $D[0, \infty)$, the case $\alpha < 1$ for stable processes cannot be deduced from Avram's work. However, we can modify his method to obtain the limit distribution for the generating function

$$\psi_n(\lambda) = \sum_{k=0}^n \lambda^k a_n^{-k} \sigma_n^k \quad \text{for all} \quad \lambda > 0, \tag{3.1}$$

where $a_n^{-1}[(X_1 + \cdots + X_n) - nb_n] \Rightarrow Y$, a stable random variable of index $\alpha$. Here $b_n = EX1(|X| \leq a_n)$ and if $\alpha < 1$ or if $EX = 0$ then $b_n = 0$. We note that, in case $b_n \neq 0$, we replace $X_i$ by $X_i - b_n$ in the definition of $\sigma_n^k$.

The next theorem is a generalization of the Dynkin and Madelbaum result for the above mentioned special case $h_k(x_1, \ldots, x_k) = \varphi(x_1), \ldots, \varphi(x_k)$. The proof is based on the so-called LePage representation introduced in [9]. We now give some details regarding this representation. Define $X_{in} = a_n^{-1}(X_i - b_n)$ and let $W_{1n} \geq W_{2n} \geq \cdots \geq W_{nn}$ denote the order statistics of $|X_{in}|$. Let $\pi_n$ denote random permutations of $1, \ldots, n$ for which $W_{in} = |X_{\pi_n(i)n}|$ and define $\delta_{in} = \text{sign}(X_{\pi_n(i)n})$. Next let $\Gamma_n$ denote the arrival times in a Poisson process with mean rate one. LePage et. al. show that if we let $W^n = (W_{1n}, \ldots, W_{nn}, 0, \ldots)$ and $\delta^n = (\delta_{1n}, \ldots, \delta_{nn}, 1, \ldots)$ the $(W^n, \delta^n) \Rightarrow (W, \delta)$ where $W = (W_1, W_2, W_3, \ldots)$ with $W_n = \Gamma_n^{-1/\alpha}$ and $\delta = (\delta_1, \delta_2, \delta_3, \ldots)$ is independent of $W$ with $\delta_n$ i.i.d. and concentrated on $\pm 1$ with $P[\delta_n = -1] = a$ and $P[\delta_n = 1] = b$.

**Theorem 2** *Suppose $X$ belongs to the domain of attraction of a stable law $Y$ on $\mathbf{R}^1$ with index $0 < \alpha < 2$. Then for all $\lambda > 0$ we have $\psi_n(\lambda) \Rightarrow \psi(\lambda)$ where*

$$\psi(\lambda) = \prod_{n=1}^{\infty} (1 + \lambda \delta_n W_n) e^{-\lambda(b-a)EW_n I(W_n \leq 1)} \tag{3.2}$$

**Proof** Notice that

$$\psi_n(\lambda) = \prod_{i=1}^{n} (1 + \lambda X_{in}) = \prod_{i=1}^{n} (1 + \lambda \delta_{in} W_{in}).$$

Now define $xA = xI\{x \in A\}$ and note that $\delta_{in}^2 = 1$. Then the above product can be written in the form

$$\prod_{i=1}^{n} (1 + \lambda \delta_{in} W_{in}(0, \lambda^{-1})) \prod_{i=1}^{n} (1 + \lambda \delta_{in} W_{in}[\lambda^{-1}, \infty))$$

$$= \exp(\sum_{i=1}^{n} \log(1 + \lambda \delta_{in} W_{in}(0, \lambda^{-1}))) \prod_{i=1}^{n} (1 + \lambda \delta_{in} W_{in}[\lambda^{-1}, \infty))$$

$$= \exp(\lambda \sum_{i=1}^{n} \delta_{in} W_{in} - \frac{\lambda^2}{2} \sum_{i=1}^{n} W_{in}^2) \tag{3.3}$$

$$\cdot \exp(\sum_{i=1}^{n} [\log(1 + \lambda \delta_{in} W_{in}(0, \lambda^{-1})) - \lambda \delta_{in} W_{in}(0, \lambda^{-1}) + \frac{\lambda^2}{2} W_{in}^2(0, \lambda^{-1})])$$

$$\cdot \prod_{i=1}^{n} [(1 + \lambda \delta_{in} W_{in}[\lambda^{-1}, \infty)) \exp(-\lambda \delta_{in} W_{in}[\lambda^{-1}, \infty) + \frac{\lambda^2}{2} W_{in}^2[\lambda^{-1}, \infty))].$$

Theorem 1 of [9] and remark 3 on p. 628 of the same paper yield that

$$\sum_{i=1}^{n} \delta_{in} W_{in} \Rightarrow \sum_{n=1}^{\infty} (\delta_n W_n - (b-a) E W_n I\{W_n \leq 1\})$$

$$\sum_{i=1}^{n} W_{in}^2 \Rightarrow \sum_{n=1}^{\infty} W_n^2$$

and so the first term in (3.3) converges in distribution to

$$\exp\left( \lambda \sum_{n=1}^{\infty} (\delta_n W_n - (b-a) E W_n I\{W_n \leq 1\} - \frac{\lambda^2}{2} \sum_{n=1}^{\infty} W_n^2 \right). \tag{3.4}$$

The second term is $\exp(\sum f(\lambda \delta_{in} W_{in}(0, \lambda^{-1})))$ where $f(x) = \log(1+x) - x + x^2/2$. Since $x^{-2} f(x) \to 0$ as $x \to 0$, for all small $a > 0$ the function $g(a) = \sup\{x^{-2} f(x) : |x| \leq a\}$ is well defined with $g(a) \to 0$ as $a \to 0$. Then for all $\varepsilon > 0$ we have

$$P\left[ \left| \sum_{i=1}^{n} f(\lambda \delta_{in} W_{in}(0, a/\lambda)) \right| > \varepsilon \right]$$

$$\leq P\left[ \sum_{i=1}^{n} (\lambda \delta_{in} W_{in}(0, a/\lambda))^2 > \varepsilon / g(a) \right]$$

$$\leq P\left[ \sum_{i=1}^{n} W_{in}^2 > \lambda^{-2} \varepsilon / g(a) \right]$$

and so, since $\sum W_{in}^2$ converges in distribution we have

$$\lim_{a \to 0} \limsup_{n \to \infty} P\left[ \left| \sum_{i=1}^{n} f(\lambda \delta_{in} W_{in}(0, a/\lambda)) \right| > \varepsilon \right] = 0.$$

Notice that the function $F(d, w) = \sum f(\lambda d_n w_n(\varepsilon, \lambda^{-1}))$ is continuous in the product topology on the set of $d = (d_1, d_2, d_3, \ldots) \in \{-1, +1\}^\infty$ and $w = (w_1, w_2, w_3, \ldots)$ positive reals with $w_1 \geq w_2 \geq w_3 \geq \cdots$ and $w_n \to 0$ as $n \to \infty$. Then the LePage representation together with the continuous mapping theorem of [3] yields that

$$\sum_{i=1}^{n} f(\lambda \delta_{in} W_{in}(\varepsilon, \lambda^{-1})) \Rightarrow \sum_{n=1}^{\infty} f(\lambda \delta_n W_n(\varepsilon, \lambda^{-1})).$$

Since $\sum W_{in}^2$ converges in distribution and $x^{-2} f(x) \to 0$ as $x \to 0$ it is easy to see that

$$\sum_{i=1}^{n} f(\lambda \delta_n W_n(\varepsilon, \lambda^{-1})) \Rightarrow \sum_{n=1}^{\infty} f(\lambda \delta_n W_n(0, \lambda^{-1}))$$

as $\varepsilon \to 0$. Now an application of Theorem 4.2 of [3] yields that the second term in (3.3) converges in distribution to

$$\exp\left(\sum_{n=1}^{\infty}[\log(1 + \lambda\delta_n W_n(0, \lambda^{-1})) - \lambda\delta_n W_n(0, \lambda^{-1}) + \frac{\lambda^2}{2}W_n^2(0, \lambda^{-1})]\right). \qquad (3.5)$$

Now let $h(x) = (1 + x)e^{-x + x^2/2}$ and note that the function

$$G(d, w) = \prod h(\lambda d_n w_n[\lambda^{-1}, \infty))$$

is also continuous in the product topology, on the same set of $(d, w)$ as before. Then another application of the continuous mapping theorem yields that the third term in (3.3) converges in distribution to

$$\prod_{n=1}^{\infty}[(1 + \lambda\delta_n W_n[\lambda^{-1}, \infty))\exp(-\lambda\delta_n W_n[\lambda^{-1}, \infty) + \frac{\lambda^2}{2}W_n^2[\lambda^{-1}, \infty))]. \qquad (3.6)$$

The product of the three terms in (3.4) through (3.6) is easily seen to reduce to (3.3).

As noted earlier, $a_n^{-k}\sigma_n^k \Rightarrow P_k(Y^{(1)}, Y^{(2)}, \ldots, Y^{(k)})$. Taking derivatives in (3.1) and using Theorem 2, we get $a_n^{-k}\sigma_n^k = \psi_n^{(k)}(0)/k! \Rightarrow \psi^{(k)}(0)/k!$. This gives two alternate forms of the limiting distribution and a series representation for them. Equating the two forms yields $P_k(Y^{(1)}, \ldots, Y^{(k)}) = \psi^{(k)}(0)/k!$. Now both the limiting distribution of the elementary symmetric statistics and the form of the polynomials $P_k$ can be recovered from (3.3) by differentiation. For example if $k = 1$ we have

$$\psi'(0) = \sum_{n=1}^{\infty}(\delta_n W_n - (b - a)EW_n I\{W_n \leq 1\}) =^d Y^{(1)}$$

and so $a_n^{-1}\sigma_n^1 \Rightarrow P_1(Y^{(1)})$ where $P_1(y_1) = y_1$, which is obvious because $\sigma_n^1 = \mu_n^1$. For $k = 2$ we have

$$\psi''(0) = \left(\sum_{n=1}^{\infty}(\delta_n W_n - (b - a)EW_n I\{W_n \leq 1\})\right)^2 - \sum_{n=1}^{\infty}W_n^2 =^d (Y^{(1)})^2 - Y^{(2)}$$

and so $a_n^{-2}\sigma_n^2 \Rightarrow P_2(Y^{(1)}, Y^{(2)})$ where $P_2(y_1, y_2) = y_1^2/2 - y_2/2$. A similar computation shows that $a_n^{-3}\sigma_n^3 \Rightarrow P_3(Y^{(1)}, Y^{(2)}, Y^{(3)})$ where $P_3(y_1, y_2, y_3) = y_1^3/6 - y_1 y_2/2 + y^3/3$.

## 4. Remarks

(1) In cases where centering is not required, we can also obtain results analogous to Theorem 2 for the noncentered elementary symmetric statistics defined by (2.6). If $\alpha < 1$ then the limit is

$$\psi(\lambda) = \prod_{n=1}^{\infty} (1 + \lambda \delta_n W_n).$$

This form differs from (3.2) only by a multiplicative constant, since in the case $\alpha < 1$ the centering terms $(b-a)EW_n I\{W_n \leq 1\}$ sum to a finite limit. If $1 < \alpha < 2$ and $EX = 0$ then the limit is the same as (3.2) whether or not we center, since in this case $a_n^{-1} n b_n \to 0$. (2) If $\alpha < 1$ then we can also obtain a formula for the characteristic function of the limiting distribution of the elementary symmetric statistics. As in (3.3) above, write the generating function in the form $\psi(\lambda) = \exp(S_\lambda) R(\lambda)$ where

$$S_\lambda = \sum_{i=1}^{n} \log(1 + \lambda \delta_n W_n(0, \lambda^{-1}))$$

and $R(\lambda) \to 1$ almost surely as $\lambda \to 0$. Then both $\psi$ and $\exp(S_\lambda)$ have the same derivatives at $\lambda = 0$. The random variable $S_\lambda$ is infinitely divisible with Lévy representation $[0, 0, M_\lambda]$ where

$$M_\lambda(r, \log 2) = a\lambda^\alpha (e^r - 1)^{-\alpha} - a\lambda^\alpha$$

$$M_\lambda(-\infty, -r) = b\lambda^\alpha (1 - e^{-r})^{-\alpha} - b\lambda^\alpha.$$

The characteristic function of the limiting distributions of the elementary symmetric statistics can be obtained by taking derivatives.

## 5. Appendix

Let $\{X_{nj}, \; j = 1, 2, \ldots, n, \; n = 1, 2, \ldots\}$ be a triangular array of row i.i.d. random variables and let

$$\mu_n^k = \sum_{j=1}^{n} X_{nj}^k.$$

If $\sum_{j=1}^{n} X_{nj} \Rightarrow Y$ (non-normal) then $(\mu_n^1, \ldots, \mu_n^m) \Rightarrow \underline{Z}$ for some random variable $\underline{Z}$. Observe that for $\underline{t} = (t_1, \ldots, t_m) \in \mathbf{R}^m$

$$\sum_{k=1}^{m} t_k \mu_n^k = \sum_{j=1}^{n} \sum_{k=1}^{m} t_k X_{nj}^k.$$

Hence

$$Ee^{i\Sigma_{k=1}^{m}t_k\mu_n^k} = (Ee^{i\Sigma_{k=1}^{m}t_k X_{n1}^k})^n.$$

We can rewrite, with $\Gamma_n = F_{X_{n1}}$, the RHS of the above equation as

$$\left[1 + \frac{1}{n}\int_{\mathbf{R}}\left(e^{i\Sigma_{k=1}^{m}t_k x^k} - 1\right)nF_n(dx)\right]^n.$$

Since $nF_n \Rightarrow M$ the Lévy measure outside the neighbourhood of zero and

$$\lim_{\varepsilon\to 0}\limsup_{n\to\infty}\int_{-\varepsilon}^{\varepsilon}|x|^k\,nF_n(dx) = 0,$$

we get that for each $\underline{t}$ the above converges to

$$e^{(\underline{a},\underline{t})} + \int_{\mathbf{R}}\left(e^{i\Sigma_{k=1}^{m}t_k x^k} - 1 - \frac{i\sum_{k=1}^{m}t_k x_k}{1 + \left\|\sum_{k=1}^{n}|x_k|^2\right\|^2}\right)dM(x)$$

which is continuous in $\underline{t}$ at zero and is a characteristic function. Using a result of Skorokhod that $\sum_{j=1}^{n}X_{nj}$ converges weakly iff $\sum_{j=1}^{[nt]}X_{nj}$ converges in $D[0,1]$ in the Skorokhod topology (see for example [1]) one can get the result of Avram and Taqqu [2].

## References


[1] Avram, F.: Weak convergence of the variations, iterated integrals and Doléans-Dade exponentials of sequences of semimartingales; *Ann. Probab.* **16** (1988) 246–250

[2] Avram, F. and Taqqu, M.: Symmetric polynomials of random variables attracted to an infinitely divisible law; *Prob. Theory and Rel. Fields* **71** (1986) 491–500

[3] Billingsley, P.: *Convergence of Probability Measures.* Wiley, New York, 1968

[4] Darling, D.: The influence of the maximal term in the addition of independent random variables; *Trans. Amer. Math. Soc.* **83** (1952) 95–107

[5] Dynkin, E. and Mandelbaum, A.: Symmetric statistics, Poisson point processes, and multiple Wiener integrals; *Ann. Stat.* **11** (1983) 739–745

[6] Hahn, M., Hudson, W., and Veeh, J.: Operator stable laws: Series representations and domains of normal attraction; *J. Theoretical Probab.* **2** (1989) 3–35

[7] Klosowska, M.: Domain of operator attraction of a Gaussian measure on $\mathbf{R}^N$; *Ann. Soc. Math. Pol., Ser. I, Comment. Math.* **XXII** (1980) 73–80

[8] Krakowiak, W.: Operator stable probability measures on Banach spaces; *Colloq. Math.* **41** (1979) 313–326

[9] LePage, R., Woodroofe, M., and Zinn, J.: Convergence to a stable distribution via order statistics; *Ann. Probab.* **4** (1981) 624–632

[10] Logan, B., Mallows, C., Rice, S., and Shepp, L.: Limit distributions of self-normalized sums; *Ann. Probab.* **1** (1973) 788–809

[11] Mandrekar, V. and Patterson, R. F.: Limit theorems for symmetric statistics of exchangeable random variables; *Statistics and Prob. Letters* **17** (1993) 157–161

[12] Mandrekar, V. and Rao, B. V.: On a limit theorem and invariance principle for symmetric statistics; *Probab. and Math. Statist.* **10** (1989) 271–276

[13] Meerschaert, M.: Domains of attraction of nonnormal operator stable laws; *J. Mult. Anal.* **19** (1986) 342–347

[14] Meerschaert, M.: Spectral decomposition for generalized domains of attraction, *Ann. Probab.* **19** (1991) 875–892

[15] Meerschaert, M.: Regular variation and generalized domains of attraction in $\mathbf{R}^k$, *Stat. Prob. Lett.* **18** (1993) 233–239

[16] Rubin, H. and Vitale, R.: Asymptotic distribution of symmetric statistics; *Ann. Stat.* **8** (1980) 165–170

[17] Seidl', L.: On the limit distributions of random symmetric functions; *Theory Probab. Appl.* **31** (1986) 590–603

[18] Siegel, G.: Exponents of operator stable distributions in Banach space; *Prob. Theory and Math. Stat.* **2** (1990) 437–445

[19] Tucker, H.: Joint limit laws of sample moments of a symmetric distribution; *Ann. Probab.* **8** (1980) 991–998

V. Mandrekar
Department of Statistics and Probability
Michigan State University
East Lansing, MI 48824, U.S.A.
*E-mail*: 20974vsm@msu.edu

Mark M. Meerschaert
Department of Mathematics
University of Nevada
Reno, NV 89557, U.S.A.
*E-mail*: mmm@math.unr.edu

R. MIKULEVICIUS & B. L. ROZOVSKII

# Soft solutions of linear parabolic SPDE's and the Wiener chaos expansion

## 1. Introduction

The Wiener chaos expansion has proven to be one very important aspect of modern stochastic analysis. It is of central importance for white noise analysis [2, 3], Malliavin calculus [12, 17, 20], the Hitsuda-Skorohod integral [14, 16], etc.

An interesting and unexpected application of Wiener chaos expansion was found by Krylov and Veretennikov [19]. They gave an "explicit" Wiener chaos formula for a solution of an Ito stochastic equation. More specifically they explicitly calculated the non-random kernels in the multiple Wiener integral expansion of the solution.

In this work we extend the result of Krylov and Veretennikov to more general Wiener functionals. In particular we develop "explicit" Wiener chaos expansions for solutions of linear parabolic SPDE's. To describe the class of solutions which allow explicit Wiener chaos expansions we introduce the notion of a soft solution.

To define the soft solution of a SPDE we first consider an associated deterministic PDE. The latter is derived (at least informally) by calculating the $S$-transform (see [3]) of the SPDE in question. (We will call this the associated deterministic PDE $S$-equation and occasionally will abbreviate the term "soft solution" as "$S$-solution".) Loosely speaking, the soft solution is defined as a random field whose $S$-transform is a solution of the associated $S$-equation.

Of course a "traditional" solution, if any, is also a soft solution but the converse is not necessarily true. We prove the existence of soft solutions under assumptions substantially less stringent than usually required for the existence of traditional solutions. Fortunately the noted "softness" of the requirements for the existence of a $S$-solution do not result in unpleasant side effects related to uniqueness or other desirable properties of a solution. Under the same assumptions which guarantee the existence of a soft solution we prove uniqueness, derive a Feynman-Kac type representation and an explicit Wiener chaos expansion, and establish a maximum principle.

## 2. Soft Solutions of SPDE's

To begin with we introduce some necessary notation and definitions.

Let $(\Omega, \mathcal{F}, P)$ be a probability space and $(E, \mathcal{E}, \kappa)$ a separable measure space. Let $W$ be a cylindrical Wiener process in $L^2(E, \mathcal{E}, \kappa)$, i.e., $\forall f, g \in L^2(E, \mathcal{E}, \kappa)$, $W_t(f) = \int_0^t \int_E f(\alpha) W(dt, d\alpha)$ is a one-dimensional Wiener process and $\langle W(f), W(g) \rangle_t = t \int_E fg\, d\kappa$.

It is assumed here that $\mathcal{F}_t = \sigma(W_s, s \in [0, t])$ and $\mathcal{F} = \mathcal{F}_1$.

We denote $H = [0, 1] \times I\!R^d$ and suppose that the following measurable functions are given: $a : I\!R^d \to R^{d^2}$, $b : I\!R^d \to I\!R^d$, $\sigma : I\!R^d \times E \to I\!R^d$, $\varphi$, and $c : I\!R^d \to I\!R$, $f : H \to I\!R$, $h : H \times E \to I\!R$ and $g : H \times E \to I\!R$ such that

$$|a| + |b| + |c| + \int_E (|\sigma|^2 + |h|^2) d\kappa \leq K < \infty.$$

It is also assumed that $f$, $\int_E |g|^2 d\kappa$ and $\varphi$ are locally integrable.

The main objective of this section is to study the equation

$$\begin{cases} -du(t, x) = \big(\mathcal{L}u(t, x) + f(t, x)\big) dt + \int_E \big(\mathcal{M}^\alpha u(t, x) + g(t, x)\big) W(\overleftarrow{dt}, d\alpha) \\ u(1, x) = \varphi(x), \end{cases} \tag{1}$$

where

$$\mathcal{L}u = a^{ij} u_{x_i x_j} + b^i u_{x_i} + cu \quad \text{and}$$

$$\mathcal{M}u = \mathcal{M}^\alpha u = \sigma Du + hu = \sigma(x, \alpha) Du + h(x, \alpha) u.$$

The stochastic integral in (1) is understood as a backward Ito integral (for details see, e.g., [15]).

Before proceeding further with the study of equation (1) we consider two particular cases of this SPDE which are of special interest.

$1^o$ *Backward diffusion equation* (see [8, 10]). Let $E = \{1, \dots, d_1\}$ and $\kappa$ be a counting measure on $E$, i.e., $\int_A d\kappa = \sum_1^{d_1} 1_A(i)$. Assume also that $c = h = f = g = 0$, $\varphi(x) = x$ and $a(x) = \frac{1}{2} \sum_{\alpha=1}^{d_1} \sigma^i(x, \alpha) \sigma^j(x, \alpha)$. Then equation (1) reduces to the backward diffusion equation

$$\begin{cases} -du(t, x) = \frac{1}{2}(\sigma\sigma^*)^{ij}(x) u_{x_i x_j}(t, x) + b^i(x) u_{x_i}(t, x) dt + \\ \qquad\qquad \sum_{\alpha=1}^{d_1} \sigma^i(x, \alpha) u_{x_i}(t, x) \overleftarrow{dw}(t, \alpha), \quad t < 1, \\ u(1, x) = x, \end{cases} \tag{2}$$

where $w(t, \alpha), \alpha = 1, 2, \dots, d_1$ are independent one dimensional Wiener processes.

It is now well known (see, e.g., [8, 10]) that under mild assumptions on the smoothness of $b(x)$ and $\sigma(x)$ the solution of equation (2) is given by

$$u(t, x) = \xi^{x,t}(1) \qquad \text{a.a.}$$

where $\xi^{x,t}(r)$ is a diffusion process defined by the ordinary Ito equation

$$\xi^{x,t}(r) = x + \int_t^r b(\xi^{x,t}(s))ds + \sum_{\alpha=1}^{d_1} \int_t^r \sigma(\xi^{x,t}(s), \alpha)dw(t, \alpha) .$$

$2^o$ *Zakai equation* (see [15, 21]). Let $E$ and $\kappa$ be the same as in $1^o$, $f = g = 0$

$$a(x) = \frac{1}{2} \sum_{\alpha=1}^{d_1} \sigma^i(x, \alpha)\sigma^j(x, \alpha) + \frac{1}{2} \sum_{\alpha=1}^{d} \hat{\sigma}^i(x, \alpha)\hat{\sigma}^j(x, \alpha) ,$$

where $\hat{\sigma} : \mathbb{R}^d \times E \to \mathbb{R}^d$ is a measurable function and $\varphi$ is the density function for some random variable $x_0$. Let us choose $c$ in such a way that $\mathcal{L}u = (a^{ij}u)_{x_i x_j} + (b^i u)_{x_i}$ for every $u \in C^2$ (the latter is always obviously possible if the coefficients $a$ and $b$ are smooth enough). It is now readily checked that $p(t, x) = u(1 - t, x)$ is a solution to the Zakai equation

$$\begin{cases} dp(t, x) = \left[\left(a^{ij}(x)p(t, x)\right)_{x_i x_j} - \left(\tilde{b}^i(x)p(t, x)\right)_{x_i}\right]dt + \\ \qquad \sum_{\alpha=1}^{d_1}\left[\left(-\tilde{\sigma}^i(x, \alpha)p(t, x)\right)_{x_i} + \tilde{h}^i(x, \alpha)p(t, x)\right]dw(t, \alpha), \quad t < 1, \\ p(0, x) = \varphi(x), \end{cases}$$

where $\tilde{b} = -b$, $\tilde{\sigma} = -\sigma$, and $\tilde{h} = h^i + \sum_{i=1}^{d} \tilde{\sigma}^i_{x_i}$. Set $\tilde{\mathcal{F}} = \mathcal{B}([0, 1] \times \mathbb{R}^d) \otimes \mathcal{F}$ and $\tilde{\mathcal{F}}_t = \mathcal{F}_t \otimes \mathcal{B}(\mathbb{R}^d)$. For $l \in L^2([0, 1] \times E, dt\, d\kappa)$, we define $q_t(l)$ as the unique solution of the equation

$$\begin{cases} -dq_t(l) = q_t(l) \int_E l(t, \alpha)W(\overleftarrow{dt}, d\alpha), \quad t < 1 \\ q_1(l) = 1. \end{cases}$$

Let $\mathcal{D}$ be the class of all $l \in L^2([0, 1] \times E, dt\, d\kappa)$ such that $\int l(t, \alpha)^2 d\kappa$ is bounded. For an $\tilde{\mathcal{F}}$-measurable $(\tilde{\mathcal{F}}_t)$-adapted function $u(t, x)$ such that $E|u(t, x)|^2$ is $dx$-locally integrable and $l \in \mathcal{D}$ we denote the $S^l$-transform by

$$S^l u(t, x) = Eu(t, x)q_t(l) .$$

**Definition** An $\tilde{\mathcal{F}}$-measurable, $(\tilde{\mathcal{F}}_t)$-adapted function $u$ is a *soft solution* to (1) if for each $l \in \mathcal{D}$, $u^l(t,x) = S^l u(t,x)$ is a generalized solution of the deterministic PDE:

$$\begin{cases} \partial_t u^l + \mathcal{L}u^l + f + \int_E l(\mathcal{M}u^l + g)d\kappa = 0, & t < 1 \\ u^l(1,x) = \varphi(x) . \end{cases} \tag{3}$$

Obviously one could derive (3) informally making use of the equations for $u(t,x)$ and $q_t(l)$ and the Ito formula. Note that the definition does not single out the class of functions where the solution of equation (3) belongs. This has to be specified separately in accordance with assumptions on the coefficients of the equations. Below we consider two different sets of assumptions on the coefficients and free terms of equation (1) and show that each one of them guarantees existence and uniqueness of the soft solution in one of the two classes: $(C^{2+\alpha}(H), W^{2,p}(H)$, respectively). $C^{2+\beta}(H)$ is the space of continuous function $u$ on $H$ having finite norm $|u|_{2,\beta} = \sup_{t,x} |u(t,x)| + \sup_{t,x} |D_x^2 u(t,x)| + \sup_{t,x \neq y} |D_x^2 u(t,x) - D_y^2 u(t,y)|/|x-y|^\beta$, $W^{2,p}$ is the space of measurable functions $u$ on $H$ having the finite norm $|u|_{2,p,H} = |u|_{p,H} + |D_x^2 u|_{p,H}$, where $|g|_{p,H} = \left( \int_0^T \int_{\mathbb{R}^d} |g(t,x)|^p dt\, dx \right)^{1/p}$.

First of all we shall further specify the structure of the matrix $a$. For this purpose let us introduce a measurable function $\hat{\sigma} : \mathbb{R}^d \times E \to E$. Everywhere below we assume that

$$a^{ij}(x) = \frac{1}{2}\left( \int_E \sigma^i(x,\alpha)\sigma^j(x,\alpha)d\kappa + \int_E \hat{\sigma}^i(x,\alpha)\hat{\sigma}^j(x,\alpha)d\kappa \right). \tag{P}$$

This assumption assures that equation (1) is parabolic in the sense of [15].

In addition we will require one of the following

(C1) There exists $\beta \in (0,1)$ and $\delta > 0$ such that

a) $|f|_\beta + |a|_\beta + |b|_\beta + |c|_\beta \leq K$, where $|v|_\beta = \sup |v(x) - v(y)|/|x-y|^\beta$.

b) $a^{ij}(x)\xi_i\xi_j \geq \delta |\xi|^2$; for all $x$.

c) $\int_E \{ |\sigma(x,\alpha) - \sigma(y,\alpha)|^2 + |h(x,\alpha) - h(y,\alpha)|^2 + |g(t,x,\alpha) - g(t,y,\alpha)|^2 \}\kappa(d\alpha) \leq K|x-y|^{2\beta}$ and $\int_E |g(t,x,\alpha)|^2 \kappa(d\alpha) \leq K$.

d) $\varphi \in C^{2,\beta}$.

(W1)

a) $a$ is uniformly continuous and condition b) of (C1) holds.

b) For $p > d$, $f \in L^p(H)$, $\left( \int_E |g(\cdot,\alpha)|^2 \kappa(d\alpha) \right)^{1/2} \in L^p(H)$ and $\varphi \in W^{2,p}(\mathbb{R}^d)$.

**Theorem 1** *Assume that the assumption (C1) (resp. (W1)) holds. Then there exists a unique soft solution of equation (1) such that $u^l \in C^{2+\beta}(H)$ (resp. $u^l \in W^{2,p}(H)$) $\forall l \in \mathcal{D}$.*

Before we proceed with the proof of the theorem, we shall introduce some additional notation and auxiliary results.

Let $\hat{W}$ be a cylindrical Wiener process in $L^2(E, \mathcal{E}, \kappa)$ independent of $W$. Denote $\mathcal{G}_t = \sigma(W_s, \hat{W}_s, s \in [0, t])$ and $\mathcal{G} = \mathcal{G}_1$. Let $\mathcal{X} = C_{[0,1]}(I\!\!R^d)$ be the space of continuous $I\!\!R^d$-valued trajectories on $[0, 1]$, equipped with the canonical process $X_t = X_t(w) = w_t, w \in \mathcal{X}$ and the canonical $\sigma$-algebra $\mathcal{C} = \sigma(X_s, s \geq 0)$ and the filtration $\mathcal{C}_t = \sigma(X_s, s \leq t)$. We define the product space $\bar{\Omega} = \Omega \times \mathcal{X}$ and $\sigma$-algebras $\bar{\mathcal{F}} = \mathcal{G} \otimes \mathcal{C}, \bar{\mathcal{F}}_t = \cap_{s>t} \mathcal{G}_s \otimes \mathcal{C}_s, \mathcal{F}_t^W = \sigma(W_s - W_u, u, s \in [T, 1])$.

Denote $B_{mc}(\Omega)$ as the set of all bounded measurable functions $f : \bar{\Omega} \to I\!\!R$ such that $f(w, \cdot)$ is continuous on $\mathcal{X}$ for all $w \in \Omega$. Let $M^1_{mc}(\bar{\Omega})$ be the set of probability measures $\mu$ on $(\bar{\Omega}, \bar{\mathcal{F}})$ such that $\mu|_\Omega = P$ endowed with the topology generated by the maps $\mu \to \mu(f)(\mu \in M^1_{mc}(\bar{\Omega}), f \in B_{mc}(\bar{\Omega})$ (see [4]).

For $(s, x) \in H$ define

$$S(s, x, \sigma, \hat{\sigma}, B) = \{\mu \in M^1_{mc}(\bar{\Omega}) : \mu - a.e.$$

$$X_u = x \quad \forall u \leq s, dX_t = \int_E \sigma(X_t, \alpha) W(dt, d\alpha) +$$

$$\int_E \hat{\sigma}(x_t, \alpha) \hat{W}(dt, d\alpha) + b(X_t) dt\}$$

**Proposition 1** *Assume that*

a) *The functions $a^{ij}(x)$ are uniformly continuous and condition (b) from (C1) holds.*

b) $|a| + |b| \leq K.$

*Then for each $(s, x)$ there exists a unique $P_{s,x} \in \mathcal{S}(s, x, \sigma, \hat{\sigma}, b)$ and the map $(s, x) \to P_{sx}$ is continuous.*

The proof of Proposition 1 is rather long and is based on the same ideas which were used to prove similar results for Ito equations driven by a finite-dimensional Wiener process (see, e.g., [5, 18]), so we omit it.

**Remark 1** Proposition 1 holds if $b, \sigma$ and $\hat{\sigma}$ also depend on $t \in [0, 1]$ assuming that they are appropriately measurable and the above assumptions hold uniformly in $t$.

**Proof of Theorem 1** For $l \in \mathcal{D}$ set $B_l(t, x) = b(x) + \int_E \sigma(x, \alpha) l(t, \alpha) \kappa(d\alpha)$ and denote $B(t, x) = b(x) - \int_E \sigma(x, \alpha) h(x, \alpha) \kappa(d\alpha)$.

Proposition 1 and Remark 1 imply existence and uniqueness of the measures $P^B_{s,x} \in \mathcal{S}(s, x, \sigma \hat{\sigma}, B)$ and $P^{B_l}_{s,x} \in \mathcal{S}(s, x, \sigma, \hat{\sigma}, B_l)$.

**Proposition 2** (Feynman-Kac formula) *Let*

$$\rho(t,s) = \exp\{\int_t^s c(X_r)dr + \int_t^s \int_E h(X_r,\alpha)W(dr,d\alpha) - \frac{1}{2}\int_t^s \int_E |h(X_r,\alpha)|^2 \kappa(d\alpha)dr\}$$

*and assume that at least one of the assumptions* (C1) *or* (W1) *holds. Then*

$$\tilde{u}(t,x) = P_{t,x}^B\Big[\int_t^1 f(s,X_s)\rho(t,s)ds + \int_t^1 g(s,X_s)\rho(t,s)W(ds,d\alpha) +$$
$$\varphi(X_1)\rho(t,1)|\mathcal{F}_t^W\Big] \tag{4}$$

*is a soft solution of* (1).

**Proof of Proposition 2**  By Proposition 1, $\rho(t,s)$ and $\tilde{u}(t,z)$ are well-defined. Let us fix $l \in \mathcal{D}$ and set $v^l(t,x) = P_{t,x}^B[q_t(l)\tilde{u}(t,x)]$. Simple computations based on Girsanov's theorem show that

$$v^l(t,x) = P_{t,x}^{B_l}\Big[\big[\int_t^1 \exp\{\int_t^S \tilde{c}(r,X_r)dr\}\big(f(s,X_s) +$$
$$\int_E l(s,\alpha)g(s,X_s,\alpha)\kappa(d\alpha)\big)\,ds + \varphi(X_1)\exp\{\int_t^1 \tilde{c}(X_r)dr\}\big]\Big] \tag{5}$$

where $\tilde{c}(r,x) = c(x) + \int_E l(r,\alpha)h(x,\alpha)\kappa(d\alpha)$ (cf. [9] or [15]). On the other hand, it is well known (see, e.g., [5] that the right-hand side of (5) solves equation (3) so we are done.

Uniqueness of the soft solution follows from uniqueness of the solutions to equation (3) in $C^{2+\beta}(H)$ and $W^{2,p}(H)$ (see, e.g., [1, 11, 13]) and the following result.

**Lemma 1**  *Let* $\eta \in L^2(\Omega,\mathcal{F}_T^W,P)$ *where* $T \in (0,1]$ *and* $Eq_T(l)\eta = 0$   $\forall l \in \mathcal{D}$. *Then* $\eta = 0$. *P-a.s.*

Proof of Lemma 1 is based on the same idea as Theorem 4.1 §4.2 in [2]. This completes the proof of theorem.

**Remark 2**  To guarantee the existence of strong solutions in $W_p^2$ for equation (1), the differentiability of $\sigma, h, g$ is needed (see [7]).

The following Corollary is an obvious implication of (4) and uniqueness of soft solutions.

**Corollary 1** (maximum principle) *If* $g = 0$, $f \geq 0$ *(a.p.) and* $\varphi \geq 0$ *(a.e.), then the S-solution of equation* (1) *is non-negative.*

216

# 3. Wiener Chaos Expansion of Soft Solutions

Fix $T \in [0,1)$. Let $(l_n)_{n \geq 1}$ be a complete orthonormal system (CONS) in $L^2([T,1] \times E, dt\,d\kappa)$ such that $l_n \in \mathcal{D}$. Denote by $\mathcal{Z}$ the set of all sequences $(z_k)_{k \geq 1}$ such that $\sum_n z_k^2 < \infty$. For $z \in \mathcal{Z}$ we define

$$l_z = \sum_k z_k l_k \in L^2([T,1] \times E).$$

Let $p_t(z) := q_t(l_z)$. For each infinite multiindex $\mu = (i_1, i_2, i_3, \ldots)$ we denote $|\mu| = i_1 + i_2 + \ldots$, and $\mu! = i_1! i_2! \ldots$. Let $\mathcal{I}$ be the set of all multiindices of finite length. We consider the following system of elements in $L^2(\Omega, \mathcal{F}_T^W, P)$:

$$\xi_\mu = \frac{1}{\sqrt{\mu!}} \frac{\partial^{|\mu|} p_T(z)}{\partial z^\mu}\Big|_{z=0}, \quad \mu \in \mathcal{I}.$$

Recall the following standard fact (see, e.g., [2]).

**Lemma 2** $(\xi_\mu)_{\mu \in \mathcal{I}}$ *is a CONS in* $L^2(\Omega, \mathcal{F}_T^W, P)$.

Obviously Lemma 2 yields

**Corollary 2** *If* $\eta \in L^2(\Omega, \mathcal{F}_T^W, P)$, *then*

a) $\eta = \displaystyle\sum_\mu \frac{1}{\sqrt{\mu!}} \partial_z^\mu E[\eta p_T(z)]|_{z=0} \xi_\mu.$

b) $E\eta^2 = \displaystyle\sum_\mu \frac{1}{\mu!} \left( \partial_z^\mu E[\eta p_T(z)]|_{z=0} \right)^2.$

Let $u(t,x)$ be a soft solution represented by (4). For $z \in \mathcal{Z}$, we define

$$\psi(\alpha, x, t) = E[u(t,x) p_t(z)].$$

Assume that assumption (C1) or (W1) holds. Then it is readily checked that

$$\begin{cases} \partial_t \psi + \mathcal{L}\psi + f + \int_E l_z (\mathcal{M}\psi + g) d\kappa = 0 \\ \psi(1, x) = \varphi(x). \end{cases}$$

Let $\varphi^\mu(\alpha, x) = \partial_z^\mu \psi|_{z=0}$. Then

$$\begin{cases} \partial_t \varphi^0 + \mathcal{L}\varphi^0 + f = 0 \\ \varphi^0(1, x) = \varphi(x) \end{cases}$$

$$\begin{cases} \partial_t \varphi^\mu + \mathcal{L}\varphi^\mu + \int_E \partial^\mu l_z|_{z=0} (\mathcal{M}\varphi_t^0 + g) d\kappa = 0 \\ \varphi^\mu(1, x) = 0, \qquad \text{if } |\mu| = 1 \end{cases}$$

$$\begin{cases} \partial_t \varphi^\mu + \mathcal{L}\varphi^\mu + \int_E \sum_{\substack{j+\nu=\mu \\ |j|=1}} \binom{u}{j} \partial^j l_z|_{z=0} \mathcal{M}\varphi^\nu d\kappa = 0 \\ \varphi^\mu(1, x) = 0, \qquad \text{if } |\mu| > 1. \end{cases}$$

To each multiindex $\mu = (\mu_1, \mu_2, \ldots) \in \mathcal{I}$ of the length $n$ we relate a set $K_\mu = \{i_1, \ldots i_n\}$ of positive integers $k$ such that $\mu_k \neq 0$ and each $k$ is represented by $\mu_k$ copies. Denote $\mu(k) = (\mu_1, \ldots, (\mu_k - 1)^+, \mu_{k+1}, \ldots)$. Then we can rewrite the equations of $\varphi^\mu$:

$$\begin{cases} \partial_t \varphi^\mu + \mathcal{L}\varphi^\mu + \int_E \sum_{k=1}^n l_{i_k} \mathcal{M}\varphi^{\mu(k)} d\kappa = 0 \\ \varphi^\mu(1, \cdot) = 0, \qquad \text{if } |\mu| > 1 \end{cases} \tag{$6_1$}$$

$$\begin{cases} \partial_t \varphi^\mu + \mathcal{L}\varphi^\mu + \int_E l_{i_1}(\mathcal{M}\varphi^0 + g) d\kappa = 0 \\ \varphi^\mu(1, \cdot) = 0, \qquad \text{if } |\mu| = 1, \ K_\mu = \{i_1\} \end{cases} \tag{$6_2$}$$

$$\begin{cases} \partial_t \varphi^0 + \mathcal{L}\varphi^0 + f = 0 \\ \varphi^0(1, \cdot) = \varphi. \end{cases} \tag{$6_3$}$$

So we arrive at

**Theorem 2** *If $f, \int_{E^0} |g|^2 d\kappa$ are bounded, then under assumption* (C1) *or* (W1) *the coefficients in the Wiener chaos expansion of the soft solution of equation* (1) *are given by system* ($6_1$)–($6_3$).

Let $\mathcal{P}^n$ be the permutation group of the set $\{1, \ldots, n\}$.

For $\mu \in \mathcal{I}_n$ such that $K_\mu = \{i_1, \ldots, i_n\}$ we define $e^\mu = \sum_{\sigma \in \mathcal{P}^n} l_{i_{\sigma(1)}} \otimes \ldots \otimes l_{i_{\sigma(n)}} / \sqrt{\mu! |\mu|!}$. The set $(e^\mu)_{|\mu| = n}$ forms a CONS in the symmetric part of $L^2(([T, 1] \times E)^n, dt_1 \kappa(d\alpha_1) \ldots dt_n \kappa(d\alpha_n))$.

Denote by $T_s p$ a semigroup corresponding to $\mathcal{L}$. Making use of the recursive structure of system ($6_1$)–($6_3$) we arrive at the following statement.

**Proposition 3** *For $\mu \in \mathcal{I}_n$ such that $K_\mu = \{i_1, \ldots, i_n\}$, we have*

$$\frac{\varphi^\mu(t, x)}{\sqrt{\mu!}} = \sqrt{n!} \int_{E^n} \int_t^1 \int_{s_1}^1 \cdots \int_{s_{n-1}}^1 T_{s_1 - t}\mathcal{M}^{\alpha_1} T_{s_2 - s_1} \mathcal{M}_\sigma^{\alpha_2} \cdot$$

$$T_{s_{n-1} - s_n}\left[\mathcal{M}^{\alpha_n - 1}\left(T_{1 - s_n}\varphi + \int_{s_n}^1 T_{u - s_n} f \, du\right) + g(s_n, x)\right] \cdot$$

$$e^\mu(s_1, \alpha_1, \ldots, s_n, \alpha_n) ds^n \kappa(d\alpha_1) \ldots \kappa(d\alpha_n) \, .$$

By Ito's theorem we have

**Corollary 3** *For each $n$,*

$$\sum_{\mu \in \mathcal{I}_n} \frac{\varphi^\mu(t, x)}{\sqrt{\mu!}} \xi^\mu = \int_{E^n} \int_t^1 \int_{s_1}^1 \cdots \int_{s_{n-1}}^1 T_{s_1 - t}\mathcal{M}^{\alpha_1} \cdots \cdot$$

$$\cdots T_{s_{n-1} - s_n}\left[\mathcal{M}^{\alpha_n - 1}\left(T_{1 - s_n}\varphi + \int_{s_n}^1 T_{u - s_n} f\right) du + g(s_n, x)\right] \cdot$$

$$W(\overleftarrow{ds}_1, d\alpha_1) \ldots W(\overleftarrow{ds}_n, d\alpha_n) \, .$$

**Acknowledgements** This work was partially supported by ONR Grant N00014-91-J-1526

## References

[1] Eidelman, S. D.: *Parabolic Systems.* North-Holland Publishing Company, Amsterdam, (1969

[2] Hida, T.: *Brownian Motion.* Applications of Math., Vol. II. Springer-Verlag, 1980

[3] Hida, T., Kuo, H. -H., Potthoff, J., and Streit, L.: *White Noise: An Infinite Dimensional Calculus.* Kluwer Academic Publishers, 1993

[4] Jacod, J. and Mémin, J.: Sur un type de convergence intermediaire entre la convergence en loi et la convergence en probabilité; *Lecture Notes in Math.* **850** (1981) 529–560

[5] Krylov, N. V.: On Ito's stochastic integral equations; *Theory Probab. Appl.* **14** (1969) 330–336

[6] Krylov, N. V.: Some estimates of the probability density of a stochastic integral; *Math. USSR Izv.* **8** (1974) 233–254

[7] Krylov, N. V.: On $L_p$ theory of stochastic partial differential equations in the whole space; *Preprint* (1994)

[8] Krylov, N. V. and Rozovskii, B. L.: On the first integrals and Liouville equations for diffusion processes; *Lecture Notes in Control and Information Sciences* **36** (1981) 117–125

[9] Krylov, N. V. and Rozovskii, B. L.: On characteristics of degenerate second order parabolic equations; *J. Soviet Math.* **32** (1986) 226–248

[10] Kunita, H.: On backward stochastic differential equations; *Stochastics* **6** (1982) 293–313

[11] Ladyzhenskaia, O. A., Solonnikov, V. A., and Ural'ceva, N. N.: *Linear and Quasilinear Equations of Parabolic Type.* Amer. Math. Soc., Providence, 1968

[12] Malliavin, P.: $C^K$-hypoellipticity with degeneracy; in:*Stochastic Analysis.* Academic Press (1978) 199–214, 327–340

[13] Mikulevicius, R. and Pragarauskas, H.: On the Cauchy problem for certain integro-differential operators in Sobolev and Hoelder spaces; *Lithuanian Math. J.* **32** (1992) 238–264

[14] Nualart, D.: Noncausal Stochastic Integrals and Calculus; in:*Stochastic Analysis and Related Topics, Lecture Notes in Math.* **1316** (1988) 80–126

[15] Rozovskii, B. L.: *Stochastic Evolution Systems.* Kluwer Academic Publishers, Dordrecht, Boston, 1990

[16] Skorohod, A. V.: On a generalization of a stochastic integral; *Theory Prob. and Appl.* **XX** (1975) 219–233

[17] Stroock, D. W.: *Lecture Notes on the Malliavin Calculus and Applications.* IMA, Minneapolis, *Preprint* #218, 1986

[18] Stroock, D. W. and Varadhan, S. R. S.: *Multidimensional Diffusion Processes.* Springer-Verlag, 1979

[19] Veretennikov, A. Yu. and Krylov, N. V.: On explicit formulas for solutions of stochastic differential equations; *Math. USSR Sbornik* **29** (1976) 229–256

[20] Watanabe, S.: *Lectures on stochastic differential equations and Malliavin calculus.* Tata Institute of Fundamental Research, Springer-Verlag, 1984

[21] Zakai, M.: On the optimal filtering of diffusion processes; *Wahrsch II* (1969) 230–243

R. Mikulevicius
Center for Applied Mathematical Sciences
University of Southern California
Los Angeles, CA 90089-1113, U.S.A.
*E-mail*: mikulvcs@cams.usc.edu

B. L. Rozovskii
Center for Applied Mathematical Sciences
University of Southern California
Los Angeles, CA 90089-1113, U.S.A.
*E-mail*: rozovski@cams.usc.edu

P. NEY

# Large deviations via subadditivity

## 1. Introduction

Let $\{S_n; n = 0, 1, \cdots\}$ be a sequence of random variables taking values in a topological space $S$. Large deviation (LD) theory is concerned with the behavior of $P\{(S_n/n) \in \cdot\}$. If, for a l.s.c. function $I : S \to \mathbb{R}$,

$$\overline{\lim} \frac{1}{n} \log P\{\frac{S_n}{n} \in K\} \leq - \inf_{x \in K} I(x) \tag{1.1}$$

for all closed $K$, and

$$\underline{\lim} \frac{1}{n} \log P\{\frac{S_n}{n} \in U\} \geq - \inf_{x \in U} I(x) \tag{1.2}$$

for all open $U$, then one says that the "LD principle" (LDP) holds for $\{S_n\}$ with rate function $I$. The $LD$ upper bound is (1.1), the lower bound is (1.2). One wants to investigate conditions on $\{S_n\}$ and $S$ under which (1.1) and/or (1.2) hold, and to describe $I$ in some reasonable way.

In both old and recent developments in this theory, subadditivity arguments have been found to be useful in establishing the *existence* of the $LDP$ limit. Early ideas on this technique are attributed to Ruelle [16] and Lanford [12]; and for sums of independent random variables it was applied directly by Bahadur and Zabell [3]. The notes of Azencott [2] contain a nice exposition of this method, as do the monographs of Stroock [17], Deuschel and Stroock [7] and Dembo and Zeitouni [6]. The last three works go beyond sums of i.i.d. random variables, and apply this approach to prove versions of the Donsker, Varadhan [10] theorem for empirical measures of a Markov chain. However, to activate the method, they need a strong uniformity condition which eliminates some important cases (e.g. birth-death chains).

Recently Dinwoodie and this author [9] made use of a regeneration construction to make the subadditivity approach applicable to very general Markov chains. An important component of our argument is to use an adaptation of a theorem of Varadhan [18] by Dinwoodie [8] to *identify* the rate function after one has proved its existence. The main ideas of this paper are taken from [9].

This note is mostly expository, with a few small new touches. I will try to explain and illustrate the basic idea, not to give complete, detailed proofs. Thus I will quote technical facts freely from the literature.

In section two I describe the old use of subadditivity for sums of i.i.d.r.v.'s and combine it with the identification results of Varadhan and Dinwoodie. Section three discusses the Markov case, adapted from [9], and Section 4 summarizes the results of that paper for empirical measures, leading to the Donsker, Varadhan Theorem [10].

For other works on the above subjects using other techniques e.g. see [4, 5, 11, 13, 14]. My purpose here is not to obtain the most general results, but to illustrate the subadditivity/identification approach.

## 2. Sums of Independent and Identically Distributed Random Variables

Let $X_1, X_2, \cdots$ be i.i.d.r.v.'s taking values in a separable Banach space $E$, and $S_{m,n} = X_m + \cdots + X_n$, $S_{1,n} = S_n$.

Let $B(x, \epsilon) = \{y \in E : \|x - y\| < \epsilon\}, x \in E$, and

$$a_n = P\{S_n \in nB(x, \epsilon)\}.$$

Then

$$a_{n+m} \geq P\{S_n \in nB, S_{n+1,n+m} \in mB\}$$

$$\text{(2.1)}$$

$$= P\{S_n \in nB\}P\{S_m \in mB\} = a_n a_m.$$

Thus $- \log a_n$ is a subadditive sequence. If $a_{n_0} > 0$ for some $n_0 \geq 1$, then also

$$a_n > 0 \quad \text{for all sufficiently large} \quad n. \tag{2.2}$$

(See e.g. Section 3.1 of [7] or Section 6.1 of [6].) From (2.1) and (2.2) it follows by the subadditivity theorem for sequences (having nothing to do with probability) that the limit

$$\lim \frac{1}{n} \log a_n = \lim \frac{1}{n} \log P\{\frac{S_n}{n} \in B\} = s(x, \epsilon) \quad \text{exists.}$$

Let

$$I(x) = - \lim_{\epsilon \searrow 0} s(x, \epsilon).$$

222

Now by standard arguments which are amply described in [2, 3, 6, 7] and in many other places, one can conclude that the LD lower bound (1.2) holds for *open $U$*, the upper bound (1.1) holds for *compact $K$*, and that $I(\cdot)$ is l.s.c. and convex.

A somewhat less well known fact is that the upper bound also holds for closed convex sets. This fact is stated without proof in Lemma 2.6 of [3]. The proof was communicated by the authors. The following simple version was pointed out by A. de Acosta (private communication – thank you Alex).

**Lemma**

*If $\frac{S_n}{n}$ satisfies (1.1) for compact sets, then it satisfies (1.1) for closed convex sets.*

**Proof** Let $\mu_n(\Gamma) = P\{\frac{S_n}{n} \in \Gamma\}$. Then for any convex $\Gamma$, as in (2.1), $\mu_{n+m}(\Gamma) \geq \mu_n(\Gamma)\mu_m(\Gamma)$. Take $\Gamma$ to be closed, convex, and let $K$ be compact convex $\subset \Gamma$. Then for fixed $n$

$$[\mu_n(K)]^m \leq \mu_{mn}(K) \quad \text{for} \quad m \geq 1.$$

Hence

$$\frac{1}{n}\log\mu_n(K) \leq \frac{1}{mn}\log\mu_{mn}(K),$$

and

$$\frac{1}{n}\log\mu_n(K) \leq \overline{\lim_{m\to\infty}}\, \frac{1}{mn}\log\mu_{mn}(K)$$
$$\leq -I(K) \leq -I(\Gamma) \quad \text{for} \quad n \geq 1.$$

Then given any $\epsilon > 0$, there is a compact convex $K_\epsilon \subset \Gamma$ such that

$$\mu_n(\Gamma) - \epsilon \leq \mu_n(K_\epsilon) \leq e^{-nI(\Gamma)}.$$

Letting $\epsilon \searrow 0$ implies the lemma, in fact the stronger inequality

$$\mu_u(\Gamma) \leq e^{-nI(\Gamma)} \qquad n \geq 1.$$

The remaining goal is to identify $I(\cdot)$, and to extend (1.1) to closed $K$.

**Identification of $I(x)$:**

Intuitively, the idea is as follows. We quote the following

**Theorem (Varadhan [18])**

*Let $\{S_n\}$ be a sequence of r.v.'s taking values in a Polish space $X$, and satisfying (1.1) and (1.2) for open and closed sets respectively, with convex, l.s.c. $I(\cdot)$. Then for any bounded, continuous function $F$ on $X$*

$$\lim_{u \to \infty} \frac{1}{n} \log E e^{nF(\frac{S_n}{n})} = \sup_{x \in X} [F(x) - I(x)]. \tag{2.3}$$

Now take $X$ to be our space $E$, and "suppose" we were able to take $F(x) = \langle \alpha, x \rangle$ for $\alpha \in E^*$. Then (2.3) would say

$$\lim_{n \to \infty} \frac{1}{n} \log E e^{\langle \alpha, S_n \rangle} = I^*(\alpha),$$

where $I^*$ is the convex conjugate of $I$. But since $S_n$ are sums of i.i.d.r.v.'s, this says that

$$\Lambda(\alpha) \overset{(\mathrm{def})}{=} \log E e^{\langle \alpha, X_1 \rangle} = I^*(\alpha).$$

Now since $I$ is convex, $I^{**}(x) = I(x) = \Lambda^*(x)$, and we *would* have identified

$$I(x) = \Lambda^*(x) \tag{2.4}$$

*if* we could take $F(\cdot) = \langle \alpha, \cdot \rangle$ in (2.3). There are two difficulties. First $\langle \alpha, \cdot \rangle$ is not bounded, as required. Secondly we only have (1.1) for closed *and* convex sets, not for all closed sets, as required in Varadhan's theorem.

The second difficulty is easily circumvented by noting (in the proof of Varadhan's Thm.) that due to the linearity of $\langle \alpha, \cdot \rangle$, having (1.1) for closed convex $K$ is sufficient.

The difficulty with the boundedness was resolved by Dinwoodie in [8]. He observed that it was sufficient to have $F$ bounded above, and then showed (in a more general context) that

$$I(x) = \psi^*(x),$$

where $\psi$ is the limit of a truncated version of $\Lambda$, namely

$$\psi(\alpha) = \lim_{M \to \infty} \lim_{n \to \infty} \frac{1}{n} \log E e^{n[\langle \alpha, \frac{S_n}{n} \rangle \wedge M]}.$$

But this is

$$\geq \lim_{M \to \infty} \lim_{n \to \infty} \frac{1}{n} \log E e^{\sum_1^n (\langle \alpha, X_i \rangle \wedge M)}$$

$$= \lim_{M \to \infty} E e^{\langle \alpha, X_1 \rangle \wedge M} = \Lambda(\alpha)$$

224

by the dominated convergence theorem. Conversely $\psi(\alpha) \leq \Lambda(\alpha)$ trivially, and hence

$$I(x) = \psi^*(x) = \Lambda^*(x),$$

which is the identification in (2.4).

We have thus proved part (i) of the following theorem; and part (ii) again follows by an easy standard argument.

**Theorem 1**

(i) *If $X_1, X_2, \cdots$ are i.i.d.r.v.'s taking values in a separable Banach space, then $S_n = X_1 + \cdots + X_n$ satisfies the LD lower bound for open sets and the upper bound for compact sets and for closed convex sets. The rate function is $I(x) = \Lambda^*(x) = $ convex conjugate of $\log E e^{\langle \alpha, X_1 \rangle}$, and is l.s.c. and convex.*

(ii) *If $X_1$ is exponentially tight, then the upper bound holds for all closed sets.*

## 3. Markov Chains

Let $X_1, X_2, \cdots$ be an irreducible, aperiodic Markov chain taking values in a general measurable space $(S, \mathcal{S})$, with transition function $\{p(x, A); x \in S, A \in \mathcal{S}\}$. Then there exist a set $C \in \mathcal{S}$ (called a "small" set) a measure $\nu$ on $(S, \mathcal{S})$, an integer $k_0 < \infty$, and a $\delta > 0$ such that

$$\delta 1_C(x)\nu(A) \leq P^{k_0}(x, A) \quad \text{for all} \quad x, A. \tag{3.1}$$

(See e.g. Nummelin [15].) For simplicity, assume at first that $k_0 = 1$. This restriction can be removed.

Let $f : S \to E = $ a separable Banach space, and $S_n = \sum_{i=1}^{n} f(X_i)$. More generally the setup can be easily extended to additive functions like $\sum_1^n f(X_{i-1}, X_i) = S_n$, or to so called Markov-additive chains [13].

The following argument is an adaptation of that in Dinwoodie and Ney [9].

We start by recalling that a "regeneration structure" exists for $\{X_n\}$; namely a sequence of random times $T_0, T_1, \cdots$ such that

(i) $\{X_{T_i+1} \cdots, X_{T_{i+1}}\}$ are i.i.d. blocks, and

(ii) $P\{X_{T_i+1} \in \cdot | \mathcal{F}_{T_i}\} = \nu(\cdot)$.

The idea figuratively speaking, is that whenever $X_n \in C$, flip a coin with $P(\text{Head}) = \delta$. If it is head distribute $X_{n+1}$ according to $\nu$. If it is tail distribute by the compensating distribution needed to leave the overall distribution unchanged. This can be done due to (3.1). (See [1] or [15].) The "regeneration times" are the times of "heads preceded by an entry into $C$". Let $R_n$ = the event for some $n = 1, 2, \cdots, X_{n-1} \in C$, and the coin is $H$ i.e. regeneration occurs at time $n$, and let

$$c_n = P_\nu\{\frac{S_n}{n} \in B, R_n\} = \delta P_\nu\{\frac{S_n}{n} \in B, X_n \in C\} \overset{(\text{def})}{=} \delta a_n,$$

where $B = B(x, \epsilon)$ (or in fact is any convex set). Then as in (2.1)

$$c_{n+m} \geq P_\nu\{S_n \in nB, R_n, S_{n+1,n+m} \in mB, R_{n+m}\}$$

$$= P_\nu\{S_n \in nB, R_n\} P_\nu\{S_m \in mG, R_m\} = c_n c_m.$$

Thus $-\log c_n$ is a subadditive sequence. Similarly to (2.2) one now shows (see Lemma 2.2 of [9]) that if $c_{n_0} > 0$ for some $n_0$, then

$$c_n > 0, \quad \text{for } n \text{ sufficiently large.}$$

(The proof in [9] relies on the boundedness of $f$.) Therefore $\lim n^{-1} \log c_n$ exists, and

$$s(x, \epsilon) = \lim \frac{1}{n} \log P\{\frac{S_n}{n} \in B, X_n \in C\} \quad \text{exists.}$$

Now again one lets

$$I(x) = \lim_{\epsilon \searrow 0} s(x, \epsilon),$$

and shows as in the i.i.d. case that $I$ is l.s.c. and convex. Then, again repeating the argument of the i.i.d. case, find that the sequence of measures $P_\nu\{\frac{S_n}{n}\epsilon\cdot, X_n \in C\}$ satisfies the $LD$ lower bound, and the upper bound for compact sets and for closed convex sets.

**Identification of $I(\cdot)$:**

Suppose first that $f(\cdot)$ is bounded, say $\|f\| \leq A < \infty$. Then $\|\frac{S_n}{n}\| \leq A$ and $\langle\alpha, \frac{S_n}{n}\rangle \leq \|\alpha\|A$. Fix $\alpha$, and take $F_\alpha(x) = \langle\alpha, x\rangle \wedge M$ with $M > A\|\alpha\|$. Then $F_\alpha(\frac{S_n}{n}) = \langle\alpha, \frac{S_n}{n}\rangle \wedge M = \langle\alpha, \frac{S_n}{n}\rangle$. Since $F_\alpha(\cdot)$ is bounded, we can apply Varadhan's theorem, and conclude by (2.3) that

$$\Lambda(\alpha) = \lim_{n\to\infty} \frac{1}{n} \log E_\nu[e^{\langle\alpha, S_n\rangle}; X_n \in C] \quad \text{exists, and}$$

$$= \sup_{x \in E}[\langle\alpha, x\rangle \wedge M - I(x)] \quad \text{for all} \quad M > A\|\alpha\|.$$

226

Letting $M \to \infty$, this

$$= \sup[\langle \alpha, x \rangle - I(x)] = I^*(\alpha) \quad \text{(as in [8, Thm. 3.1]).}$$

Hence

$$I(x) = \Lambda^*(x),$$

as in (2.4).

Now let $P_\alpha = \{p_\alpha(x, dy)\} = \{e^{\langle \alpha, f(y) \rangle} p(x, dy)\}$. Then

$$\lim \tfrac{1}{n} \log(\nu p_\alpha^n)(C) = \lim \tfrac{1}{n} \log E_\nu \left[ e^{\langle \alpha, S_n \rangle}; X_n \in C \right] = \Lambda(\alpha).$$

Let $R(\alpha) = $ the convergence parameter of the kernel $\{p_\alpha(x, dy)\}$. Then (see Nummelin [15]) $\Lambda(\alpha) = \log[R(\alpha)]^{-1}$, and we can conclude that

$$I(x) = [\log[R(\cdot)]^{-1}]^*(x).$$

This is the required identification.

So far we have treated only $P_\nu\{\tfrac{S_n}{n} \in \cdot, X_n \in C\}$. We want to replace $\nu$ by an arbitrary initial measure $\mu$, and remove the temporary restriction $k_0 = 1$. This can be done as in Theorem 2.1 of [9]. We summarize:

## Theorem 2

*Let $\{X_n\}_0^\infty \subset \mathcal{S}$ be an irreducible M.C., $f$ bounded: $\mathcal{S} \to E = $ separable Banach space, $S_n = X_1 + \cdots + X_n$, $\mu = $ a probability measure on $\mathcal{S}$. Then the sequence of measure $P_\mu\{\tfrac{S_n}{n} \in \Gamma, X_n \in A\}$*

(i) *satisfies the LD lower bound with rate $I = [\log R^{-1}]^*$ for any open $\Gamma \subset E$ and any set $A$ containing a small set.*

(ii) *satisfies the LD upper bound for compact or closed convex $\Gamma$, and small $A \subset \mathcal{S}$, with the same rate $I$.*

(iii) *satisfies the upper bound for closed sets if it is exponentially tight.*

## Corollary

*$P_\mu\{\tfrac{S_n}{n} \in \cdot\}$ satisfies the LD lower bound with rate $(\log \tfrac{1}{R})^*$.*

**Remark (i)** The boundedness of $f$ was used only in the identification $I(\cdot) = [\log R^{-1}(\cdot)]^*$. The unbounded case might be approached via a truncation argument as in [14].

**Remark (ii)** If there is a uniform minorization in (3.1), namely

$$\delta\nu(A) \le p^{k_0}(x, A) \qquad \text{for all } x, A \in \mathcal{S},$$

then $P_\mu\{\frac{S_n}{n} \in \cdot\}$ also satisfies the $LD$ upper bound for compacts and closed convex sets, and for closed sets under exponential tightness.

**Remark (iii)** Another approach to the upper bound is via the spectral radius $r_\alpha$ of $P_\alpha$, as in [4]. The problem then is to compare $r_\alpha$ with $R^{-1}(\alpha)$.

**Example** If the state space $\mathcal{S}$ is countable then $P_x\{\frac{S_n}{n} \in \cdot, X_n = y\}$, $x, y \in \mathcal{S}$, always satisfies the $LD$ lower bound, and the upper bound for compacts and closed convex sets. If $O \in \text{In}\{\alpha : \Lambda(\alpha) < \infty\}$, then the upper bound also holds for closed sets.

For $P_x\{\frac{S_n}{n} \in \cdot\}$ the lower bound always holds as above. But the upper bound may be a problem. For example if $\{X_n\}$ is a birth-death chain with $p(i, i+1) = 1 - p(i, i-1) = p_i$, $f(i) = i$, and $S_n = \sum_{i=1}^n f(x_i)$, then

$$E_i e^{\alpha S_n} = \sum_j P_\alpha^n(i, j)$$

is unbounded (as a function of $i$). Hence the spectral radius of $P_\alpha$ is infinite for $\alpha > 0$ and may not yield the "right" upper bound (even if $\{X_n\}$ is strongly recurrent).

## 4. Empirical Measures

Let $\{X_n\}_0^\infty$ be an irreducible, aperiodic Markov chain on a Polish space $S$; take for $f$ the measure valued function $\{1_A(x), x \in S, A \in \mathcal{S}\}$; let $S_n(A) = \sum_1^n 1_A(X_i)$, and let

$$L_n(\cdot) = \tfrac{1}{n} S_n(\cdot).$$

This plays the role of $S_n/n$ in the previous section. In [9] we used a subadditivity and identification argument like that above to give a relatively simple proof of the Donsker-Varadhan Theorem, which is an $LDP$ for $L_n$. I summarize its main ingredients.

We let $E = $ the finite signed measures on $\mathcal{S}$, $L_{11} = $ Lipschitz functions with exponent 1 and bounded by 1, and for $q \in E$

$$\|q\| = \sup_{f \in L_{11}} |\int f dq|.$$

Then the dual space $E^*$ of $E$ is the bounded Lipschitz functions. Let $M_1 = $ the probability measures $\subset E$. The norm $\| \cdot \|$ agrees with the weak topology on $M_1$. Let $C$ be a small set as in (3.1), and $B(q, \epsilon) = B = \{m \in E : \|m - q\| < \epsilon\}$. Let $a_n = P_\nu\{L_n \in B, X_n \in C\}$, $\delta$ and $\nu$ as in (3.1), and $k_0 = 1$. Then $-\log \delta a_n$ is subadditive, and if $a_{n_0} > 0$ for some $n_0$, then $a_n > 0$ for $n$ sufficiently large. Hence

$$s(q, \epsilon) \doteq \lim_{n \to \infty} \tfrac{1}{n} \log P\{L_n \in B(q, \epsilon), X_n \in C\}$$

exists and we define

$$I(q) = \lim_{\epsilon \searrow 0} s(q, \epsilon).$$

Now as before, Varadhan's Theorem and Theorem 3.1 of [8] are used to identify $I$, namely one shows that for $f \in E^*$

$$\Lambda(f) \overset{(\text{def})}{=} \lim_{n \to \infty} \frac{1}{n} \log E_\nu \left[ e^{\langle f, S_n \rangle}; X_n \in C \right] = I^*(f)$$

and

$$I(q) = \Lambda^*(q).$$

Also

$$\Lambda(f) = \log[R(f)]^{-1}, \quad f \in E^*,$$

where $R(f)$ is the convergence parameter of $\{p(x, dy)e^{f(y)}\}$.

From this we obtain the lower bound for $P_\mu\{\frac{S_n}{n} \in U, X_n \in C\}$, hence also for $P_\mu\{\frac{S_n}{n} \in U\}$, with $U$ open; and the upper bound for $P_\mu\{\frac{S_n}{n} \in K, X_n \in C\}$, $K$ compact or closed, bounded. The argument parallels that in section 3. The upper bound for $P_\mu\{\frac{S_n}{n} \in K\}$ is obtained in terms of the spectral radius by de Acosta's theorem [4], and the upper and lower rates are shown to be equal, and equal to the variational formula of Donsker and Varadhan. Namely

$$I(q) = \sup_u \int \log \left( \frac{u}{Pu} \right) dq,$$

where the sup is taken over bounded Lipschitz functions that are bounded away from 0.

## References

[1] Athreya, K. and Ney, P.: A new approach to the limit theory of recurrent Markov chains; *Trans. Amer. Math. Soc.* **245** (1978) 493–501

[2] Azencott, R.: Grandes deviations et applications; *Lecture Notes in Math.*, Springer, Berlin **774** (1980) 1–176

[3] Bahadur, R. and Zabell, S.: Large deviations of the sample mean in general vector spaces; *Ann. Prob.* **7** (1979) 587–621

[4] de Acosta, A.: Upper bounds for large deviations of dependent random vectors; *Zeitschrift für Wahrschlichkeitstheorie* **69** (1985) 551–565

[5] de Acosta, A.: Large deviations for vector-valued functionals of a Markov chain: lower bounds; *Annals of Prob.* **16** (1988) 925–960

[6] Dembo, A. and Zeitouni, O.: *Large Deviation Techniques.* Boston: Jones & Bartlett (1993)

[7] Deuschel, J. -D. and Stroock, D. W.: *Large Deviations.* San Diego: Academic Press (1989)

[8] Dinwoodie, I. H.: Identifying a large deviation rate function; *Ann. Probab.* **21** (1993) 216–231

[9] Dinwoodie, I. H. and Ney, P.: Occupation measures for Markov Chains; *J. of Th. Prob.* (1994) to appear

[10] Donsker, M. D. and Varadhan, S. R. S.: Asymptotic evaluation of certain Markov process expectations for large time III; *Comm. Pure Appl. Math.* **29** (1976) 389–461

[11] Jain, N. C.: Large deviation lower bounds for additive functionals of Markov processes; *Ann. Probab.* **18** (1990) 1071–1098

[12] Lanford, D. E.: Entropy and equilibrium states in classical stat. mechanics; *Lecture Notes in Physics*, Berlin: Springer **20** (1973) 1–113

[13] Ney, P. and Nummelin, E.: Markov additive processes I; *Ann. Probab.* **15** (1987) 561–592

[14] Ney, P. and Nummelin, E.: Markov additive processes II; *Ann. Probab.* **15** (1987) 593–609

[15] Nummelin, E.: *General Irreducible Markov Chains and Non-negative Operators.* New York: Cambridge University Press (1984)

[16] Ruelle, D.: A variational formulation of stat. mechanics; *Comm. Math. Phys.* **5** (1967) 324–329

[17] Stroock, D. W.: *An Introduction to the Theory of Large Deviations.* New York: Springer (1984)

[18] Varadhan, S. R. S.: *Large Deviations and Appl's.* Philadelphia: SIAM (1984)

Peter Ney
Department of Mathematics
University of Wisconsin
Madison, WI 53706, U.S.A.
*E-mail*: ney@math.wisc.edu

M. NISIO

# On sensitive control for stochastic partial differential equations

## 1. Introduction

Recently Fleming and McEneaney [2] investigated a connection between risk sensitive control for the systems governed by stochastic differential equations and differential games. In the present note, we will deal with the same problems for infinite dimensional state systems governed by stochastic partial differential equations.

Let $W_k$, $k = 1, 2, \ldots$, be independent 1-dimensional Brownian motions, defined on a probability space $(\Omega, \mathcal{F}, P)$. $\mathcal{F}_t$ denotes the $\sigma$-field generated by $\{W_k(s), s \le t, k = 1, 2, \ldots\}$. Let $D$ be a bounded open set of $R^n$ with smooth boundary and $\Gamma$ a compact convex subset of $R^q$. Putting $\gamma = L^2(D; \Gamma)$, we call $\gamma$ a *control region*. A mapping $U\colon [0, T] \times \Omega \to \gamma$ is called an *admissible control* if it is $\mathcal{F}_t$-progressively measurable. Let us put $H = L^2(D)$, $\| \ \|$ = its norm and

$$A\zeta = \sum_{i,j=1}^{n} \frac{\partial}{\partial x_i}\left(a^{ij}(x)\frac{\partial \zeta}{\partial x_j}\right) + \sum_{i=1}^{n} r^i(x)\frac{\partial \zeta}{\partial x_i} - c(x)\zeta.$$

For an admissible control $U$, we will consider a nearly parabolic system $\xi^\varepsilon(t) = \xi^\varepsilon(t, \eta, U)$, governed by the following stochastic partial differential equation on a fixed time interval $[0, T]$,

$$d\xi^\varepsilon(t, x) = (A\xi^\varepsilon(t, x) + b(x, \xi^\varepsilon(t, x), U(t)(x)))dt + \sqrt{\varepsilon}dM(t, x), \tag{1}$$

$$x \in D, \ 0 < t < T$$

with boundary condition $\xi^\varepsilon(t, x) = 0$ on $\partial D$, and

initial condition $\xi^\varepsilon(0, \cdot) = \eta(\in H)$,

where $\varepsilon(\in (0, 1))$ is a small parameter and $M$ (a random force) is the $H$-valued colored noise defined by

$$M(t, x) = \sum_{k=1}^{\infty} m_k e_k(x) W_k(t)$$

with bounded smooth functions $e_k, k = 1, 2, \ldots$, of an orthonormal base of $H$ and $\sum m_k^2 < \infty$.

We assume some regularity conditions on A (see Section 2), which provide nice solution of (1). Suppose that $h\colon H \to R^1$ is bounded and Lipschitz continuous. Now, putting $Q_\varepsilon(s) = \exp(s/\varepsilon)$, the controller wishes to choose an admissible control which minimizes an exponential criterion $J^\varepsilon$,

$$J^\varepsilon(t,\eta,U) = EQ_\varepsilon\left(\int_0^t h(\xi^\varepsilon(s,\eta,U))ds\right), \quad t \in [0,T].$$

We define the *value function* $v^\varepsilon$ and (certainty) *equivalent value function* $V^\varepsilon$ as follows:

$$v^\varepsilon(t,\eta) = \inf_{U \in \alpha} J^\varepsilon(t,\eta,U)$$

and

$$V^\varepsilon(t,\eta) = Q_\varepsilon^{-1}(v^\varepsilon(t,\eta)),$$

where $\alpha =$ the set of all admissible controls.

In Section 2, we will study some regularity properties of $\xi^\varepsilon$ and a priori estimate of $V^\varepsilon$, which yields compactness of $\{V^\varepsilon,\ \varepsilon > 0\}$. Section 3 is devoted to dynamic programming equations of $v^\varepsilon$ and $V^\varepsilon$. We will prove that $v^\varepsilon$ is a unique viscosity solution of Bellman equation (Theorem 5) and $V^\varepsilon$ is a unique viscosity solution of the min-max equation (Theorem 6). The asymptotic behavior of $V^\varepsilon$, as $\varepsilon \to 0$, will be considered in Section 4.

## 2. Preliminaries

We assume the following four conditions on the elliptic differential operator A:

    (A1) $a^{ij}$ and $r^i$ are in $C^3(\bar{D})$.

    (A2) $n \times n$ matrix $(a^{ij}(x))$ is uniformly positive definite, say $(a^{ij}(x)) \geq \tilde{\mu}I$, where $\tilde{\mu} > 0$ is a constant and $I =$ identity matrix.

    (A3) $c$ is non-negative and continuous in $\bar{D}$.

    (A4) $b\colon \bar{D} \times R^1 \times \Gamma \to R^1$ is bounded and Lipschitz continuous.

Hence, putting $H^k =$ Sobolev space $H_0^k(D)$ and $\| \ \|_k = H^k$-norm, we can easily see that A satisfies the coercive condition,

$$\langle -A\zeta, \zeta \rangle + \mu\|\zeta\|^2 \geq \lambda\|\zeta\|_1^2 \quad \text{for} \quad \zeta \in H^1 \tag{2}$$

with constants $\mu \geq 0$ and $\lambda > 0$, where $\langle\ ,\ \rangle$ stands for duality pair between $H^{-1}$ and $H^1$. Moreover, $A$ generates a $C_0$ semigroup $e^{tA}$ such that

$$\|e^{tA}\varphi\| \leq K\|\varphi\| \quad \text{for} \quad \varphi \in H \quad \text{and} \quad t \in [0,T]$$

with a constant K.

Define $\beta$: $H \times \gamma \to H$ by

$$\beta(\zeta, u)(x) = b(x, \zeta(x), u(x)).$$

It follows from $(A4)$ that $\beta$ is bounded and Lipschitz continuous, say,

$$|\beta(\zeta, u)| \le \bar{\beta} \quad \text{and} \quad \|\beta(\zeta, u) - \beta(\psi, w)\| \le \ell(\|\zeta - \psi\| + \|u - w\|).$$

By a solution $\xi^\varepsilon$ of (1), we mean a mild solution, namely

$$\xi^\varepsilon(t) = e^{tA}\eta + \int_0^t e^{(t-s)A}\beta(\xi^\varepsilon(s), U(s))ds + \sqrt{\varepsilon}\sum_{k=1}^{\infty} m_k \int_0^t e^{(t-s)A}e_k dW_k(s).$$

From $(A1) \sim (A4)$, we can obtain Theorem 1.

**Theorem 1** *There is a unique solution $\xi^\varepsilon$ with $H$-valued continuous path and it satisfies the following properties,*

$$E\left( \sup_{t \le T} \|\xi^\varepsilon(t, \eta, U)\|^2 + \int_0^T \|\xi^\varepsilon(t, \eta, U)\|_1^2 dt \right) \le K_1(1 + \|\eta\|^2),$$

$$\|\xi^\varepsilon(t, \eta_1, U) - \xi^\varepsilon(t, \eta_2, U)\| \le K_2\|\eta_1 - \eta_2\| \tag{3}$$

*with constants $K_1$ and $K_2$ independent of $U$, $\varepsilon$ and $\omega$. (Hereafter $K_i$ stands for a constant independent of $U$, $\varepsilon$ and $\omega$.)*

Let us set $\hat{A} = A - \sum_{i=1}^{n} r^i \dfrac{\partial}{\partial x_i}$ and $B =$ the resolvent $(I - \hat{A})^{-1}$ with boundary value 0. Then $B\varphi \in C_0(\bar{D}) \cap H^2$ and $B$ is a compact operator on H. Moreover, (4) and (5) below hold,

$$\|\varphi\|^2 \ge |\varphi|_B^2 \ge \|B\varphi\|^2 + \tilde{\mu}\|\nabla B\varphi\|^2, \tag{4}$$

$$\text{Structural condition} \ : \ \langle -A^*B\varphi, \varphi \rangle \ge \frac{1}{2}\|\varphi\|^2 - p\langle B\varphi, \varphi \rangle \tag{5}$$

with a constant $p \ge 0$, where $A^* =$ dual operator of $A$ and $|\varphi|_B^2 = \langle B\varphi, \varphi \rangle$.

**Theorem 2**

$$\int_0^T \|\xi^\varepsilon(t, \eta_1, U) - \xi^\varepsilon(t, \eta_2, U)\|^2 dt \le K_3|\eta_1 - \eta_2|_B^2.$$

**Proof** Putting $\xi_i = \xi^\varepsilon(,\eta_i,U)$ and $\zeta = \xi_1 - \xi_2$, we calculate the dynamics of $|\zeta(t)|_B^2$ as follows:

$$
\begin{aligned}
\frac{1}{2}\frac{d}{dt}|\zeta(t)|_B^2 &= \langle B\zeta(t), \frac{d\zeta}{dt}(t)\rangle \\
&= \langle A^*B\zeta(t),\zeta(t)\rangle + \langle B\zeta(t),\beta(\xi_1(t),U(t)) - \beta(\xi_2(t),U(t))\rangle \\
&\le -\frac{1}{2}\|\zeta(t)\|^2 + p|\zeta(t)|_B^2 + \ell\|B\zeta(t)\|\|\zeta(t)\| \\
&\le -\frac{1}{4}\|\zeta(t)\|^2 + (p + 2\ell^2)|\zeta(t)|_B^2
\end{aligned}
\tag{6}
$$

by virtue of (4) and (5). So, we have

$$
\frac{d}{dt}|\zeta(t)|_B^2 \le (2p + 4\ell^2)|\zeta(t)|_B^2.
$$

Hence, putting $k = \exp\{(2p + 4\ell^2)T\}$, we get

$$
|\zeta(t)|_B^2 \le k|\eta_1 - \eta_2|_B^2.
\tag{7}
$$

Again noting (6), we get

$$
\int_0^T \|\zeta(t)\|^2 dt \le 4(p + 2\ell^2)\int_0^T |\zeta(t)|_B^2 dt - 2|\zeta(T)|_B^2 + 2|\eta_1 - \eta_2|_B^2.
\tag{8}
$$

Inserting (7) into (8), we complete the proof. $\qquad\square$

Next, putting $|h(\zeta)| \le \bar{h}$ and $\bar{\ell}$ = Lipschitz constant of h, we evaluate the equivalent value function $V^\varepsilon$:

$$
\int_0^{t_1} h(\xi_1(s))ds = \int_0^{t_2} h(\xi_2(s))ds + \int_0^{t_2}\Big(h(\xi_1(s)) - h(\xi_2(s))\Big)ds + \int_{t_2}^{t_1} h(\xi_1(s))ds
$$

$$
\le \int_0^{t_2} h(\xi_2(s))ds + \bar{\ell}\int_0^{t_2}\|\xi_1(s) - \xi_2(s)\|ds + \bar{h}|t_1 - t_2|,
$$

where $\xi_i = \xi^\varepsilon(,\eta_i,U)$. Appealing to Theorem 2, we get

$$
Q_\varepsilon\Big(\int_0^{t_1} h(\xi_1(s))ds\Big) \le Q_\varepsilon\Big(\int_0^{t_2} h(\xi_2(s))ds\Big) Q_\varepsilon(K_4(|\eta_1 - \eta_2|_B + |t_1 - t_2|))
$$

with $K_4 = \bar{\ell}\sqrt{K_3 T} + \bar{h}$. Hence, taking the expectation, we have

$$
J^\varepsilon(t_1,\eta_1,U) \le J^\varepsilon(t_2,\eta_2,U)Q_\varepsilon(K_4(|\eta_1 - \eta_2|_B + |t_1 - t_2|)).
$$

Therefore,

$$v^\varepsilon(t_1, \eta_1) \le v^\varepsilon(t_2, \eta_2) Q_\varepsilon(K_4(|\eta_1 - \eta_2|_B + |t_1 - t_2|)). \tag{9}$$

In the same way, we have

$$v^\varepsilon(t_1, \eta_1) \ge v^\varepsilon(t_2, \eta_2) Q_\varepsilon(-K_4(|\eta_1 - \eta_2|_B + |t_1 - t_2|)). \tag{10}$$

Applying $Q_\varepsilon^{-1}$ to (9) and (10), we obtain the following theorem.

**Theorem 3**

$$|V^\varepsilon(t_1, \eta_1) - V^\varepsilon(t_2, \eta_2)| \le K_4(|\eta_1 - \eta_2|_B + |t_1 - t_2|) \tag{11}$$

*and*

$$|V^\varepsilon(t, \eta)| \le \bar{h}T. \tag{12}$$

Since $B$ is a compact operator, the following theorem holds.

**Theorem 4** *$V^\varepsilon$ and $v^\varepsilon$ are bounded and weakly continuous.*

## 3. Nonlinear Equations Related to $v^\varepsilon$ and $V^\varepsilon$

The value function $v^\varepsilon$ satisfies the dynamic programming principle (13), since it is bounded and continuous,

$$v^\varepsilon(t + \tau, \eta) = \inf_{U \in \alpha} E\left[v^\varepsilon(\tau, \xi^\varepsilon(t, \eta, U)) Q_\varepsilon\left(\int_0^t h(\xi^\varepsilon(s, \eta, U)) ds\right)\right]. \tag{13}$$

If we assume $v^\varepsilon \in C^{12}((0, T) \times H)$ and its Fréchet derivative $\partial v^\varepsilon(t, \eta) \in H^2$, then $v^\varepsilon$ satisfies the dynamic programming equation,

$$0 = \frac{\partial v^\varepsilon}{\partial t} - \langle A^* \partial v^\varepsilon, \eta \rangle - \inf_{u \in \gamma} \langle \partial v^\varepsilon, \beta(\eta, u) \rangle - \frac{\varepsilon}{2} Tr(S \partial^2 v^\varepsilon) - \frac{1}{\varepsilon} h(\eta) v^\varepsilon, \tag{14}$$

$$0 < t < T, \ \eta \in H$$

with initial condition $v^\varepsilon(0, \eta) = 1$,

where $Tr(X) = $ trace of $X$ and $S$ is a trace class operator on $H$ defined by $S\zeta = \sum m_k^2 \langle \zeta, e_k \rangle e_k$. Furthermore, $V^\varepsilon \in C^{12}((0, T) \times H)$, $\partial V^\varepsilon(t, \eta) \in H^2$ and the equation (15) holds, under the above assumption on $v^\varepsilon$,

$$0 = \frac{\partial V^\varepsilon}{\partial t} - \langle A^* \partial V^\varepsilon, \eta \rangle - \inf_{u \in \gamma} \langle \partial V^\varepsilon, \beta(\eta, u) \rangle - \frac{\varepsilon}{2} Tr(S \partial^2 V^\varepsilon) - h(\eta)$$

$$- \frac{1}{2} \langle S \partial V^\varepsilon, \partial V^\varepsilon \rangle, \quad 0 < t < T, \ \eta \in H \tag{15}$$

with initial condition $V^\varepsilon(0,\eta) = 0$.

By the Legendre transformation, $\|\sqrt{S}\xi\|^2 = \langle S\xi, \xi \rangle = \sup_{\zeta \in H}(2\langle S\zeta, \xi \rangle - \langle S\zeta, \zeta \rangle)$, (15) turns out to be the min-max equation,

$$0 = \frac{\partial V^\varepsilon}{\partial t} - \langle A^*\partial V^\varepsilon, \eta \rangle - \inf_{u \in \gamma}\langle \partial V^\varepsilon, \beta(\eta, u) \rangle - \frac{\varepsilon}{2}Tr(S\partial^2 V^\varepsilon) - h(\eta)$$

$$- \sup_{\zeta \in H}(\langle S\zeta, \partial V^\varepsilon \rangle - \frac{1}{2}\langle S\zeta, \zeta \rangle), \quad 0 < t < T, \ \eta \in H. \tag{16}$$

Now, we will show that $v^\varepsilon$ and $V^\varepsilon$ still satisfy (14) and (15), respectively, in the viscosity sense without the above restrictive assumptions. According to Crandall and Lions (see [1], [6]), we will define the viscosity solution of (17) below, which is more general than equations (14) and (16). $\Phi \in C^{12}((0,T) \times H)$ is called a *test function* if (i) $\Phi$ is weakly lower semicontinuous and bounded from below, (ii) $\partial\Phi(t,\eta) \in H^2$ and $A^*\partial\Phi$ is continuous. A function $g : H \to [0,\infty)$ is called *radial* if $g(\zeta) = \tilde{g}(\|\zeta\|)$ with a strictly increasing function $\tilde{g} \in C^2[0,\infty)$ such that $\tilde{g}(0) = 0$ and $\lim_{\theta \to \infty} \tilde{g}(\theta) = \infty$. We consider the following nonlinear equation

$$0 = \frac{\partial v}{\partial t}(t,\eta) - \langle A^*\partial v(t,\eta), \eta \rangle + F(t, \eta, v(t,\eta), \partial v(t,\eta)) - \lambda Tr(S\partial^2 v(t,\eta)) \tag{17}$$

$$0 < t < T, \ \eta \in H$$

with initial condition $v(0,\eta) = \Psi(\eta)$,

where $\Psi$ is bounded and uniformly continuous, $\lambda$ is a non-negative constant, and $F$: $[0,T] \times H \times R^1 \times H \to R^1$ is locally uniformly continuous.

**Definition**  A function $v \in C([0,T] \times H)$ is called a *subsolution* (respectively, *supersolution*) of (17) if $v(0,\eta) = \Psi(\eta)$ and the following condition (i) (respectively, (ii)) holds for any test function $\Psi$ and any radial function $g$:

(i)  If $v - \Psi - g$ has a global maximum at $(\hat{t}, \hat{\eta})$ with $\hat{t} \in (0,T)$, then

$$\frac{\partial\Phi}{\partial t}(\hat{t}, \hat{\eta}) - \langle A^*\partial\Phi(\hat{t}, \hat{\eta}), \hat{\eta} \rangle + F(\hat{t}, \hat{\eta}, v(\hat{t}, \hat{\eta}), \partial(\Phi + g)(\hat{t}, \hat{\eta}))$$

$$- \lambda Tr(S\partial^2(\Phi + g)(\hat{t}, \hat{\eta})) \leq \mu\tilde{g}'(\|\hat{\eta}\|)\|\hat{\eta}\|.$$

(ii)  If $v + \Psi + g$ has a global minimum at $(\hat{t}, \hat{\eta})$ with $\hat{t} \in (0,T)$, then

$$-\frac{\partial\Phi}{\partial t}(\hat{t}, \hat{\eta}) + \langle A^*\partial\Phi(\hat{t}, \hat{\eta}), \hat{\eta} \rangle + F(\hat{t}, \hat{\eta}, v(\hat{t}, \hat{\eta}), -\partial(\Phi + g)(\hat{t}, \hat{\eta}))$$

$$+ \lambda Tr(S\partial^2(\Phi + g)(\hat{t}, \hat{\eta})) \geq -\mu\tilde{g}'(\|\hat{\eta}\|)\|\hat{\eta}\|,$$

236

where $\mu$ is the constant in (2). We call $v$ a *viscosity solution* if it is both a subsolution and a supersolution.

**Remark** In the definition above, we can replace "global" by "strictly global".

Recently, Świech [6] proved the following comparison theorem for more general nonlinear equations:

**Comparison theorem** *Let $\underline{v}$ and $\bar{v}$ be a subsolution and a supersolution, respectively, of (17). If $\underline{v}$ and $\bar{v}$ are weakly continuous, then $\underline{v} \leq \bar{v}$.*

Now, we state our theorems.

**Theorem 5** *$v^\varepsilon$ is a unique viscosity solution of (14).*

**Proof** Suppose that $v^\varepsilon - \Phi - g$ has a global maximum at $(\hat{t}, \hat{\eta})$ with $\hat{t} \in (0, T)$, namely

$$v^\varepsilon(t, \eta) - v^\varepsilon(\hat{t}, \hat{\eta}) \leq \Phi(t, \eta) - \Phi(\hat{t}, \hat{\eta}) + g(\eta) - g(\hat{\eta}).$$

Then, putting $v = v^\varepsilon$ and $\xi = \xi^\varepsilon(\ , \hat{\eta}, U)$ and using the dynamic programming principle and Itô's formula, we have

$$0 = \inf_{U \in \alpha} E\left[ v(\hat{t} - \theta, \xi(\theta)) Q_\varepsilon \left( \int_0^\theta h(\xi(s)) ds \right) \right] - v(\hat{t}, \hat{\eta})$$

$$= \inf_{U \in \alpha} E\left[ v(\hat{t} - \theta, \xi(\theta)) \left\{ Q_\varepsilon \left( \int_0^\theta h(\xi(s)) ds \right) - 1 \right\} + v(\hat{t} - \theta, \xi(\theta)) - v(\hat{t}, \hat{\eta}) \right]$$

$$\leq \inf_{U \in \alpha} E\left[ v(\hat{t} - \theta, \xi(\theta)) \left\{ Q_\varepsilon \left( \int_0^\theta h(\xi(s)) ds \right) - 1 \right\} \right.$$

$$\left. + \Phi(\hat{t} - \theta, \xi(\theta)) - \Phi(\hat{t}, \hat{\eta}) + g(\xi(\theta)) - g(\hat{\eta}) \right] \tag{18}$$

and

$$dg(\xi(\theta)) = [\langle \partial g(\xi(\theta)), A\xi(\theta) + \beta(\xi(\theta), U(\theta)) \rangle + \frac{\varepsilon}{2} Tr(S\partial^2 g(\xi(\theta)))] d\theta$$

$$+ \sqrt{\varepsilon} \sum_{k=1}^\infty m_k \langle \partial g(\xi(\theta)), e_k \rangle dW_k$$

$$\leq [\mu \tilde{g}'(\|\xi(\theta)\|) \|\xi(\theta)\| + \langle \partial g(\xi(\theta)), \beta(\xi(\theta), U(\theta)) \rangle$$

$$+ \frac{\varepsilon}{2} Tr(S\partial^2 g(\xi(\theta)))] d\theta + \sqrt{\varepsilon} \sum_{k=1}^\infty m_k \langle \partial g(\xi(\theta)), e_k \rangle dW_k \tag{19}$$

by the fact that $\xi(\theta) \in H^1$. Hence recalling (3), we get

$$0 \leq \left[ \frac{1}{\varepsilon} v(\hat{t}, \hat{\eta}) h(\hat{\eta}) - \frac{\partial \Phi}{\partial t}(\hat{t}, \hat{\eta}) + \langle A^* \partial \Phi(\hat{t}, \hat{\eta}), \hat{\eta} \rangle + \mu \tilde{g}'(\|\hat{\eta}\|) \|\hat{\eta}\| \right.$$

$$\left. + \frac{\varepsilon}{2} Tr(S\partial^2 (\Phi + g)(\hat{t}, \hat{\eta})) \right] \theta + \inf_{U \in \alpha} E \int_0^\theta \langle \partial(\Phi + g)(\hat{t}, \hat{\eta}), \beta(\hat{\eta}, U(s)) \rangle ds + o(\theta).$$

So, noting the evaluation (20) below, we can conclude that $v$ is a subsolution of (14),

$$\inf_{U\in\alpha} E\int_0^\theta G(U(s))ds \geq \int_0^\theta \inf_{u\in\gamma} G(u)ds = \inf_{u\in\gamma}\int_0^\theta G(u)ds \geq \inf_{U\in\alpha} E\int_0^\theta G(U(s))ds. \tag{20}$$

By the same arguments, we can show that $v$ is a supersolution. Hence Theorem 4 and the comparison theorem complete the proof. $\qquad\square$

**Theorem 6** $V^\varepsilon$ *is a unique viscosity solution of* (15) *and* (16).

**Proof** For the proof, we will only show that $V^\varepsilon$ is a subsolution. Suppose that $V^\varepsilon - \Phi - g$ has a global maximum at $(\hat{t},\hat{\eta})$ with $\hat{t}\in(0,T)$, namely

$$V^\varepsilon(t,\eta) \leq \Phi(t,\eta) - \Phi(\hat{t},\hat{\eta}) + g(\eta) - g(\hat{\eta}) + V^\varepsilon(\hat{t},\hat{\eta}). \tag{21}$$

Putting $v = v^\varepsilon$, $V = V^\varepsilon$, $\xi = \xi^\varepsilon(\hat{t},\hat{\eta},U)$, and $F(\theta,\xi(\theta)) = Q_\varepsilon\{\Phi(\hat{t}-\theta,\xi(\theta)) + g(\xi(\theta)) - \Phi(\hat{t},\hat{\eta}) - g(\hat{\eta}) + V(\hat{t},\hat{\eta})\} - v(\hat{t},\hat{\eta})$, we see from (21)

$$v(t-\theta,\xi(\theta)) \leq F(\theta,\xi(\theta)) + v(\hat{t},\hat{\eta}) \tag{22}$$

and

$$\lim_{\theta\to 0} F(\theta,\xi(\theta)) = 0 \quad \text{uniformly in} \quad U.$$

By the same argument as (19), we have

$$\begin{aligned}
EF(\theta,\xi(\theta)) \leq \frac{1}{\varepsilon}E\int_0^\theta \Big\{ &F(s,\xi(s)) + v(\hat{t},\hat{\eta})\Big\}\Big\{ -\frac{\partial\Phi}{\partial t}(\hat{t}-s,\xi(s)) \\
&+ \langle A^*\partial\Phi(\hat{t}-s,\xi(s)),\xi(s)\rangle + \mu\tilde{g}'(\|\xi(s)\|)\|\xi(s)\| \\
&+ \langle\partial(\Phi+g)(\hat{t}-s,\xi(s)),\beta(\xi(s),U(s))\rangle + \frac{\varepsilon}{2}Tr(S\partial^2(\Phi+g)(\hat{t}-s,\xi(s))) \\
&+ \frac{1}{2}\langle S\partial(\Phi+g)(\hat{t}-s,\xi(s)),\partial(\Phi+g)(\hat{t}-s,\xi(s))\rangle\Big\}ds.
\end{aligned}$$

Recalling (18) and (22), we get

$$\begin{aligned}
0 \leq \inf_{U\in\alpha} E\Big[&v(\hat{t}-\theta,\xi(\theta))\Big\{Q_\varepsilon\Big(\int_0^\theta h(\xi(s))ds\Big) - 1\Big\} + F(\theta,\xi(\theta))\Big] \\
= \frac{v(\hat{t},\hat{\eta})}{\epsilon}\Big\{&h(\hat{\eta}) - \frac{\partial\Phi}{\partial t}(\hat{t},\hat{\eta}) + \langle A^*\partial\Phi(\hat{t},\hat{\eta}),\hat{\eta}\rangle + \mu\tilde{g}'(\|\hat{\eta}\|)\|\hat{\eta}\| \\
&+ \frac{\varepsilon}{2}Tr[S(\partial^2(\Phi+g)(\hat{t},\hat{\eta}))] + \langle S(\partial(\Phi+g)(\hat{t},\hat{\eta})),\partial(\Phi+g)(\hat{t},\hat{\eta})\rangle \\
&+ \inf_{u\in\gamma}\langle\partial(\Phi+g)(\hat{t},\hat{\eta}),\beta(\hat{\eta},u)\rangle\Big\}\theta + o(\theta).
\end{aligned}$$

This concludes that $V$ is a subsolution of (15).

Using the Legendre transformation for $\langle S(\partial(\Phi+g)(\hat{t},\hat{\eta})), \partial(\Phi+g)(\hat{t},\hat{\eta})\rangle$, we can conclude that $V$ is a subsolution of (16). $\qquad\qquad\square$

## 4. Limit Function of $V^\varepsilon$

As $\varepsilon_n \to 0$, we can choose a subsequence $n_j$ such that $V_j \equiv V^{\varepsilon_{n_j}}$ converges uniformly on any bounded set of $[0,T] \times H$ by virtue of Theorem 3. Let us set $V = \lim V_j$. Then $V$ satisfies (11) and (12). So, $V$ is weakly continuous.

**Theorem 7** *$V$ is a unique viscosity solution of the following min-max equation*

$$0 = \frac{\partial V}{\partial t} - \langle A^*\partial V(t,\eta), \eta\rangle - \inf_{u\in\gamma}\langle \partial V(t,\eta), \beta(\eta,u)\rangle$$

$$- \sup_{\zeta\in H}(\langle \partial V(t,\eta), S\zeta\rangle - \frac{1}{2}\langle\zeta, S\zeta\rangle) - h(\eta), \quad 0 < t < T, \; \eta \in H \qquad (23)$$

*with initial condition $V(0,\eta) = 0$.*

**Proof** Suppose that $V - \Phi - g$ has a strictly global maximum at $(\hat{t}, \hat{\eta})$ with $\hat{t} \in (0,T)$. By Theorem 4, $V_j - \Phi - g$ has a global maximum point $(t_j, \eta_j)$ and there is a ball $\Delta = \{\|\zeta\| \leq q\}$ such that $\eta_j \in \Delta$, $j = 1, 2, \ldots$. Since $\Delta$ is compact in $|\,|_B$-topology, we can choose a convergent subsequence, say $\eta_{j_k} \to \eta^*$ in $|\,|_B$ and $t_{j_k} \to t^*$. Now we have

$$(V - \Phi - g)(\hat{t},\hat{\eta}) = \lim(V_{j_k} - \Phi - g)(\hat{t},\hat{\eta}) \leq \overline{\lim}(V_{j_k} - \Phi - g)(t_{j_k}, \eta_{j_k})$$

$$\leq (V - \Phi - g)(t^*, \eta^*) \leq (V - \Phi - g)(\hat{t},\hat{\eta}), \qquad (24)$$

because $V_{j_k} \to V$ uniformly on $[0,T] \times \Delta$, $\eta_{j_k} \to \eta^*$ weakly, $t_{j_k} \to t^*$ and $(\hat{t},\hat{\eta})$ is the unique maximum point of $V - \Phi - g$.

It follows from (24) that $(\hat{t},\hat{\eta}) = (t^*, \eta^*)$ and $\tilde{g}(\|\eta_{j_k}\|) \to \tilde{g}(\|\eta^*\|)$. Hence, $\eta_{j_k} \to \hat{\eta}$ in $H$, since $\tilde{g}$ is strictly increasing. Noting Theorem 6, we can conclude that $V$ is a subsolution of (23). By routine, we complete the proof. $\qquad\square$

Since any convergent subsequence of $V^{\varepsilon_n}$ has the same limit, which is a unique viscosity solution of (23), $V^{\varepsilon_n}$ itself converges to $V$ uniformly on any bounded set. Moreover, using the same arguments as in [4], we can prove the following theorem.

**Theorem 8** *As $\varepsilon \to 0$, $V^\varepsilon$ converges to $V$ uniformly on any bounded set.*

**Proof** Put $\Delta_n = \{\zeta \in H; \|\zeta\| \leq n\}$, $\|w\|_n = \sup\{|w(t,\zeta)|; (t,\zeta) \in [0,T] \times \Delta_n\}$ and $m(w) = \sum_{n=1}^\infty 2^{-n}\|w\|_n(1 + \|w\|_n)^{-1}$. Then $m$ provides the topology of uniform convergence on any bounded sets. On the other hand, we can choose $\varepsilon_n \to 0$ such that

$$\overline{\lim_{\varepsilon \to 0}}\ m(V^\varepsilon - V) = \lim_{n \to \infty} m(V^{\varepsilon_n} - V). \tag{25}$$

Since the right hand side of (25) is 0, we complete the proof. $\qquad\square$

The min-max equation (23) is related to differential game. The relationship between $V$ and the value function of game is the following. Let us set

$$\mathbf{Y} = \{Y : [0,T] \to H,\ \text{measurable and bounded image set } Y[0,T]\},$$
$$\mathbf{Z} = \{Z : [0,T] \to \gamma,\ \text{measurable}\},$$
$$\mathbf{M} = \{M : \mathbf{Z} \to \mathbf{Y},\ \text{non-anticipative}\},$$
$$\mathbf{N} = \{N : \mathbf{Y} \to \mathbf{Z},\ \text{non-anticipative}\},$$

where $M$ is said to be *non-anticipative* if $M(Z)(\cdot) = M(Z')(\cdot)$ on $[0,t]$, whenever $Z = Z'$ on $[0,t]$. $Y$ (respectively, $Z$) is called a *control for nature* (respectively, *player*) and $M$ (respectively, $N$) a *strategy for nature* (respectively, *player*). For a partition $\pi = \{0 = t_0 < t_1 < \cdots < t_p = T\}$, $Y$ is called $\pi$-*admissible* if $Y(t) = Y(t_j)$ for $t \in [t_j, t_{j+1})$ and $\mathbf{Y}_\pi =$ the set of $\pi$-admissible controls. $M$ is called $\pi$-*admissible* if $M \colon \mathbf{Z} \to \mathbf{Y}_\pi$ and $M(Z)(s)$ is independent of $Z$ for $s \in [0,t_1)$. $\mathbf{M}_\pi$ denotes the set of $\pi$-admissible strategies and $\mathbf{M}^* = \cup_\pi \mathbf{M}_\pi$. $\mathbf{Z}_\pi, \mathbf{N}_\pi$ and $\mathbf{N}^*$ are defined in the same way.

For $Y \in \mathbf{Y}$ and $Z \in \mathbf{Z}$, we consider the system $X$ in $H$ governed by

$$\frac{dX}{dt} = AX(t) + \beta(X(t), Z(t)) + SY(t), 0 < t < T, \tag{26}$$

with initial condition $X(0) = \eta$.

The equation (26) has a unique mild solution $X(\cdot, \eta, Y, Z)$ and a criterion $\mathcal{J}$ is given by

$$\mathcal{J}(t, \eta, Y, Z) = \int_0^t h(X(s, \eta, Y, Z)) - \frac{1}{2}\langle SY(s), Y(s)\rangle ds.$$

The *upper* (respectively, *lower*) *value of this game* $\overline{v}$ (respectively, $\underline{v}$) $: [0,T] \times H \to R^1$ is defined by

$$\overline{v}(t, \eta) = \inf_{N \in \mathbf{N}^*}\ \sup_{Y \in \mathbf{Y}}\ \mathcal{J}(t, \eta, Y, N(Y))$$

respectively,

$$\underline{v}(t, \eta) = \sup_{M \in \mathbf{M}^*}\ \inf_{Z \in \mathbf{Z}}\ \mathcal{J}(t, \eta, M(Z), Z).$$

240

If $\underline{\nu} = \overline{\nu}$, then we call it the *value of game*. Suppose that $\mathcal{J}(\cdot, \cdot, Y, Z)$ is continuous in $R^1 \times |\ |_B$-topology uniformly in $(Y, Z)$. Then both $\overline{\nu}$ and $\underline{\nu}$ are viscosity solutions of (23). Again noting uniqueness of viscosity solution, we see

$$\overline{\nu} = \underline{\nu} = V.$$

Namely, the limit function $V$ turns out to be the value of game.

## References


[1] Crandall, M. G. and Lions, P. L.: Hamilton-Jacobi equations in infinite dimensions. Part 4; *J. Funct. Anal.* **90** (1990) 237–283

[2] Fleming, W. H. and McEneaney, W. M.: Risk sensitive control and differential games; *Lecture Notes in Control and Infor. Sci.* **184** (1992) 185–197

[3] Itô, K.: *Foundations of Stochastic Differential Equations in Infinite Dimensional Spaces.* SIAM Reg. Conf. Series **47**, 1984

[4] Nisio, M.: On total risk aversion and differential games for controlled parabolic equations; *Proc. Jap. Acad.* **69** (1993) 119–124

[5] Rozovskii, B. L.: *Stochastic Evolution Systems.* Kluwer, 1990

[6] Świech, A.: Viscosity solutions of fully nonlinear partial differential equations with unbounded terms in infinite dimensions; *Ph.D. dissertation, Univ. of Calif.* (1993)



Makiko Nisio
Department of Mathematics
Osaka Elect. Comm. Univ.
Hatsu-cho, Neyagawa 572, JAPAN

L. D. PITT & R. S. ROBEVA

# On the sharp Markov property for the Whittle field in 2-dimensions

This paper presents a general discussion of the sharp Markov property for Gaussian fields on Euclidean space. A detailed discussion is given in the particular case of the two dimensional Whittle field. While our arguments are, in principle, completely general their application requires detailed information concerning the particular associated reproducing kernel Hilbert space. These calculations are only presented in the case of the two dimensional Whittle field, which is the simplest meaty example.

The general case will be distinguished notationaly from the Whittle field as follows: $\Psi = \{\psi(x) : x \in R^n\}$ will always denote a general stochastically continuous, separable, real valued and mean zero Gaussian field on $R^n$. The Whittle field on $R^2$ will be denoted with $\Phi = \{\phi(x) : x = (x_1, x_2) \in R^2\}$. $\Phi$ is a continuous, stationary solution of the elliptic equation

$$(I - \Delta)\phi(x_1, x_2) = \dot{W}(x_1, x_2) \tag{1}$$

with $\dot{W}$ representing the two dimensional Gaussian white noise. All fields will be defined on a complete probability space $(\Omega, \Sigma, P)$, and $N$ will denote the $\sigma$-field generated by the null sets.

We introduce the two customary families of sub $\sigma$-fields: For $A \subseteq R^n$ the *sharp field* $F_0(\Psi, A)$ is given by

$$F_0(\Psi, A) = \sigma\{\psi(x) : x \in A\} \cup N,$$

while the *germ field* $F(\Psi, A)$ is defined as

$$F(\Psi, A) = \bigcap F_0(\Psi, O),$$

where the intersection is over all open sets $O$ containing the closure $\bar{A}$ of $A$.

In the present context the fields $F(\Psi, A)$ are analogous to the right continuous filtrations familiar from the theory of Markov processes, and the question of when $F(\Psi, \Gamma) = F_0(\Psi, \Gamma)$ is of comparable interest to the analogous question there.

To motivate the following discussion consider a surface $\Gamma \subseteq R^n$ that divides $R^n$ into complementary open sets $D_+$ and $D_-$. We say that $(\Gamma, D_+, D_-)$ (or simply $\Gamma$) has the *sharp Markov property* relative to $\Psi$ provided that $F(\Psi, D_+)$ and

$F(\Psi, D_-)$ are conditionally independent given the sharp field $F_0(\Psi, \Gamma)$. If $F(\Psi, D_+)$ and $F(\Psi, D_-)$ are conditionally independent given the germ field $F(\Psi, \Gamma)$, we say that $\Gamma$ has the *germ-field Markov property* relative to $\Psi$. Since the field $F(\Psi, \Gamma)$ is contained in both $F(\Psi, D_+)$ and $F(\Psi, D_-)$ it follows that, modulo null sets, $F(\Psi, \Gamma)$ contains no proper sub $\sigma$-fields for which $F(\Psi, D_+)$ and $F(\Psi, D_-)$ are conditionally independent. Thus, since $F_0(\Psi, \Gamma) \subseteq F(\Psi, \Gamma)$, the sharp Markov property for $\Gamma$ relative to $\Psi$ would, of necessity, implies that

$$F_0(\Psi, \Gamma) = F(\Psi, \Gamma). \tag{2}$$

The Whittle field $\Phi$ is known to have the germ field Markov property for any bounded separating curve $\Gamma$ in $R^2$, (see [8], [10], or [12]) and hence, for any such $\Gamma$, (2) is equivalent to $\Gamma$ having the sharp Markov property for $\Phi$. When (2) holds is the focus point of this paper.

There are also non-Markovian settings where the issue of "right continuity" of the fields $F_0(\Psi, \Gamma)$ is of considerable interest. For example, special case solutions to the following general problem:

*Given a stochastically continuous field $\Psi$ and a closed set $\Gamma$, find conditions on $\Gamma$ (necessary and sufficient) for (2) to occur,*

include the Blumenthal $0 - 1$ law for Markov processes $\Psi$ when $\Gamma$ contains only the single element 0. Zero-one laws for non-Markov Gaussian processes and fields have been discussed in [7], [11, sec.2], and [16]. Section 3 of [10] contains a proof that (2) holds for bounded convex $\Gamma$ and certain stationary Gaussian fields.

Our results and arguments here turn on the reproducing kernel Hilbert space $H(\rho)$ (defined in section 1) associated with $\Psi$. For a closed set $\Gamma$ in $R^n$ we write $H_{00}(\rho, R^n \setminus \Gamma)$ to denote the closure of functions $u$ in $H(\rho)$ that vanish on some open neighborhood of $\Gamma$. For the two dimensional Whittle field the differential equation (1) implies that $H(\rho)$ is the classical second order Sobolev space $W^2(R^2)$ with squared norm

$$\|u\|_2^2 = \int_{R^2} \{u(x) - \Delta u(x)\}^2 dx.$$

Each $u$ in $W^2(R^2)$ is continuous and is shown to be differentiable off a set of logarithmic capacity 0. It thus is possible to speak of the gradient $\nabla u(x)$ as being well defined modulo sets of capacity 0. We will write $Cap_l(.)$ for logarithmic capacity.

Our principle general result is,

**Theorem 1**

(a) *For a general stochastically continuous Gaussian field* $\Psi = \{\psi(x) : x \in R^n\}$ *and a closed set* $\Gamma \subseteq R^n$, *(2) holds iff for each* $u \in H(\rho)$ *with* $u\mid_\Gamma = 0$ *it follows that* $u \in H_{00}(\rho, R^n \setminus \Gamma)$.

(b) *For the Whittle field* $\Phi = \{\phi(x) : x \in R^2\}$ *and a closed* $\Gamma \subseteq R^2$, *(2) holds iff for each* $u \in W^2(R^2)$ *with* $u\mid_\Gamma = 0$ *it follows that* $Cap_l(\Gamma \cap \{x : \nabla u(x) \neq 0\}) = 0$.

**Remark a** The notation $W_0^2(A)$ is used to denote the closure in $W^2(A)$ of smooth functions $f$ for which $supp(f) \subseteq A$. Thus, if $\Gamma$ is closed, $W_0^2(R^2 \setminus \Gamma) = H_{00}(\rho, R^2 \setminus \Gamma)$. The general results on spectral synthesis in Sobolev spaces of Hedberg [6] imply that $u \in W_0^2(R^2 \setminus \Gamma)$ iff $u\mid_\Gamma = 0$ and $\nabla u\mid_\Gamma = 0$ in the sense that the exceptional set of $x$ in $\Gamma$ for which $\nabla u(x) \neq 0$ must have logarithmic capacity 0. Thus part (b) follows directly from the identification $H_{00}(\Phi, R^2 \setminus \Gamma) = W_0^2(R^2 \setminus \Gamma)$ and part (a).

For Theorem 1 to be an effective tool for determining when $F_0(\Phi, \Gamma) = F(\Phi, \Gamma)$ holds, it is necessary to gain a detailed understanding of the sets $\{x : u(x) = 0, \nabla u(x) \neq 0\}$ for $u$ in $W^2(R^2)$. This is quite delicate because functions in $W^2(R^2)$ will not, in general, be continuously differentiable. Thus familiar tools from multi-variable calculus, such as the implicit function theorem, are not directly applicable.

We require the following definitions. For points $x \neq y \in R^2$ we let $l(x, y)$ denote the line in $R^2$ that contains $x$ and $y$. If $x$ is an accumulation point of a set $\Gamma \subseteq R^2$ and if $l$ is a line through $x$, we will say that $l$ is a *contingent of* $\Gamma$ *at* $x$ provided there is a sequence of points $y_n \neq x$ in $\Gamma$ that converges to $x$ with $\lim_{n \to \infty} l(x, y_n) = l$. We let $C(\Gamma, x)$ denote the set of all contingents to $\Gamma$ at $x$. For a subset $\Gamma$ of $R^2$ we say that $\Gamma$ has a tangent at $x \in \Gamma$ provided that $C(\Gamma, x)$ contains an unique line. We write $T(\Gamma)$ for the set of points on $\Gamma$ at which $\Gamma$ has a tangent. A discussion of contingents may be found in [13, ch.9].

**Theorem 2** *If* $Cap_l(T(\Gamma)) = 0$, *then* $F_0(\Phi, \Gamma) = F(\Phi, \Gamma)$.

**Theorem 3** *If* $C \subseteq R^2$ *is a twice continuously differentiable smooth curve with* $\Gamma \subseteq C$ *and* $Cap_l(\Gamma) > 0$, *then* $F_0(\Phi, \Gamma) \neq F(\Phi, \Gamma)$.

Theorems 2 and 3 are relatively elementary consequences of Theorem 1 and for most standard examples they are sufficient to decide if $F_0(\Phi, \Gamma) = F(\Phi, \Gamma)$ . In particular, by Theorem 2 we have

**Corollary 4** *If* $F$ *is the Whittle field and* $\Gamma$ *is the Sierpinski gasket, the Sierpinski carpet, or the von Koch Snow Flake then* $F_0(\Phi, \Gamma) = F(\Phi, \Gamma)$ *and* $\Gamma$ *has the sharp Markov property for* $\Phi$.

**Remark b**  Since a Brownian graph is not a bounded separating set in $R^2$, it was excluded from the statement of Corollary 4, but $F_0(\Phi, \Gamma) = F(\Phi, \Gamma)$ holds in this case as well, and with appropriate modifications in the statement it could be included in the statement of Corollary 4.

**Remark c**  The classical Cantor set on the line, when considered as a subset of the plane satisfies the conditions of Theorem 3; in particular $Cap_l(\Gamma) > 0$, and hence

**Corollary 5**  *If $\Phi$ is the Whittle field and $\Gamma$ is the Cantor set, then $F_0(\Phi, \Gamma) \neq F(\Phi, \Gamma)$.*

**Remark d**  The product of two Cantor sets has tangents nowhere and satisfies $Cap_l(\Gamma) > 0$ and hence

**Corollary 6**  *If $\Phi$ is the Whittle field and $C = \Gamma \times \Gamma \subseteq R^2$, where $\Gamma$ is the Cantor set, then $F_0(\Phi, C) = F(\Phi, C)$.*

**Corollary 7**  *If $\Phi$ is the Whittle field and $Cap_l(\Gamma) = 0$, then $F_0(\Phi, \Gamma) = F(\Phi, \Gamma)$.*

**Remark e**  The set $T(\Gamma)$ where $\Gamma$ has tangents can be contained in a countable union of rectifiable curves, [13, ch.9]. When this fact is compared with Theorem 3 it is reasonable, but incorrect, to think that the correct condition for $F_0(\Phi, \Gamma) \neq F(\Phi, \Gamma)$ might be that $Cap_l(T(\Gamma)) > 0$. In fact, the actual condition for $F_0(\Phi, \Gamma) \neq F(\Phi, \Gamma)$ requires more than $Cap_l(T(\Gamma)) > 0$ but less than the conditions of Theorem 3. Precise conditions on $\Gamma$ require a closer look at Theorem 1 and functions in $W^2(R^2)$. In a subsequent work, with more space to explore these technical questions, we will show that for $u$ in $W^2(R^2)$, the set $\{x : u(x) = 0 \text{ and } \nabla u(x) \neq 0\}$ must be approximately a $C^1$ curve whose derivative is Holder continuous with exponent $1/2$. In particular we will derive the following refinements of Theorems 2 and 3.

**Theorem 2′**  *Let $\Phi$ be the Whittle field on $R^2$ and let $\delta < 1/2$ be fixed. Then there is a curve $\Gamma$ of class $C^{1+\delta}$ with $F_0(\Phi, \Gamma) = F(\Phi, \Gamma)$.*

**Theorem 3′**  *Let $\Phi$ be the Whittle field on $R^2$ and let $\delta > 1/2$ be fixed. Then for each curve $C$ of class $C^{1+\delta}$, and each closed set $\Gamma \subseteq C$ with $Cap_l(\Gamma) > 0$ $F_0(\Phi, \Gamma) \neq F(\Phi, \Gamma)$.*

Finally, we state

**Theorem 8**  *Let $\Phi$ be the Whittle field on $R^2$. Then $F_0(\Phi, D) = F(\Phi, D)$ for each bounded connected open set $D$.*

However, we give an example due, in a different context, to Hedberg [5]:

**Example 9** There exists a disconnected bounded open set for which (2) fails.

**Remark f** Our primary reason for stating Theorem 8 and Example 9 is to clarify the particular definition of the germ field Markov property that we have chosen to work with. Our choice, that $F(\Psi, D_+)$ and $F(\Psi, D_-)$ are conditionally independent when conditioned on $F(\Psi, \Gamma)$, differs from the following alternative germ field Markov property: $F_0(\Psi, D_+)$ and $F_0(\Psi, D_-)$ are conditionally independent when conditioned on $F(\Psi, \Gamma)$. This is the definition used by Dalang and Walsh in [3, 4]. When the sets $D_+$ and $D_-$ are sufficiently regular (for example, if they are finitely connected and $\Psi = \Phi$) Theorem 8 would imply that $F_0(\Psi, D_+) = F(\Psi, D_+)$ and $F_0(\Psi, D_-) = F(\Psi, D_-)$. And, in this case, the two Markov properties will coincide. However, when $F_0(\Psi, D_+) \neq F(\Psi, D_+)$ or $F_0(\Psi, D_-) \neq F(\Psi, D_-)$ the properties will be distinct. Example 9 shows that for the Whittle field $\Phi$ they are distinct for general open sets.

We have chosen to stress this point in part because of the work [3] that presents a very complete discussion of the sharp Markov property for the Brownian sheet (see also the earlier works [2] and [17]) using the alternative germ field Markov property. Their work is closely related to the present work, but care must be taken when comparing results because of the two different Markov properties. (The example 3.5 in [3] shows that $F_0(\Psi, D) \neq F(\Psi, D)$ can also occur for the Brownian sheet when $D$ is disconnected.) To assist comparison with [3] we next present an alternative statement of one of the principle results in [3]. We let $N(\Gamma)$ denote the set of points on $\Gamma$ for which the contingent set $C(\Gamma, x)$ contains neither a vertical nor a horizontal line, and we let $\Lambda_1$ denote the linear Hausdorff measure in the plane.

**Theorem 10** (See [3, Theorem 6.1] and [2, Theorem 3.12]) *Let $B = \{B(\tau) : \tau = (s, t) \text{ with } s, t > 0\}$ denote the Brownian sheet defined on $R_+^2$ and let $\Gamma \subseteq R_+^2 \cup \{\infty\}$ be a Jordan curve with complementary open sets $D_-$ and $D_+$. Then given $F_0(B, \Gamma)$, $F_0(B, D_-)$ and $F_0(B, D_+)$ are conditionally independent iff*

$$\Lambda_1(N(\Gamma)) = 0.$$

**Remark g** Rather than the elliptic equation (1) satisfied by the Whittle field, the Brownian sheet $B$ satisfies the hyperbolic equation

$$\partial_s \partial_t B(s, t) = \dot{W}(s, t).$$

246

The characteristic directions of the operator $\partial_s \partial_t$ play a special role in describing the reproducing kernel space associated with $B$, and this fact is reflected by the special role played by the set $N(\Gamma)$ in Theorem 10.

The remainder of the paper is arranged as follows. In section 1 the reproducing kernel Hilbert space machinery is set up and Theorem 1 is proved. Proofs of Theorems 2 and 3 are given in Sections 2 and 3. Theorem 8 and Example 9 are discussed in Section 4.

## 1. Generalities and Notation

Let $\Psi = \{\psi(x) : x \in R^n\}$ be a stochastically continuous separable real valued random field defined over a complete probability space $(\Omega, \Sigma, P)$ and let $F_0(\Psi, A)$ and $F(\Psi, A)$ be defined as in section 0. By the stochastic continuity of $\Psi$, it follows that any set $A$ in $R^n$ with closure $\bar{A}$ satisfies the equality $F_0(\Psi, A) = F_0(\Psi, \bar{A})$. Thus $A \mapsto F_0(\Psi, A)$ is continuous from below. The condition for continuity from above is

$$F_0(\Psi, A) = F(\Psi, A) \tag{3}$$

and can be rephrased in terms of strong continuity of $L^2$ projection operators. In fact, if $\mathbf{E}_A$ is the orthogonal projection (conditional expectation operator) in $L^2(\Omega, \Sigma, P)$ onto the space of $F_0(\Psi, A)$ measurable random variables, then (3) is known to be equivalent to the condition

$$\text{strong} \lim_{O \downarrow \bar{A}} \mathbf{E}_O = \mathbf{E}_A. \tag{4}$$

When $\Psi$ is a mean zero Gaussian process this may be reduced further using standard Wiener-Hermite expansions. In fact, (4) is easily seen to be equivalent to

$$H_A(\Psi) = \bigcap_{O \supseteq \bar{A}} H_O(\Psi), \tag{5}$$

where $H_A(\Psi)$ is defined as the closed $L^2$ span of the variables $\{\psi(x) : x \in A\}$. Letting $H(\Psi) = H_{R^n}(\Psi)$, and taking orthogonal complements, condition (5) can be rephrased as the

**Criterion 1** *Let $\Psi$ be Gaussian and $A \subseteq R^n$ be bounded and closed. Then (3) holds iff for each $X \in H(\Psi)$ which satisfies $\mathbf{E}X\psi(x) = 0$ for all $x$ in $A$, it follows that there exists a sequence $X_n$ in $H(\Psi)$ and a sequence $O_n$ of neighborhoods of $A$ with $\mathbf{E}X_n\psi(x) = 0$ for all $x \in O_n$, and $\lim_{n \to \infty} X_n = X$ in $H(\Psi)$.*

We next translate the Criterion 1 into the language of reproducing kernel Hilbert spaces. Let $\rho(x, y) = \mathbf{E}\psi(x)\psi(y)$ be the covariance kernel of $\Psi$. The reproducing kernel Hilbert space $H = H(\rho)$ associated with $\rho$ is the completion of the space of finite linear combinations of the functions $\rho(x,.)$ with the inner product $< .,. >$ that is given by $< \rho(x,.), \rho(y,.) > = \rho(x, y)$. The correspondence $\psi(x) \leftrightarrow \rho(x,.)$ determines a linear isometry between the two spaces $H(\Psi)$ and $H(\rho)$ that takes the form $X \leftrightarrow \mathbf{E}X\psi(x)$, and maps the spaces $H_A(\Psi)$ isometrically onto spaces $H_A(\rho)$ that are defined as:

$$H_A(\rho) = closed\ span\{\rho(x,.) : x \in A\},$$

and

$$\bar{H}_A(\rho) = \bigcap_{O \supseteq \bar{A}} H_O(\rho).$$

Functions $u$ in $H(\rho)$ satisfy the reproducing equation

$$u(x) = < u, \rho(x,.) >$$

and this gives an easy formal identification of the orthogonal complements of $H_A(\rho)$ and $\bar{H}_A(\rho)$. Namely,

$$H(\rho) \ominus H_A(\rho) \stackrel{\text{def}}{=} H_0(\rho, R^n \setminus A) = \{u \in H(\rho) : u(x) = 0 \quad \text{for } x \text{ in } A\},$$

and

$$H(\rho) \ominus \bar{H}_A(\rho) \stackrel{\text{def}}{=} H_{00}(\rho, R^n \setminus A) = \overline{\bigcup\{H_0(\rho, R^n \setminus O) : O \supseteq A\}}.$$

The direct translation of Criterion 1 into the language of $H(\rho)$ and the spaces $H(\rho) \ominus H_A(\rho)$ is:

**Criterion 2**  *Let $\Psi$ be Gaussian with covariance function $\rho$. For closed $\Gamma \subseteq R^n$, (2) holds iff*

$$H_0(\rho, R^n \setminus \Gamma) = H_{00}(\rho, R^n \setminus \Gamma).$$

Criterion 2 is equivalent to part (a) of Theorem 1. For part (b) we need to identify the reproducing kernel space for $\Phi$, and this may be done using Fourier transforms. If $\rho(x, y) = \mathbf{E}\phi(x)\phi(y)$, then it follows from (1) (see e.g. [18] that $\rho$ has the representation

$$\rho(x,y) = \int_{R^2} e^{i<x-y,\lambda>}(1+|\lambda|^2)^{-2}d\lambda,$$

and that there is a unique correspondence $u \leftrightarrow f$ between $u$ in $H(\rho)$ and $f$ satisfying $\int_{R^2}|f(\lambda)|^2(1+|\lambda|^2)^{-2}d\lambda < \infty$, that is given by

$$u(x) = \int_{R^2} e^{i<x,\lambda>}f(\lambda)(1+|\lambda|^2)^{-2}d\lambda. \tag{6}$$

Setting $F(\lambda) = f(\lambda)(1+|\lambda|^2)^{-2}$, we see that $u$ is the Fourier transform of $F$, and that $F$ satisfies

$$\int_{R^2}|F(\lambda)|^2(1+|\lambda|^2)^2d\lambda < \infty.$$

Thus $H(\rho)$ is exactly the space of Bessel potential representation of the Sobolev space $W^2(R^2)$, (see [14, sec. V.3]) with the squared norm

$$\|u\|_2^2 = \int_{R^2}\{u(x) - \Delta u(x)\}^2 dx.$$

The condition on $\Gamma$ from Theorem 1(a), that

*each $u \in H(\rho)$ for which $u\mid_\Gamma = 0$ satisfies $u \in H_{00}(\rho, R^n \setminus \Gamma)$,*

when restated for the Whittle field thus becomes

*each $u \in W^2(R^2)$ for which $u\mid_\Gamma = 0$ satisfies $u \in W_0^2(R^2 \setminus \Gamma)$.*

Finally, to complete the proof of Theorem 1 we require Hedberg's characterization of $W_0^2(R^2 \setminus \Gamma)$ given in terms of the Bessel capacity

$$B_{1,2}(A) = \inf\{\|u\|_{1,2}^2 : u(x) \geq 1 \quad \text{for } x \text{ in } A\},$$

where

$$\|u\|_{1,2}^2 = \int_{R^2}(|u(x)|^2 + |\nabla u(x)|^2)dx,$$

is the Sobolev norm. It is known [19, Chap. 3] that

(i) the $L^2$ gradient $\nabla u(x)$ exists except for a set $E$ of $x$ with $B_{1,2}(E) = 0$;

(ii) $B_{1,2}(A)$ is equivalent to the classical logarithmic capacity $Cap_l(A)$.

(i) and (ii) may be used to interpret $\nabla u\mid_\Gamma = 0$ as $Cap_l(\Gamma \cap \{x : \nabla u(x) \neq 0\}) = 0$, and with this interpretation we can state Hedberg's Theorem 1.1 [6] that $u \in W_0^2(R^2 \setminus \Gamma)$ iff $u\mid_\Gamma = 0$ and $\nabla u\mid_\Gamma = 0$. Part (b) follows directly.

## 2. Proofs of Theorems 2 and 3

**Proof of Theorem 2** Suppose that $x_0 \in \Gamma \setminus T(\Gamma)$, and that $u$ vanishes on $\Gamma$ and is differentiable at $x_0$. We set $v = \nabla u(x_0)$ and will show that $v = 0$. Because of the assumed differentiability

$$u(y) = u(x_0) + <y - x_0, v> + o(|y - x_0|) \qquad \text{as } |y - x_0| \to 0,$$

and hence, if $y \in \Gamma$ approaches $x_0$, the unit vectors $(y - x_0)/|y - x_0|$ will be asymptotically orthogonal to $v$. Since $\Gamma$ does not have a tangent at $x_0$, the set of cluster points of the vectors $(y - x_0)/|y - x_0|$ must contain two linearly independent vectors, and these must both be orthogonal to $v$, which implies that $v = 0$.

It thus suffices to show that each u in $W^2(R^2)$ is differentiable except on an exceptional set $E$ with $Cap_l(E) = 0$. When the classical derivatives are replaced with $L^p$ derivatives this is a result of Bagby and Ziemer [1]. Our extension to the present setting uses the following elementary result.

**Lemma 2.1** $W^2(R^2) \subseteq \lambda^*(R^2)$, where $\lambda^*(R^2)$ is the Zygmund space of bounded smooth functions u which satisfy the condition

$$\|u(x + h) + u(x - h) - 2u(x)\|_\infty = o(|h|) \quad \text{as } h \to 0. \tag{7}$$

**Proof** Each $u \in W^2(R^2)$ has a Fourier representation of the form (6) with $f$ satisfying $\int_{R^2} |f(\lambda)|^2 (1 + |\lambda|^2)^{-2} d\lambda < \infty$. We write

$$u(x + h) + u(x - h) - 2u(x)$$
$$= 2 \int_{R^2} \left(\cos <h, \lambda> -1\right) e^{i<x,\lambda>} f(\lambda) \frac{1}{(1 + |\lambda|^2)^2} d\lambda,$$

and break this integral into pieces as $I_1 + I_2$ where

$$I_1 = 2 \int_{\{|\lambda||h|<1\}} \left(\cos <h, \lambda> -1\right) e^{i<x,\lambda>} \frac{f(\lambda)}{(1 + |\lambda|^2)^2} d\lambda$$
$$= O\left(|h|^2 \int_{\{|\lambda||h|<1\}} |\lambda|^2 \frac{|f(\lambda)|}{(1 + |\lambda|^2)^2} d\lambda\right).$$

Choosing a function $g(\lambda) > 1$, with $g(\lambda) \to \infty$ as $|\lambda| \to \infty$ and satisfying

$$\int_{R^2} |f(\lambda) g(\lambda)|^2 \frac{1}{(1 + |\lambda|^2)^2} d\lambda < \infty,$$

we have by Schwarz's inequality,

$$(I_1)^2 = O\Big(|h|^4 \int_{\{|\lambda||h|<1\}} |\lambda|^4 \frac{1}{|g(\lambda)|^2(1+|\lambda|^2)^2} d\lambda$$
$$\times \int_{\{|\lambda||h|<1\}} |f(\lambda)g(\lambda)|^2 \frac{1}{(1+|\lambda|^2)^2} d\lambda\Big) = o(|h|^2).$$

For $I_2$ we have

$$(I_2)^2 = O\Big(\big(\int_{\{|\lambda||h|\geq 1\}} |f(\lambda)| \frac{1}{(1+|\lambda|^2)^2} d\lambda\big)^2\Big)$$
$$= O\Big(\int_{\{|\lambda||h|\geq 1\}} |f(\lambda)|^2 \frac{1}{(1+|\lambda|^2)^2} d\lambda \int_{\{|\lambda||h|\geq 1\}} \frac{1}{(1+|\lambda|^2)^2} d\lambda\Big)$$
$$= o(|h|^2)$$

We now invoke the result of Bagby and Ziemer [1] to the effect that there is a set $E$ with $Cap_l(E) = 0$, and such that for each $x$ not in $E$ there is a linear polynomial $P_x(y)$ with

$$\Big(r^{-2} \int_{\{|y|<r\}} |u(x+y) - P_x(y)|^2 dy\Big)^{1/2} = o(r) \quad \text{as } r \to 0.$$

By the defining condition (7) of $\lambda^*(R^2)$, $u \in \lambda^*(R^2)$ implies that if $T_{r,x}(y)$ is the linear function that interpolates the three values: $u(x)$ at $y = 0$, $u(x + (r,0))$ at $y = (r,0)$, and $u(x + (0,r))$ at $y = (0,r)$, then

$$|u(x+y) - T_{r,x}(y)| = o(r) \quad \text{uniformly in } |y| < r \text{ as } r \to 0. \tag{8}$$

This, in turn, gives

$$\Big(r^{-2} \int_{\{|y|<r\}} |u(x+y) - T_{x,r}(y)|^2 dy\Big)^{1/2} = o(r) \quad \text{as } r \to 0.$$

Thus

$$\Big(r^{-2} \int_{\{|y|<r\}} |P_x(y) - T_{r,x}|^2 dy\Big)^{1/2} = o(r) \quad \text{as } r \to 0. \tag{9}$$

but $P_x(y)$ and $T_{r,x(y)}$ are linear functions that agree at $y = 0$, and this allows the exact calculation (9) as $|w_r|(\pi/4)^{1/2}r$ , where $w_r = \nabla P_x(0) - \nabla T_{r,x}(0)$. So (9) implies that $|w_r| \to 0$ as $r \to 0$, or that

$$\lim_{r\to 0} T_{r,x}(y) = P_x(y).$$

Together with (8) this shows that $u$ is differentiable at $x$.

**Proof of Theorem 3** If $C$ is a $C^2$ curve and if $\Gamma \subseteq C$ satisfies $Cap_l(\Gamma) > 0$ there will exist an $x$ in $\Gamma$ so that $Cap_l(A \cap \Gamma) > 0$ for each open subarc $A$ of $C$ that contains $x$. If $A$ is short enough, it is possible to find a $W^2(R^2)$ function $u(x_1, x_2)$ that is defined in a neighborhood of $A$, and which satisfies $u\big|_A = 0$, but $\nabla u(x_1, x_2) \neq 0$ for all $(x_1, x_2)$ in $A$. By multiplying $u$ by a $C^2$ function $h$ with compact support satisfying $h(x) = 1$ near $A$ we obtain a function in $W_2(R^2)$ that vanishes on $\Gamma$, but whose gradient does not vanish on $\Gamma$. By Theorem 1(b) we see that $F_0(\Phi, \Gamma) \neq F(\Phi, \Gamma)$.

## 3. Proof of Theorem 8 and Example 9

**Proof of Theorem 8** By Hedberg's Theorem 1.1 [6] it suffices to show that for each $u$ in $W^2(R^2)$ which vanishes identically on $D$, will satisfy $\nabla u(x) = 0$ on $D$ except for a set of $x$ of capacity 0. But $D$ is open so $\nabla u(x) = 0$ on $D$, and off a set of capacity zero, $\nabla u(x)$ is finely continuous. Thus it suffices to show that $D$ is thick at each point in $\Gamma = \bar{D} \setminus D$, see [6]. But it follows from the Beurling criterion , (see e.g. [15, p.105]) for regular points of the Dirichlet problem, that each point $x$ in the boundary of a connected domain in $R^2$ is a regular point for the Dirichlet problem on the complementary open set. This is simply a restatement that $D$ is thick at $x$, and completes the proof.

Example 9 is due to Hedberg [5, Th 3.14] and is a close cousin of Example 3.5 in [3]. For this example, it is necessary to construct an open set $D$ that is not thick in its boundary $\Gamma$. This is done by starting with a sequence of closed disjoint disks $\{B(x_n, R_n)\}_{n=1}^{\infty}$ with centers $\{x_n\}$ which lie on the real line. We assume that $\cup_{n=1}^{\infty} B(x_n, R_n)$ is dense in the real line while $\sum_{n=1}^{\infty} R_n < \infty$. The set $D$ is

$$D = \cup_{n=1}^{\infty} B(x_n, r_n) \text{ where } 0 < r_n < R_n \text{ for all } n \geq 1, \text{ and } \sum_{n=1}^{\infty} \left(\log \frac{R_n}{r_n}\right)^{-1} < \infty.$$

A function $u(x)$ in $W^2(R^2)$ is now constructed that vanishes on the boundary $\Gamma$ of $D$, but for which $\nabla u(x)$ does not vanish on $\Gamma$. This begins with a sequence $\{v_n(r)\}_{n=1}^{\infty}$ of $C^2$ functions of $r \geq 0$, which satisfy

$$0 \leq v_n(r) \leq 1 \quad \text{for all } r, \ v_n(r) = 1 \text{ for } r < r_n, \text{ and } v_n(r) = 0 \text{ for } r > R_n .$$

The $\{v_n\}_{n=1}^{\infty}$ can be chosen so as to satisfy the conditions

$$|v_n{}'(r)| \leq \frac{const}{r} \left(\log \frac{R_n}{r_n}\right)^{-1} ; \quad |v_n{}''(r)| \leq \frac{const}{r^2} \left(\log \frac{R_n}{r_n}\right)^{-1} .$$

252

The desired function is

$$u(x) = g(x_1, x_2) \frac{x_2}{(1 + x_1^2 + x_2^2)^3},$$

where $g(x) = 1 - \sum_{n=1}^{\infty} v_n(|x - x_n|)$. A calculation shows that $u$ is in $W^2(R^2)$ and vanishes on $D$. Moreover, $u(x)$ agrees with $\dfrac{x_2}{(1 + x_1^2 + x_2^2)^3}$ off $\cup_{n=1}^{\infty} B(x_n, R_n)$ and since $\partial_{x_2} u(x)$ is finely continuous except at a set of capacity 0, $\partial_{x_2} u(x) = 1$ on that part of $\Gamma$ that is not on the boundary of one of the circles $B(x_n, R_n)$. But each real $x$ that is not in $\cup_{n=1}^{\infty} B(x_n, R_n)$ is in $\Gamma$. Since $\sum_{n=1}^{\infty} R_n < \infty$ it is clear that $\nabla u(x) \neq 0$ on $\Gamma$ and by Hedberg's Theorem that $u$ is not in $W_0^2(R^2 \setminus D)$, so $F_0(\Phi, D) \neq F(\Phi, D)$.

**Acknowledgements** Research supported in part by Office of Naval Research Contract No. N00014-90-J-1639.

# References


[1] Bagby, T. and Ziemer, W.: Pointwise differentiability and absolute continuity; *Trans. Amer. Math. Soc.* **191** (1974) 129–148

[2] Dalang, R. C. and Russo, F.: A prediction problem for the Brownian sheet; *J. Multivariate Anal.* **26** (1988) 16–47

[3] Dalang, R. C. and Walsh, J. B.: The sharp Markov property of the Brownian sheet and related processes; *Acta Math.* **168** (1992) 153–218

[4] Dalang, R. C. and Walsh, J.B.: The sharp Markov property of Levy sheets; *Ann. Probab.* **20** (1992) 591–628

[5] Hedberg, L. I.: Spectral synthesis and stability in Sobolev spaces; in:*Euclidean Harmonic Analysis, Lecture Notes in Math.* **779**, Springer-Verlag, New York (1980)

[6] Hedberg, L. I.: Spectral synthesis in Sobolev spaces, and uniqueness of solutions of the Dirichlet problem; *Acta Math.* **147** (1981) 237–264

[7] McKean, H.P.: Brownian motion with a several dimensional time; *Theor. Probability Appl.* **8** (1963) 335–354

[8] Pitt, L. D.: A Markov property for Gaussian processes with a multidimensional parameter; *Arch. Rat. Mech. Anal.* **43** (1971) 367–391

[9] Pitt L. D.: Some problems in the spectral theory of stationary processes on $R^d$;*Indiana Univ. Math. J.* **23** (1973) 343–365

[10] Pitt L. D.: Stationary Gaussian Markov fields on $R^d$ with a deterministic component; *J. Multivariate Anal.* **5** (1975) 300–311

[11] Pitt, L. D. and Tran, L. T.: Local sample path properties of Gaussian fields; *Ann. Probab.* **7** (1979) 477–493

[12] Rozanov, Yu. A.: *Markov Random Fields.* Springer-Verlag, New York, 1982

[13] Saks, S.: *Theory of the Integral.* (2nd ed.) Monografie Matematyczne, Vol vii., Warszawa, 1937

[14] Stein, E. M.: *Singular Integrals and Differentiability Properties of Functions.* Princeton University Press, Princeton, 1970

[15] Tsuji, M.: *Potential Theory in Modern Function Theory* (2nd ed.), Chelsea, New York, 1975

[16] Tutubalin, V. N. and Friedlin, M. I.: On the structure of the infinitesimal $\sigma$-algebra of a Gaussian process; *Theor. Probability Appl.* **7** (1962) 196–199

[17] Walsh, J. B.: Cours de troisieme cycle; *Universite de Paris* **6** (1976-77)

[18] Whittle, P.: On stationary processes in the plane; *Biometrika* **41** (1954) 434–449

[19] Ziemer, W. P.: *Weakly Differentiable Functions.* Springer-Verlag, New York, 1989

Loren D. Pitt
Department of Mathematics
University of Virginia
Charlottesville, VA 22903, USA
*E-mail*: ldp@feller.math.virginia.edu

Raina S. Robeva
Department of Mathematics
University of Virginia
Charlottesville, VA 22903, USA
*E-mail*: rsr5k@virginia.edu

M. REDFERN

# Stochastic integration via white noise and the fundamental theorem of calculus

In [3] the author and Betounes defined the stochastic integral $\int_a^b X(t) \circ \dot{B}(t)dt$ as a generalized white noise functional. We wanted to formulate the Stratonovich integral in the white noise setting, independently of the dimension of the time parameter, and then use this to study the analogous idea in the multiparameter case. While we have proved that in some cases $\int_a^b X(t) \circ \dot{B}(t)dt$ and the Stratonovich integral $\int_a^b X(t) \circ dB(t)$ are the same, we do not as yet have a general theorem. Our integral does, however, exhibit some of the same properties. For instance, if $p(x)$ is a polynomial and $B(t) = I_1(1_{[0,t)}), t > 0$, in $L^2(\mathcal{S}^*)$, it is easily seen that

$$\int_a^b p(B(t)) \circ \dot{B}(t)dt = q(B(b)) - q(B(a)),$$

where $q(x) = \int p(x)dx$. The setup here is more general than what one usually finds in white noise analysis, see [4, 5]. However, our work has its origins in the standard theory.

In this paper we consider the problem of finding a larger class of functions $F : \mathbf{R} \to \mathbf{R}$ for which

$$\int_a^b F'(B(t)) \circ \dot{B}(t)dt = F(B(b)) - F(B(a)).$$

In Section 1 we review the basic ideas needed to understand the definition of $\int_a^b X(t) \circ \dot{B}(t)dt$ and describe some of its properties obtained in [3]. In Section 2 we present our results.

## 1. Preliminaries

The setting for this work is the space $(\mathcal{D}^*)$ of Wiener distributions. This is a space of generalized functionals containing $(L^2) \equiv L^2(\mathcal{S}^*)$ where $\mathcal{S}^*$ is the space of tempered distributions endowed with Gaussian measure $\mu$ given by Minlos' theorem. We will indicate the basic structure of this space but for more details see [2].

The space $(L^2)$ is isomorphic to the Fock space $\hat{L}^2 = \oplus_{n=0}^\infty \sqrt{n!}\hat{L}_n^2$ and we denote this isomorphism: $I : \hat{L}^2 \to (L^2)$. Here $L_n^2 \equiv L^2(\mathbf{R}^n)$ and $\hat{\ }$ indicates symmetrization.

For $f \in L_n^2$, $I_n(f) \equiv I(f)$ indicates the n-tuple Wiener integral of $f$. Densely embedded in $L_n^2$ is the space $\mathcal{D}_n = C_c^\infty(\mathbf{R}^n)$ and by restriction of $I_n$ to $\hat{\mathcal{D}}_n$ and extension to $\hat{\mathcal{D}}_n^*$ by duality we get the spaces $(\mathcal{D}_n) \equiv I_n(\mathcal{D}_n)$ and $(\mathcal{D}_n^*) \equiv I_n(\mathcal{D}_n^*)$. Letting $\mathcal{D} = \coprod_{n=0}^\infty \mathcal{D}_n$ with the inductive limit topology we have that $\mathcal{D}^* = \prod_{n=0}^\infty \mathcal{D}_n^*$. We call $(\mathcal{D}) \equiv \coprod_{n=0}^\infty (\mathcal{D}_n)$ the space of Wiener test functionals and $(\mathcal{D}^*) \equiv \prod_{n=0}^\infty (\mathcal{D}_n^*)$ is then the space of Wiener distributions. The pairing is the natural one induced by the isomorphism $I$. There is a natural product $\hat{\otimes}$ on $(\mathcal{D}^*)$ which generalizes the Wick product on $L^2(\mathcal{S}^*)$. More precisely, for $X = \sum_n^\infty I_n(T_n)$ and $Y = \sum_{n=0}^\infty I_n(U_n)$ in $(\mathcal{D}^*)$,

$$X \hat{\otimes} Y = \sum_{n=0}^\infty \sum_{r=0}^n I_n(T_r \hat{\otimes} U_{n-r}).$$

Here $T_r \hat{\otimes} U_{n-r}$ is the symmetric tensor product of the distributions $T_r$ and $U_{n-r}$.

Given $X : \mathbf{R} \to (\mathcal{D}^*)$, the derivative $\frac{dX(t)}{dt}$ and the integral $\int X(t)dt$ are elements of $(\mathcal{D}^*)$ which satisfy

$$\left\langle \frac{dX(t)}{dt}, \Phi \right\rangle = \frac{d}{dt}\langle X(t), \Phi \rangle, \qquad \left\langle \int X(t)dt, \Phi \right\rangle = \int \langle X(t), \Phi \rangle dt$$

for all $\Phi$ in $(\mathcal{D})$. For example, for $t > 0$, $B(t) = I_1(1_{[0,t)})$ is a Brownian motion with $\dot{B}(t) \equiv \frac{dB}{dt}(t) = I_1(\delta_t)$. For a process $X(t)$ which is Skorokhod integrable, its Skorokhod integral can be expressed as $\int_a^b X(t)\hat{\otimes}\dot{B}(t)dt$. Another way to view the Skorokhod integral is as

$$I \int_a^b I^{-1}(X(t))\hat{\otimes}\delta_t dt$$

with this integral being understood as a Pettis integral.

The operators which we use to define our stochastic integral $\int_a^b X(t) \circ \dot{B}(t)dt$ are the renormalization operator $R$ and the trace operators $T_k, k = 1, 2, \ldots$. They were defined in [1] and further studied in [3]. These operators are initially defined on $(L_1^2)^{\otimes_a n}$, the n-fold algebraic tensor product of $L_1^2$, and then extended by continuity to $L_{n,\pi}^2$ which is the completion of $(L_1^2)^{\otimes_a n}$ with respect to the projective tensor product norm, $\| \cdot \|_{n,\pi}$, see [6]. By restriction we get $R$ and $T_k$ defined and continuous on $\hat{L}_{n,\pi}^2$. $R$ is then extended to $\coprod_{n=0}^\infty L_{n,\pi}^2$ by defining $R(\sum_{n=0}^\infty f_n) = \sum_{n=0}^\infty R(f_n)$ where $R(f_0) = f_0$ and $R(f_1) = f_1$ for $f_0 \in L_{0,\pi}^2 \equiv \mathbf{R}$ and $f_1 \in L_{1,\pi}^2 \equiv L_1^2$. The renormalization operator $R$ has an inverse and $\overset{\circ}{I}(f_n) \equiv IR^{-1}(f_n)$ defines the multiple Stratonovich integral of $f_n \in \hat{L}_{n,\pi}^2$. $\overset{\circ}{I}$ extends to $\coprod_{n=0}^\infty \hat{L}_{n,\pi}^2$ componentwise.

For $\mathcal{A} = [a, b]$ and $\alpha > 0$, let

$$\mathcal{M}_\alpha \equiv \{f = \sum_{n=0}^{\infty} f_n \in \prod_{n=0}^{\infty} L^2(\mathcal{A}, \hat{L}^2_{n,\pi}) \mid \sum_{n=0}^{\infty} \alpha^n n! \|f_n\|^2_{n,\pi,\mathcal{A}} < \infty\},$$

where $\|f_n\|^2_{n,\pi,\mathcal{A}} \equiv \int_a^b \|f_n(t)\|^2_{n,\pi} dt$. If $X(t) = I(f(t))$ where $f \in \mathcal{M}_\alpha, \alpha > 2$, the integral $\int_a^b X(t) \circ \dot{B}(t) dt$ is defined by

$$\int_a^b X(t) \circ \dot{B}(t) dt = I(\mathcal{S}(f)). \tag{1}$$

Here the operator $\mathcal{S} : \mathcal{M}_\alpha \to \prod_{n=0}^{\infty} L^2_{n,\pi}$ is given by

$$\mathcal{S}(f) = \mathcal{S}(\sum_{k=0}^{\infty}(f_n)) = \sum_{n=1}^{\infty} \left( \sum_{r=0}^{\infty} \sum_{k=0}^{r} (-1)^k T_{r-k} \int_a^b T_k(f_{2r+n-1}(t)) \hat{\otimes} \delta_t dt \right).$$

For $f_n \in L^2(\mathcal{A}, \hat{L}^2_{n,\pi})$, we have shown that

$$\mathcal{S}(f_n) = R^{-1} \left( \int_a^b R(f_n(t)) \hat{\otimes} \delta_t dt \right)$$

and that for $f \in \coprod_{n=0}^{\infty} L^2(\mathcal{A}, \hat{L}^2_{n,\pi})$ with $X = I(f) = I(\sum_{n=0}^{\infty} f_n)$,

$$\int_a^b X(t) \circ \dot{B} dt = \overset{\circ}{I} \int_a^b \overset{\circ}{I}^{-1}(X(t)) \hat{\otimes} \delta_t dt = \sum_{n=0}^{\infty} I(\mathcal{S}(f_n)). \tag{2}$$

For $c > 6$ and $f \in \mathcal{M}_c^2$ :

$$\mathcal{M}_c^2 \equiv \left\{ f = \sum_{n=0}^{\infty} f_n \in \prod_{n=0}^{\infty} L^2(\mathcal{A}, \hat{L}^2_{n,\pi}) \mid \sum_{n=0}^{\infty} c^n(n+1)! \| f_n \|^2_{n,\pi,\mathcal{A}} < \infty \right\},$$

and $X = \sum_{n=0}^{\infty} I_n(f_n)$, $\int_a^b X(t) \circ \dot{B}(t) dt$ is actually an element of $(L^2)$.

The definition in (1) is natural in that it allows us to view $\int_a^b X(t) \circ \dot{B}(t) dt$ in terms of multiple Stratonovich integrals via (2). There is a simplification of (1) which provides an alternate view. If $f \in L^2(\mathcal{A}, \hat{L}^2_{n,\pi})$, then $\mathcal{T} f$ is defined by

$$\mathcal{T} f = T_1 \left( \int_a^b f(t) \hat{\otimes} \delta_t dt \right) - \int_a^b T_1(f(t)) \hat{\otimes} \delta_t dt.$$

For an elementary function $f = \sum_{n=0}^{N} 1_{A_i}\tau_i$ with $\{A_i\}_{i=1}^{N}$ being a partition of $[a,b]$ and $\tau_i \in (L_1^2)^{\hat{\otimes}_a n}$, $\mathcal{T}f$ is given by

$$\mathcal{T}f = n\sum_i a(1_{A_i})\tau_i,$$

where $a(1_{A_i})$ is the lowering operator. For $X(t) = \sum_{n=0}^{\infty} I_n(f_n)$ with $f = \sum_{n=0}^{\infty} f_n$ in $\mathcal{M}_\alpha, \alpha > 2$, we have shown that

$$\int_a^b X(t) \circ \dot{B}(t)dt = \int_a^b X(t)\hat{\otimes}\dot{B}(t)dt + \sum_{n=1}^{\infty} I_{n-1}(\mathcal{T}f_n). \tag{3}$$

## 2. The Fundamental Theorem

In order to evaluate $\int_a^b F'(B(t)) \circ \dot{B}(t)dt$, where $F : \mathbf{R} \to \mathbf{R}$ is a Borel measurable function, we need to understand $F(B(t))$ and $F'(B(t))$ as elements of $(\mathcal{D}^*)$. For $t > 0$, let $g_t(x) = \exp(-x^2/(2t))/\sqrt{2\pi t}$. Assume

$$c_n(t) \equiv \langle F^{(n)}, g_t \rangle \equiv \langle F, (-1)^n g_t^{(n)} \rangle$$

exists for $n = 0, 1, \ldots$. Then $F(B(t))$, as a measurable function, defines the following Wiener distribution:

$$F(B(t)) = \sum_{n=0}^{\infty} \frac{c_n(t)}{n!} B(t)^{\hat{\otimes}n} \tag{4}$$

(see [2, Theorem 2]). For $t \leq 0$, let $B(t) = I_1(1_{[t,0)})$. If $F$ is analytic on $(-\infty, \infty)$ and $F(x) = \sum_{n=0}^{\infty} \frac{a_n}{n!} x^n$ where $a_n = F^{(n)}(0)$, then pointwise in $\mathcal{S}^*$,

$$F(B(t)) = \sum_{n=0}^{\infty} \frac{a_n}{n!} B(t)^n. \tag{5}$$

If $F(B(t))$ is an element of $(L^2)$, then one can show that its Wiener-Ito decomposition is

$$F(B(t)) = \sum_{n=0}^{\infty} \frac{1}{n!} \left( \sum_{k=0}^{\infty} \frac{a_{n+2k}}{k!2^k} t^k \right) B(t)^{\hat{\otimes}n}$$

and also that for $t > 0$,

$$\sum_{k=0}^{\infty} \frac{a_{n+2k}}{k!2^k} t^k = c_n(t).$$

We will consider formulations (4) and (5) in order to highlight some of the features of stochastic integration in our framework .

First we will look at $F(B(t))$ as defined by (5). In this case we will see that $\int_a^b F'(B(t)) \circ \dot{B}(t)dt$ can be given in terms of multiple Stratonovich integrals and the resulting evaluation is very simple, but first we need to extend the domain of $\overset{\circ}{I}$. Let us adopt the following notation:

$$L_{\infty,\pi}^{2\,*} = \prod_{n=0}^{\infty} L_{n,\pi}^2, \qquad K_{\infty,\pi}^* = \prod_{n=0}^{\infty} K_{n,\pi},$$

where $K_{n,\pi} = I(L_{n,\pi}^2)$. If on the right hand side the symmetrization of the component spaces is taken, then we write $\hat{L}_{n,\pi}^{2\,*}$ and $\hat{K}_{\infty,\pi}^*$.

For $\alpha > 0$, let

$$L_\pi^2(\alpha) = \left\{ f = \sum_{n=0}^{\infty} f_n \in L_{\infty,\pi}^{2*} \; ; \; \|f\|_{L_\pi^2(\alpha)}^2 \equiv \sum_{n=0}^{\infty} \alpha^n n! \|f_n\|_{n,\pi}^2 < \infty \right\}.$$

The proofs of the following two theorems are, with only slight modifications, those of Theorems 4 and 5 of [1].

**Proposition 2.1** *If $\alpha > 1$ and $f \in L_\pi^2(\alpha)$, then for every $n$ the series*

$$\sum_{k=0}^{\infty} \epsilon_k T_k(f_{n+2k})$$

*converges absolutely in $L_{n,\pi}^2$. Here $\epsilon_k = -1$ for every $k$ or $\epsilon_k = (-1)^k$ for every $k$. Hence*

$$\sum_{n=0}^{\infty} R^{-1}(f_n) = \lim_{M \to \infty} \sum_{n=0}^{M} R^{-1}(f_n) = \lim_{M \to \infty} \sum_{n=0}^{M} \sum_{k=0}^{[\frac{M-n}{2}]} T_k(f_{n+2k})$$

*converges in $L_{\infty,\pi}^{2*}$, and we denote the element it converges to by $R^{-1}(f)$. Similarly, $R(f)$ is defined.*

**Proposition 2.2** *If $\alpha > 2$ and $f \in L_\pi^2(\alpha)$, then $R^{-1}(f)$ and $R(f)$ are elements of $(L^2)$ and*

$$\| R^{-1}(f) \|_{(L^2)} \leq \beta_0(c+1) \left( \frac{\alpha}{\alpha - (c+1)} \right)^{\frac{1}{2}} \| f \|_{L_\pi^2(\alpha)} .$$

*Here $c$ is any number in $(1, \alpha - 1)$ and $\beta_0(c+1) = \left( \sum_{k=0}^{\infty} \frac{(2k)!}{(k!)^2} \frac{1}{(2(c+1))^{2k}} \right)^{\frac{1}{2}}$.*

For $\alpha > 1$ and $f \in \hat{L}^2_\pi(\alpha)$, let $\overset{\circ}{I}(f) = IR^{-1}(f)$. Thus for $\alpha > 1$ we have the continuous extensions:

$$R, R^{-1} : L^2_\pi(\alpha) \to L^{2*}_{\infty,\pi}, \qquad \overset{\circ}{I}: \hat{L}^2_\pi(\alpha) \to K^*_{\infty,\pi}$$

and for $\alpha > 2$, the ranges are in $(L^2)$.

Now, suppose that $F(x) = \sum_{n=0}^\infty \frac{a_n}{n!} x^n$ is analytic on $(-\infty, \infty)$ with $a_n = F^{(n)}(0)$ for all $n$. For $t \in \mathbf{R}$, let

$$R_t = \begin{cases} [0, t) & if \ \ t \geq 0 \\ [t, 0) & if \ \ t < 0 \end{cases}$$

and define $f : \mathbf{R} \to L^{2\,*}_{\infty,\pi}$ by

$$f(t) = \sum_{n=0}^\infty \frac{a_n}{n!} 1_{R_t}^{\hat{\otimes}n}$$

and set $f_n(t) = \frac{a_n}{n!} 1_{R_t}^{\hat{\otimes}n}$. If $|a_n| \leq M$ for all $n$, then for every $\alpha > 0$

$$\sum_{n=0}^\infty \alpha^n n! \, \| f_n(t) \|^2_{n,\pi} < \sum_{n=0}^\infty \alpha^n \frac{M^2}{n!} |t|^n < \infty$$

so that $f(t) \in \hat{L}^2_\pi(\alpha)$. Hence, in $(L^2)$

$$\overset{\circ}{I}(f(t)) = \overset{\circ}{I}\left( \sum_{n=0}^\infty \frac{a_n}{n!} 1_{R_t}^{\hat{\otimes}n} \right) = \sum_{n=0}^\infty \frac{a_n}{n!} \overset{\circ}{I}(1_{R_t}^{\hat{\otimes}n})$$

$$= \sum_{n=0}^\infty \frac{a_n}{n!} B(t)^n = F(B(t)).$$

**Theorem 2.1** *Let $F$ be analytic on $(-\infty, \infty)$. Suppose that there is an $M$ such that $|F^{(n)}(0)| \leq M$ for all $n$. Then in $(L^2)$*

$$\int_a^b F'(B(t)) \circ \dot{B}(t) dt = F(B(b)) - F(B(a)).$$

**Proof** Let $a_n = F^{(n)}(0)$. Without loss of generality we will assume that $t \geq 0$ and $a \geq 0$. Then

$$\overset{\circ}{I}^{-1}(F'(B(t)) = \overset{\circ}{I}^{-1}\left( \sum_{n=0}^\infty \frac{a_{n+1}}{n!} B(t)^n \right) = \sum_{n=0}^\infty \frac{a_{n+1}}{n!} 1_{[0,t)}^{\hat{\otimes}n}. \tag{6}$$

Thus we have

$$\int_a^b \overset{\circ}{I}{}^{-1}(F'(B(t))\hat{\otimes}\delta_t dt = \sum_{n=0}^{\infty} \frac{a_{n+1}}{n!} \int_a^b 1_{[0,t)}^{\hat{\otimes}n} \hat{\otimes}\delta_t dt$$

$$= \sum_{n=0}^{\infty} \frac{a_{n+1}}{n!} \cdot \frac{1}{n+1}[1_{[0,b)}^{\hat{\otimes}n+1} - 1_{[0,a)}^{\hat{\otimes}n+1}].$$

Since $|a_n| \leq M$ for all $n$, one easily sees that this is an element of $\hat{L}_\pi^2(\alpha)$ for all $\alpha > 0$. Therefore, in $(L^2)$

$$\int_a^b F'(B(t)) \circ \dot{B}(t)dt = \overset{\circ}{I} \int_a^b \overset{\circ}{I}{}^{-1}(F'(B(t))\hat{\otimes}\delta_t dt$$

$$= \overset{\circ}{I} \sum_{n=0}^{\infty} \frac{a_{n+1}}{(n+1)!}[1_{[0,b)}^{\hat{\otimes}n+1} - 1_{[0,a)}^{\hat{\otimes}n+1}]$$

$$= F(B(b)) - F(B(a)).$$

If $F$ is not analytic but is defined by (4), we might not be able to obtain $\overset{\circ}{I}{}^{-1}(F(B(t)))$ explicitly as in (6) above. By means of the reduction formula (3), however, we are able to get the following result.

**Theorem 2.2** *Let $F : \mathbf{R} \to \mathbf{R}$ be a continuous function such that $F \cdot g_t^{(n)}$ is integrable for every $n$. Let $c_n(t) = \int_{-\infty}^{\infty} F(x)(-1)^n g_t^{(n)}(x)dx$ and suppose that $|c_n(t)| \leq M$ for all $t \in [a, b], a > 0$. Then in $(L^2)$*

$$\int_a^b F'(B(t)) \circ \dot{B}(t)dt = F(B(b)) - F(B(a)).$$

**Proof** Since

$$\frac{d}{dt}g_t(x) = \frac{1}{2}\frac{d^2}{dx^2}g_t(x) = \frac{1}{2}g_t^{(2)}(x),$$

it is easy to see that $c_n'(t) = \frac{1}{2}c_{n+2}(t)$. Also, as an element of $(\mathcal{D}^*)$

$$F(B(t)) = \sum_{n=0}^{\infty} I_n\left(\frac{c_n(t)}{n!}1_{[0,t)}^{\hat{\otimes}n}\right)$$

and since $|c_n(t)| \leq M$ for all $n$, it is easily checked that $F(B(t))$ is in $(L^2)$. Also,

$$\sum_{n=0}^{\infty} \alpha^n(n+1)!||\frac{c_{n+1}(t)}{n!}1_{[0,t)}^{\hat{\otimes}n}||_{n,\pi,\mathcal{A}}^2 \leq \sum_{n=0}^{\infty} \frac{\alpha^n(n+1)!}{(n!)^2}M^2\int_a^b t^n dt$$

is finite for all $\alpha > 0$. Hence we have that $\int_a^b F'(B(t)) \circ \dot{B}(t)dt$ is an element of $(L^2)$ and that formula (3) applies :

$$\int_a^b F'(B(t)) \circ \dot{B}(t)dt = \int_a^b F'(B(t))\hat{\otimes}\dot{B}(t)dt + \sum_{n=1}^{\infty} \frac{1}{n!}I_{n-1}(\mathcal{T}(c_{n+1}(t)1_{[0,t)}^{\hat{\otimes}n})). \quad (7)$$

Now,

$$\int_a^b F'(B(t))\hat{\otimes}\dot{B}(t)dt = I\left[\sum_{n=0}^{\infty}\int_a^b \frac{c_{n+1}(t)}{n!}1_{[0,t)}^{\hat{\otimes}n}\hat{\otimes}\delta_t dt\right]$$

and we can write the right hand side of (7) as

$$I\left[\sum_{n=0}^{\infty}\frac{1}{n!}\int_a^b c_{n+1}(t)1_{[0,t)}^{\hat{\otimes}n}\hat{\otimes}\delta_t dt + \sum_{n=1}^{\infty}\frac{1}{n!}\mathcal{T}(c_{n+1}(t)1_{[0,t)}^{\hat{\otimes}n})\right].$$

Let

$$g(t, u_1, \ldots, u_n) = 1_{[a,b]}(t)c_{n+1}(t)1_{[0,t)}(u_1)1_{[0,t)}^{\hat{\otimes}n-1}(u_2, \ldots, u_n)$$

$$= 1_{[a,b]}(t)c_{n+1}(t)1_{[0,t)}(u_1)1_{[u_2 \vee \cdots \vee u_n, b]}(t).$$

Let $\tilde{g}$ represent the symmetrization of $g$ in the variables $t$ and $u_1$. Then in $L^2_{n+1,\pi}$

$$\tilde{g}(t_1, u_1, \ldots, u_n) = \frac{1}{2}1_{[a,b]^2}(t, u_1)c_{n+1}(t \vee u_1)1_{[u_2 \vee \cdots \vee u_n, b]}(t \vee u_1)$$

and since $F$ is continuous

$$\mathcal{T}g = \mathcal{T}\tilde{g} = \frac{n}{2}\int_a^b c_{n+1}(t)1_{[0,t)}^{\hat{\otimes}n-1}dt.$$

Therefore,

$$\int_a^b F'(B(t)) \circ \dot{B}(t)dt$$

$$= I\left[\sum_{n=0}^{\infty}\frac{1}{n!}\int_a^b c_{n+1}(t)1_{[0,t)}^{\hat{\otimes}n}\hat{\otimes}\delta_t dt + \sum_{n=1}^{\infty}\frac{1}{n!}\frac{n}{2}\int_a^b c_{n+1}(t)1_{[0,t)}^{\hat{\otimes}n-1}dt\right]$$

$$= I\left[\sum_{n=0}^{\infty}\frac{1}{n!}\int_a^b c_{n+1}(t)1_{[0,t)}^{\hat{\otimes}n}\hat{\otimes}\delta_t dt + \sum_{n=0}^{\infty}\frac{1}{n!}\int_a^b \frac{1}{2}c_{n+2}(t)1_{[0,t)}^{\hat{\otimes}n}dt\right].$$

Now observe that for $n \geq 1$,

$$c_n(b)1_{[0,b)}^{\hat{\otimes}n} - c_n(a)1_{[0,a)}^{\hat{\otimes}n} = \int_a^b \frac{d}{dt}[c_n(t)1_{[0,t)}^{\hat{\otimes}n}]dt$$

$$= n\int_a^b c_n(t)1_{[0,t)}^{\hat{\otimes}n-1}\hat{\otimes}\delta_t dt + \int_a^b c'_n(t)1_{[0,t)}^{\hat{\otimes}n}dt$$

$$= n\int_a^b c_n(t)1_{[0,t)}^{\hat{\otimes}n-1}\hat{\otimes}\delta_t dt + \int_a^b \frac{1}{2}c_{n+2}(t)1_{[0,t)}^{\hat{\otimes}n}dt.$$

Summing on $n$ we get

$$F(B(b)) - F(B(a)) = I\left[\sum_{n=0}^{\infty} \frac{1}{n!} \int_a^b \frac{d}{dt}[c_n(t)1_{[0,t)}^{\hat{\otimes}n}]dt\right]$$

$$= I\left[\sum_{n=1}^{\infty} \frac{n}{n!} \int_a^b c_n(t)1_{[0,t)}^{\hat{\otimes}n-1}\hat{\otimes}\delta_t dt + \int_a^b c_0'(t)dt + \sum_{n=1}^{\infty} \frac{1}{n!} \int_a^b \frac{1}{2}c_{n+2}(t)1_{[0,t)}^{\hat{\otimes}n} dt\right]$$

$$= I\left[\sum_{n=0}^{\infty} \frac{1}{n!} \int_a^b c_{n+1}(t)1_{[0,t)}^{\hat{\otimes}n}\hat{\otimes}\delta_t dt + \sum_{n=0}^{\infty} \frac{1}{n!} \int_a^b \frac{1}{2}c_{n+2}(t)1_{[0,t)}^{\hat{\otimes}n} dt\right]$$

and the result follows.

## References


[1] Betounes, D.: Trace operators, Feynman distributions and multiparameter white noise; *J. of Theoretical Prob.* (1993) to appear.

[2] Betounes, D. and Redfern, M.: Wiener distributions and white noise analysis; *Appl. Math. Optim.* **26** (1992) 63–93

[3] Betounes, D. and Redfern, M.: The Stratonovich integral via the renormalization on Fock space; *Preprint* (1994)

[4] Hida, T., Kuo, H. -H., Potthoff, J. and Streit, L.: *White Noise: An Infinite Dimensional Calculus.* Kluwer Academic Publishers, 1993

[5] Kuo, H. -H.: Lectures on white noise analysis; *Soochow J. Math.* **18** (1992) 229–300

[6] Treves, F.: *Topological Vector Spaces, Distributions, and Kernels.* Academic Press, New York, 1967



Mylan Redfern
Department of Mathematics
University of Southern Mississippi
Hattiesburg, MS 39406-5045, U.S.A.
*E-mail*: mylan redfern@bull.cc.usm.edu

J. ROSINSKI

# Uniqueness of spectral representations of skewed stable processes and stationarity

## 1. Introduction and Preliminaries

In this paper we study the uniqueness of spectral representations of general stable processes extending certain results from [3, 6] obtained for symmetric processes. Uniqueness is the key property which allows us to characterize spectral representations of stationary stable processes. Our proof of uniqueness given in Section 2 employs a certain variation of Rudin's Theorem (Proposition 2.3) which may be of independent interest. In Section 3 we show that every stationary stable process can be decomposed into two mutually independent stable processes - symmetric and skewed. Since the symmetric case has been investigated in detail in [5], we characterize here the skewed case. Our work on uniqueness of (nonstationary) stable processes is based on certain ideas of Hardin [1, 2]. We should also mention that in this work we do not treat skewed 1–stable processes due to the form and nonlinearity of centering in the 1–stable case (see, e.g., [4, 8]) which require different methods.

Univariate non–Gaussian stable distributions are characterized by four parameters. These are the index of stability $\alpha \in (0, 2)$, the scale parameter $\sigma \geq 0$, the skewness parameter $\beta \in [-1, 1]$, and the shift parameter $b \in \mathbf{R}$. From now on we will assume that $\alpha \in (0, 2)$ and $\alpha \neq 1$. We will write

$$X \sim S_\alpha(\sigma, \beta, b) \tag{1.1}$$

to indicate that a random variable $X$ has a stable distribution with parameters $\alpha$, $\sigma$, $\beta$, and $b$. Specifically, (1.1) says that the characteristic function of $X$ has the following form:

$$E \exp i\theta X = \exp\left[-\sigma^\alpha |\theta|^\alpha (1 - i\beta(\mathrm{sign}(\theta)) \tan \frac{\pi\alpha}{2}) + ib\theta\right]. \tag{1.2}$$

In the sequel we will need some basic facts and properties of stable distributions, processes, and stable stochastic integrals. They can be found, for instance, in a recent monograph by Samorodnitsky and Taqqu [8] and we refer the reader to this book for more information and the proofs of these basic facts.

264

Let $T$ be an arbitrary set. A stochastic process $\{X_t\}_{t\in T}$ is said to be a stable process if all its finite dimensional distributions are stable. The latter condition is equivalent to that all linear combinations $\sum a_j X_{t_j}$ have stable univariate distributions of index $\alpha$ ($\alpha \neq 1$).

For every stable process $\{X_t\}_{t\in T}$ there exists a nonrandom function $c : T \to \mathbf{R}$ such that all finite dimensional distributions of $\{X_t - c(t)\}_{t\in T}$ are strictly stable, i.e., the shift parameter $b$ in (1.2) equals zero for all linear combinations $\sum a_j(X_{t_j} - c(t_j))$. Therefore, without loss of generality, we may restrict our study to strictly stable processes.

Let $S$ be a Borel subset of a Polish space equipped with its Borel $\sigma$–field $\mathcal{B}$, and let $\mu$ be a $\sigma$–finite measure on $\mathcal{B}$. Put $\mathcal{B}_0 = \{A \in \mathcal{B} : \mu(A) < \infty\}$. Let $\beta : S \to [-1,1]$ be a measurable function. Let $(\Omega, \mathcal{F}, P)$ be a probability space. An independently scattered $\sigma$–additive set function $M : \mathcal{B}_0 \to L^0(\Omega, \mathcal{F}, P)$ is said to be an $\alpha$–stable random measure on $(S, \mathcal{B})$ with control measure $\mu$ and skewness intensity $\beta$ if for each $A \in \mathcal{B}_0$

$$M(A) \sim S_\alpha((\mu(A))^{1/\alpha}, (\mu(A))^{-1} \int_A \beta(s)\,\mu(ds), 0).$$

Then, for every $f \in L^\alpha(S, \mathcal{B}, \mu)$, the stochastic integral $\int_S f\, dM$ exists ($\alpha \neq 1$ is important here) and

$$\int_S f\, dM \sim S_\alpha(\sigma_f, \beta_f, 0), \tag{1.3}$$

where

$$\sigma_f = \left( \int_S |f(s)|^\alpha\, \mu(ds) \right)^{1/\alpha}, \tag{1.4}$$

and

$$\beta_f = \sigma_f^{-\alpha} \int_S \beta(s)[f(s)]^{<\alpha>}\, \mu(ds). \tag{1.5}$$

By a convention, $x^{<\alpha>} = \mathrm{sign}(x)|x|^\alpha$. Let $\{X_t\}_{t\in T}$ be a strictly stable process and let $M$ be a stable random measure as above. A family of functions $\{f_t\}_{t\in T} \subset L^\alpha(S, \mathcal{B}, \mu)$ is said to be a *spectral representation* of $\{X_t\}_{t\in T}$ with respect to $M$ if

$$\{X_t\}_{t\in T} \overset{\mathcal{D}}{=} \left\{ \int_S f_t\, dM \right\}_{t\in T}. \tag{1.6}$$

Every separable in probability strictly stable process of order $\alpha \neq 1$ has a spectral representation with respect to a $\alpha$–stable random measure $M$ on $([0,1], \mathcal{B}_{[0,1]})$ with the Lebesgue control measure and constant skewness intensity $\beta \equiv 1$ (see [4, 8]).

(We recall that a stochastic process $\{X_t\}_{t\in T}$ is said to be separable in probability if there exists a countable subset $T_0$ of $T$ such that $\{X_t\}_{t\in T_0}$ is dense in $\{X_t\}_{t\in T}$ in the $L^0(\Omega, \mathcal{F}, P)$–topology.) Throughout this paper we will assume that all processes under consideration are separable in probability.

## 2. Uniqueness of Minimal Representations

Let us recall the notion of a minimal representation due to Hardin [3].

**Definition 2.1** A spectral representation $\{f_t\}_{t\in T} \subset L^\alpha(S, \mathcal{B}, \mu)$ is said to be *minimal* if the following conditions hold

(i) $\sigma(\{\frac{g}{h} : g, h \in F\}) = \mathcal{B}$ modulo $\mu$, where $F$ is the $L^\alpha$–closure of $\mathrm{lin}\{f_t : t \in T\}$. ($\frac{g}{h}$ is naturally defined as an extended–valued function taking values in the one–point compactification of $\mathbf{R}$.)

(ii) $\mathrm{supp}\{g : g \in F\} = S$ modulo $\mu$.

Hardin [3] proved that every symmetric stable process has a minimal representation. We will show that the same is true for more general strictly stable processes. It is interesting to notice that while ordinary spectral representations of strictly stable processes can be constructed with respect to stationary totally skewed random measures ($\beta \equiv 1$), minimal representations require nonstationary random measures (i.e., varying $\beta$'s).

**Theorem 2.2** *Every strictly stable process has a minimal representation.*

**Proof** The proof essentially follows arguments in [3] with one modification. Let $\{g_t\}_{t\in T} \subset L^\alpha(Y, \mathcal{G}, \nu)$ be a spectral representation of a process $\{X_t\}_{t\in T}$ with respect to a random measure $N$ having control measure $\nu$ and skewness intensity $\gamma$. Let $G$ be the $L^\alpha$–closure of $\mathrm{lin}\{g_t : t \in T\}$ and let $g \in G$ be a function with full support in $G$, i.e., $\mathrm{supp}\{g\} = \mathrm{supp}\{g_t : t \in T\}$, and such that $\|g\|_\alpha = 1$ (see [2], Lemma 3.2 for the existence of functions with full support). Define a probability measure $\nu_0$ on $(Y, \mathcal{G})$ by $\nu_0(dy) = |g(y)|^\alpha \nu(dy)$. Let

$$\mathcal{A} = \sigma(\{\frac{g_t}{g} : t \in T\}), \qquad \beta_0 = \int_Y \mathrm{sign}(g(y))\gamma(y)\,\nu_0(dy|\mathcal{A}).$$

Clearly $|\beta_0| \le 1$. From the assumption that $\{X_t\}_{t\in T}$ is separable in probability we infer that the Boolean measure $\sigma$–algebra $(Y, \mathcal{A}, \nu_0)$ is separable. Hence $(Y, \mathcal{A}, \nu_0)$ is $\sigma$–isomorphic to some $(S, \mathcal{B}, \mu)$, where $S$ is a Polish space equipped with the Borel $\sigma$–field and a probability measure $\mu$. Call this isomorphism $\Psi$, it acts from $\mathcal{A}$ modulo

$\nu_0$ onto $\mathcal{B}$ modulo $\mu$. Since $\beta_0$ is $\mathcal{A}$–measurable, $\beta := \Psi(\beta_0)$ is well–defined and takes values in $[-1, 1]$ (see [1, pp. 452-454]). Set $f_t = \Psi(g_t/g)$, $t \in T$. For every $a_1, \ldots, a_n \in \mathbf{R}$ and $t_1, \ldots, t_n \in T$ we have

$$\int_S |\sum a_j f_{t_j}|^\alpha \, d\mu = \int_Y |\sum a_j g_{t_j}/g|^\alpha \, d\nu_0 = \int_Y |\sum a_j g_{t_j}|^\alpha \, d\nu,$$

and, since $\beta_0$ is a conditional expectation,

$$\int_S \beta(\sum a_j f_{t_j})^{<\alpha>} \, d\mu = \int_Y \beta_0(\sum a_j g_{t_j}/g)^{<\alpha>} \, d\nu_0$$

$$= \int_Y \text{sign}(g)\gamma(\sum a_j g_{t_j}/g)^{<\alpha>} \, d\nu_0 = \int_Y \gamma(\sum a_j g_{t_j})^{<\alpha>} \, d\nu.$$

These equations in conjunction with (1.3)–(1.5) show that $\{f_t\}_{t \in T}$ is a spectral representation of $\{X_t\}_{t \in T}$ with respect to a random measure $M$ having control measure $\mu$ and skewness intensity $\beta$. The minimality of $\{f_t\}_{t \in T}$ follows from the construction and the fact that $1 = \Psi(g/g)$ belongs to the $L^\alpha$–closure of $\text{lin}\{f_t : t \in T\}$. Q.E.D.

The following proposition can be viewed as a variation of Rudin's Theorem (see [2, 7, 9]). The difference between this and Rudin's Theorem is that in Rudin's Theorem one assumes the equality of integrals with respect to the $|.|^p$–function and $p \notin 2\mathbf{Z}$ while here we consider $(.)^{<p>}$ and $p \notin 2\mathbf{Z} - 1$. Let $x^{<p>} = \bar{x}|x|^{p-1}$, if $x \in \mathbf{C}$.

**Proposition 2.3** *Let $p > 0$ be not an odd integer. Let $\nu_1$ and $\nu_2$ be finite measures on $Y_1$ and $Y_2$, respectively, and let $f_1, \ldots, f_n \in L^p(Y_1, \nu_1)$ and $g_1, \ldots, g_n \in L^p(Y_2, \nu_2)$ be real–valued (complex–valued, resp.) functions such that for every $\lambda_1, \ldots, \lambda_n \in \mathbf{R}$ ($\mathbf{C}$, resp.)*

$$\int_{Y_1} (1 + \sum \lambda_k f_k(y))^{<p>} \nu_1(dy) = \int_{Y_2} (1 + \sum \lambda_k g_k(y))^{<p>} \nu_2(dy).$$

*Then $\nu_1 \circ (f_1, \ldots, f_n)^{-1} = \nu_2 \circ (g_1, \ldots, g_n)^{-1}$.*

**Proof** The method of proof of Rudin's Theorem from [2] can be adapted to this case if one can show that there exists a nonzero real function $h$ of the form

$$h(x) = \sum_{n=1}^{N} a_n (1 + nx)^{<p>} \tag{2.1}$$

that is bounded and Lebesgue integrable on $\mathbf{R}$ ($\mathbf{C}$, resp.). First we observe that any function of this form, with at least one $a_n \neq 0$, does not vanish. Indeed, this follows from the fact that the $\lceil p\rceil$–th derivative of $h$ does not exist at $x = -n^{-1}$ (the assumption that $p$ is not an odd integer is crucial here). Next we observe that the function $\phi(x) = (1+x)^{<p>}$ is analytic in $\{|x| < 1\}$ and has power series expansion $\phi(x) = \sum_{k=0}^{\infty} c_k x^k$ with $\limsup_{k\to\infty} \sqrt[k]{|c_k|} \leq 1$. Moreover, $\phi(x) = x^{<p>}\phi(x^{-1})$. Therefore, similar arguments as in the proof of Lemma 1.3 [2] can be used to construct $h$ in (2.1). Indeed, fix $N > p + 3$ and let $a_1, \ldots, a_N$ be a nontrivial solution of the system of $N - 1$ linear equations

$$\sum_{n=1}^{N} a_n n^{p-k} = 0, \qquad k = 0, \ldots, N - 2.$$

Then, if $h$ is defined with such $a_n$'s, we get for $|x| \geq 2$

$$|h(x)| = |\sum_{n=1}^{N} a_n n^p x^{<p>}\phi((nx)^{-1})|$$

$$= |x|^p | \sum_{k=N-1}^{\infty} c_k (\sum_{n=1}^{N} a_n n^{p-k}) x^{-k}|$$

$$\leq const \, |x|^{p-N+1}.$$

This shows that $h$ is bounded and integrable on $\mathbf{R}$ ($\mathbf{C}$, resp.). Q.E.D.

**Theorem 2.4** *Let $\{f_t^{(i)}\}_{t\in T} \subset L^\alpha(S_i, \mathcal{B}_i, \mu_i)$ be a minimal spectral representation of a $\alpha$–stable process $\{X_t\}_{t\in T}$ with respect to a random measure $M_i$ having control measure $\mu_i$ and skewness intensity $\beta_i$, $i = 1, 2$. Then there exists a unique bijective nonsingular map $\phi : S_2 \to S_1$ such that $|\beta_2| = |\beta_1 \circ \phi|$ $\mu_2$–a.e. and for every $t \in T$*

$$f_t^{(2)} = \varepsilon \left\{ \frac{d(\mu_1 \circ \phi)}{d\mu_2} \right\}^{1/\alpha} f_t^{(1)} \circ \phi \qquad \mu_2 - a.e.,$$

*where $\varepsilon : S_2 \to \{-1, 1\}$ is a measurable function given by*

$$\varepsilon(s) = \begin{cases} \frac{\beta_1(\phi(s))}{\beta_2(s)}, & \text{if } \beta_2(s) \neq 0; \\ \\ arbitrary, & \text{if } \beta_2(s) = 0. \end{cases}$$

**Proof** By (1.3)–(1.4)

$$\int_{S_1} |\sum a_k f_{t_k}^{(1)}|^\alpha \, d\mu_1 = \int_{S_2} |\sum a_k f_{t_k}^{(2)}|^\alpha \, d\mu_2$$

for every $a_1, \ldots, a_n \in \mathbf{R}$ and $t_1, \ldots, t_n \in T$. Hence $\{f_t^{(i)}\}_{t \in T}$, $i = 1, 2$ are minimal spectral representations of the symmetrization $\{\tilde{X}_t\}_{t \in T}$ of $\{X_t\}_{t \in T}$. By a result of Hardin [3] (see also Remark 2.3 in [5]) there exists a unique bijective nonsingular mapping $\phi : S_2 \to S_1$ such that for every $t \in T$

$$f_t^{(2)} = h f_t^{(1)} \circ \phi, \tag{2.2}$$

and

$$|h|^\alpha = \frac{d(\mu_1 \circ \phi)}{d\mu_2} \tag{2.3}$$

$\mu_2$–a.e. Put $\varepsilon(s) = \operatorname{sign}(h(s))$. To complete the proof it is enough to show that $\beta_2 \varepsilon = \beta_1 \circ \phi$ $\mu_2$–a.e. By (1.3)–(1.5), for every $a_1, \ldots, a_m \in \mathbf{R}$ and $t_1, \ldots, t_m \in T$, we have

$$\int_{S_1} \beta_1 \Big(\sum a_k f_{t_k}^{(1)}\Big)^{<\alpha>} d\mu_1 = \int_{S_2} \beta_2 \Big(\sum a_k f_{t_k}^{(2)}\Big)^{<\alpha>} d\mu_2. \tag{2.4}$$

Let $F_i$ be the $L^\alpha$–closure of $\operatorname{lin}\{f_t^{(i)} : t \in T\}$ and let $g_i \in F_i$ be a function with full support in $F_i$, $i = 1, 2$. In view of (2.2) we may assume that $g_2 = h g_1 \circ \phi$. Since $g_i \in F_i$ and $|\beta_i| \leq 1$, we get from (2.4) by the $L^\alpha$–continuity

$$\int_{S_1} \beta_1 \Big(g_1 + \sum \lambda_k f_{t_k}^{(1)}\Big)^{<\alpha>} d\mu_1 = \int_{S_2} \beta_2 \Big(g_2 + \sum \lambda_k f_{t_k}^{(2)}\Big)^{<\alpha>} d\mu_2$$

for every $\lambda_1, \ldots, \lambda_n \in \mathbf{R}$, $t_1, \ldots, t_n \in T$, and $n \geq 1$. Combining the above equality with (2.2) and (2.3) we have

$$\begin{aligned}
\int_{S_1} \beta_1 \Big(g_1 + \sum \lambda_k f_{t_k}^{(1)}\Big)^{<\alpha>} d\mu_1 &= \int_{S_2} \beta_2 \Big(h g_1 \circ \phi + \sum \lambda_k h f_{t_k}^{(1)} \circ \phi\Big)^{<\alpha>} d\mu_2 \\
&= \int_{S_2} \beta_2 \varepsilon \Big(g_1 \circ \phi + \sum \lambda_k f_{t_k}^{(1)} \circ \phi\Big)^{<\alpha>} d(\mu_1 \circ \phi) \\
&= \int_{S_1} (\beta_2 \circ \phi^{-1})(\varepsilon \circ \phi^{-1})\Big(g_1 + \sum \lambda_k f_{t_k}^{(1)}\Big)^{<\alpha>} d\mu_1.
\end{aligned}$$

Put $\psi = (\beta_2 \circ \phi^{-1})(\varepsilon \circ \phi^{-1}) - \beta_1$. We have shown that

$$\int_{S_1} \Big(g_1 + \sum \lambda_k f_{t_k}^{(1)}\Big)^{<\alpha>} \psi \, d\mu_1 = 0 \tag{2.5}$$

for every $\lambda_1, \ldots, \lambda_n \in \mathbf{R}$ and $t_1, \ldots, t_n \in T$, $n \geq 1$. Define the following finite measures on $(S_1, \mathcal{B}_1)$

$$\nu(ds) = (\psi(s)\operatorname{sign}(g_1(s)))^+ |g_1|^\alpha \, \mu_1(ds),$$

$$\rho(ds) = (\psi(s)\mathrm{sign}(g_1(s)))^- |g_1|^\alpha \, \mu_1(ds).$$

Then by (2.5)

$$\int_{S_1} (1 + \sum \lambda_k f_{t_k}^{(1)}/g_1)^{<\alpha>} \, d\nu = \int_{S_1} (1 + \sum \lambda_k f_{t_k}^{(1)}/g_1)^{<\alpha>} \, d\rho.$$

By Proposition 2.3 and the minimality of $\{f_t^{(1)}\}_{t \in T}$ we get $\nu = \rho$ on $\mathcal{B}_1$. Since $\nu$ and $\rho$ are also mutually singular, $\nu = \rho = 0$. This proves that $\psi = 0$ $\mu_1$–a.e., i.e., $(\beta_2 \circ \phi^{-1})(\varepsilon \circ \phi^{-1}) = \beta_1$ $\mu_1$–a.e. Since $\phi$ is nonsingular, $\beta_2 \varepsilon = \beta_1 \circ \phi$ $\mu_2$–a.e. which completes the proof. Q.E.D.

Theorem 2.4 suggests the following definition of a skewed stable process.

**Definition 2.5** A stochastic process $\{X_t\}_{t \in T}$ is said to be a *skewed $\alpha$–stable process* if it possesses a minimal spectral representation with respect to a random measure with nonvanishing skewness intensity.

This definition is consistent with the notion of Lévy skewed stable processes. Indeed, in the latter case, the family of functions $f_t(s) = 1_{[0,t]}(s)$, $t \geq 0$ forms a minimal representation with respect to a $\alpha$–stable random measure having the Lebesgue control measure and constant skewness intensity $\beta \neq 0$.

By Theorem 2.4, for any minimal representation of a skewed process, skewness intensity of the associated random measure differs from zero a.e. Moreover, that theorem also says that minimal spectral representations of skewed stable processes are unique up to nonsingular bijective transformations of their domains. In this sense minimal spectral representations of skewed stable processes are more rigid than the corresponding representations of symmetric stable processes. The latter are uniquely determined up to nonsingular bijective transformations and sign multipliers. Another corollary from Theorem 2.4 is that the range of $|\beta|$ is essentially determined by a stable process. This proves our earlier remark that minimal representations require nonstationary random measures (varying $\beta$'s).

## 3. Stationary Stable Processes

Now let $T = \mathbf{Z}$ or $\mathbf{R}$. Consider a stationary strictly $\alpha$–stable process $\{X_t\}_{t \in T}$. Let $\{f_t\}_{t \in T} \subset L^\alpha(S, \mathcal{B}, \mu)$ be its minimal spectral representation with respect to a random measure $M$ having control measure $\mu$ and skewness intensity $\beta$. Such a representation exists by Theorem 2.2. Since $\{X_{t+\tau}\}_{t \in T} =^{\mathcal{D}} \{X_t\}_{t \in T}$ for each $\tau \in T$, the family $\{f_{t+\tau}\}_{t \in T}$ is also a minimal representation of $\{X_t\}_{t \in T}$. By Theorem

2.4 there exist a nonsingular bijective map $\phi_\tau : S \to S$ and a measurable function $\varepsilon_\tau : S \to \{-1, 1\}$ such that

$$f_\tau = \varepsilon_\tau \left\{ \frac{d(\mu \circ \phi_\tau)}{d\mu} \right\}^{1/\alpha} f_0 \circ \phi_\tau \qquad \mu - a.e. \tag{3.1}$$

Similarly to [5], proof of Theorem 3.1, one can show that $\{\phi_\tau\}_{\tau \in T}$ can be selected as a nonsingular flow on $(S, \mathcal{B}, \mu)$. Then $\{\varepsilon_\tau\}_{\tau \in T}$ is a cocycle for that flow (see [5]). By Theorem 2.4, $|\beta \circ \phi_\tau| = |\beta|$ $\mu$–a.e. for each $\tau \in T$ which shows that $|\beta|$ is $\{\phi_\tau\}_{\tau \in T}$–invariant. In particular, the set $\{s : \beta(s) = 0\}$ is $\{\phi_\tau\}_{\tau \in T}$–invariant. On its complement one has $\varepsilon_\tau = (\beta \circ \phi_\tau)/\beta$. Hence the stochastic processes

$$X_t^0 := \int_{\{\beta=0\}} f_t \, dM \quad \text{and} \quad X_t^1 := \int_{\{\beta \neq 0\}} f_t \, dM, \qquad t \in T,$$

are both stationary and obviously independent. Applying (1.3)–(1.5) to linear combinations of $f_\tau$'s given by (3.1), and making use of above listed properties of $\beta$, $\phi_\tau$, and $\varepsilon_\tau$, it is easy to prove the following:

**Proposition 3.1** *Let $\{X_t\}_{t \in T}$ be a stationary strictly $\alpha$–stable process. Then there exist mutually independent stationary $\alpha$–stable processes $\{X_t^0\}_{t \in T}$ and $\{X_t^1\}_{t \in T}$ such that $\{X_t^0\}_{t \in T}$ is symmetric, $\{X_t^1\}_{t \in T}$ is skewed, and $X \overset{\mathcal{D}}{=} X^0 + X^1$.*

Stationary symmetric $\alpha$–stable processes have been characterized in [2, 4]. Therefore, in order to obtain a complete picture of the structure of general stationary stable processes, we need to characterize the skewed case.

**Theorem 3.2** *Let $T = \mathbf{Z}$ or $\mathbf{R}$. Let $\{X_t\}_{t \in T}$ be a stationary skewed $\alpha$–stable process, $\alpha \neq 1$. Then there exist a finite Lebesgue measure space $(S, \mathcal{B}, \mu)$, a nonsingular flow $\{\phi_t\}_{t \in T}$ on $(S, \mathcal{B}, \mu)$, and an $f \in L^\alpha(S, \mathcal{B}, \mu)$ such that*

$$f_t = \left\{ \frac{d(\mu \circ \phi_t)}{d\mu} \right\}^{1/\alpha} f \circ \phi_t, \qquad t \in T, \tag{3.2}$$

*is a minimal spectral representation of $\{X_t\}_{t \in T}$ with respect to a random measure $M$ having control measure $\mu$ and a positive $\{\phi_t\}_{t \in T}$–invariant skewness intensity.*

**Remark 3.3** In view of Kuratowski's Isomorphism Theorem, the measure space $(S, \mathcal{B}, \mu)$ in Theorem 3.3 can always be taken as the unit interval equipped with the Lebesgue measure or a countable discrete set equipped with the counting measure or the union of both the unit interval and a disjoint discrete set with the associated measures.

**Remark 3.4** In view of (3.2), we can relate with every skewed stable process a group of *positive* isometries on $L^\alpha(S, \mathcal{B}, \mu)$ given by

$$U_t g := \left\{ \frac{d(\mu \circ \phi_t)}{d\mu} \right\}^{1/\alpha} g \circ \phi_t, \qquad t \in T, \quad g \in L^\alpha(S, \mathcal{B}, \mu).$$

In the symmetric case Hardin [3] showed that every symmetric stationary stable process is determined by a group of (arbitrary) isometries on an $L^\alpha$–space (in fact, one can take $L^\alpha[0, 1]$). The positivity of the associated group of isometries is the property that one gains in the skewed case.

**Proof of Theorem 3.2** Let $\{g_t\}_{t \in T} \subset L^\alpha(S, \mathcal{B}, \mu)$ be a minimal spectral representation of $\{X_t\}_{t \in T}$. By Theorem 2.4 and (3.1) we have, for every $t \in T$, $|\beta \circ \phi_t| = |\beta|$ and

$$g_t = \frac{\beta \circ \phi_t}{\beta} \left\{ \frac{d(\mu \circ \phi_t)}{d\mu} \right\}^{1/\alpha} g_0 \circ \phi_t, \qquad \mu - \text{a.e.},$$

where $\{\phi_t\}_{t \in T}$ is a nonsingular flow on $(S, \mathcal{B}, \mu)$. Define $f = \beta g_0 / |\beta|$ and let $\{f_t\}_{t \in T} \subset L^\alpha(S, \mathcal{B}, \mu)$ be given by (3.2). We have for every $a_1, \ldots, a_n \in \mathbf{R}$ and $t_1, \ldots, t_n \in T$

$$\int_S \left| \sum a_k f_{t_k} \right|^\alpha d\mu = \int_S \left| \sum a_k g_{t_k} \beta / |\beta \circ \phi_{t_k}| \right|^\alpha d\mu$$

$$= \int_S \left| \sum a_k g_{t_k} \right|^\alpha d\mu$$

and

$$\int_S |\beta| \left( \sum a_k f_{t_k} \right)^{<\alpha>} d\mu = \int_S |\beta| \left( \sum a_k g_{t_k} \beta / |\beta \circ \phi_{t_k}| \right)^{<\alpha>} d\mu$$

$$= \int_S \beta \left( \sum a_k g_{t_k} \right)^{<\alpha>} d\mu.$$

Hence $\{f_t\}_{t \in T}$ is a spectral representation of $\{X_t\}_{t \in T}$ with respect to a random measure $M$ with control measure $\mu$ and skewness intensity $|\beta|$. Its minimality follows from the fact that for every $t, \tau \in T$, $f_t / f_\tau = (g_t | \beta \circ \phi_\tau |) / (g_\tau | \beta \circ \phi_t |) = g_t / g_\tau$. Q.E.D.

It is easy to verify that given a nonsingular flow $\{\phi_t\}_{t \in T}$ on a Lebesgue measure space $(S, \mathcal{B}, \mu)$ and a $\{\phi_t\}_{t \in T}$–invariant function $\beta : S \to (0, 1]$, a process $\{X_t\}_{t \in T}$ given by

$$X_t := \int_S \left\{ \frac{d(\mu \circ \phi_t)}{d\mu} \right\}^{1/\alpha} f \circ \phi_t, \, dM \qquad t \in T, \tag{3.3}$$

with $f \in L^{\alpha}(S, \mathcal{B}, \mu)$, is stationary and strictly stable. (As before, $M$ is a $\alpha$–stable random measure with control measure $\mu$ and skewness intensity $\beta$.) By Theorem 3.2 every stationary skewed stable process is equal in distribution to a process given by (3.3), therefore, in this sense, (3.3) is the most general form of stationary skewed stable processes. Every such a process is generated by a nonsingular flow and the flow–invariant function $\beta \in (0,1]$. Depending on general ergodic theory types of flows one obtains various classes of stationary stable processes. For example, if $\phi_t : \mathcal{X} \times \mathbf{R} \to \mathcal{X} \times \mathbf{R}$ is given by $\phi_t(x,s) = (x, s + t)$, where $(\mathcal{X}, \mathcal{A}, \nu)$ is a Lebesgue space, and if $\beta(x,s)$ depends only on $x \in \mathcal{X}$ but not on $s$, then the process in (3.3) is a mixed moving average process (see [10] for the symmetric case). In general, if a flow is dissipative, then the resulting process is a mixed moving average (see [5, Theorem 4.4]).

## References


[1] Doob, J. L.: *Stochastic Processes*. John-Wiley & Sons, Inc. New York, 1953

[2] Hardin, C. D. Jr.: Isometries on subspaces of $L^p$; *Indiana Univ. Math. J.* **30** (1981) 449–465

[3] Hardin, C. D. Jr.: On the spectral representation of symmetric stable processes; *J. Multivariate Analysis* **12** (1982) 385–401

[4] Hardin, C. D. Jr.: Skewed stable variables and processes; *Center for Stoch. Processes Tech. Rept.* **65** , University of North Carolina (1984)

[5] Rosinski, J.: On the structure of stationary stable processes; *Preprint* (1993)

[6] Rosinski, J.: On uniqueness of the spectral representation of stable processes; *J. Theor. Probability* (1994) to appear

[7] Rudin, W.: $L^p$ isometries and equimeasurability; *Indiana Univ. Math. J.* **25** (1976) 215–228

[8] Samorodnitsky, G. and Taqqu, M. S.: *Non–Gaussian Stable Processes*. (1994) to appear

[9] Stephenson, K.: Certain integral equations which imply equimeasurability of functions; *Can. J. Math.* **29** (1977) 827–844

[10] Surgailis, D., Rosinski, J., Mandrekar, V., and Cambanis, S.: Stable mixed moving averages; *Probab. Theory Relat. Fields* **97** (1993) 543–558



Jan Rosinski
Department of Mathematics
University of Tennessee
Knoxville, TN 37922, U.S.A.
*E-mail*: rosinski@utkvx.utk.edu

K. SAITÔ

# A group generated by the Lévy Laplacian and the Fourier-Mehler transform

## 1. Introduction

An infinite dimensional Laplacian was introduced by P. Lévy in his book [23]. For any real-valued twice differentiable function $f$ on $L^2[0,1]$ with the second derivative $f''$, the Lévy Laplacian $\Delta_L f$ is defined by

$$\Delta_L f(x) = \lim_{n \to \infty} \frac{1}{n} \sum_{k=1}^{n} f''(x)(\xi_k, \xi_k)$$

if the limit exists, where $\{\xi_n; n \geq 1\}$ is a complete orthonormal system for $L^2[0,1]$. He suggested that $\Delta_L$ comes from the singular part $f_s''$ of the second derivative $f''$, i.e.,

$$\Delta_L f(x) = \int_0^1 f_s''(x; u)\, du.$$

The Lévy Laplacian has been studied by many authors. In 1975, T. Hida introduced $\Delta_L$ into the theory of generalized white noise functionals [4]. H. -H. Kuo [16, 18, 20, 21] defined the Fourier-Mehler transform on the space $(\mathcal{S})^*$ of generalized white noise functionals and gave a relationship between this transform and $\Delta_L$. An interesting characterization of $\Delta_L$ in terms of rotation groups was obtained by N. Obata [25]. M. N. Feller [1] and E. M. Polishchuck [27] also obtained good results about $\Delta_L$. Recently, T. Hida [8] applied $\Delta_L$ to S. Tomonaga's many time theory in quantum physics. Y. Zhang [34] gave a probabilistic interpretation of $\Delta_L$ by using the ideas in [22].

It is important to construct a semi-group generated by $\Delta_L$ for solving the heat equation associated with $\Delta_L$. The purpose of this paper is to construct a semi-group and a group generated by $\Delta_L$.

In Section 2 we will explain a construction of the space of generalized white noise functionals and define the Lévy Laplacian in that space. Moreover, we introduce an operator $\Delta$ and prove that $\Delta$ coincides with $\widetilde{\Delta_L} \equiv U \circ \Delta_L \circ U^{-1}$ on a domain in the space $U[(\mathcal{S})^*]$ of $U$-functionals of $(\mathcal{S})^*$. Using a semi-group generated by $\Delta$, we will construct in Section 3 a semi-group generated by $\Delta_L$ on a domain $\mathbf{D}$. In Section

4 we will construct a group generated by $\Delta_L$ and give a relationship between this group and the adjoint operator of Kuo's Fourier-Mehler transform.

## 2. Preliminaries

In this section, we introduce a space of Hida distributions following [5, 14, 28] (See also, [10, 12, 26]) and the Lévy Laplacian defined on a domain in this space.

(A) Let $L^2(\mathbf{R})$ be the Hilbert space of real square-integrable functions on $\mathbf{R}$ with the inner product $(\cdot, \cdot)_0$. Consider a Gel'fand triple

$$\mathcal{S} = \mathcal{S}(\mathbf{R}) \subset L^2(\mathbf{R}) \subset \mathcal{S}^* = \mathcal{S}^*(\mathbf{R}),$$

where $\mathcal{S}(\mathbf{R})$ is the Schwartz space consisting of rapidly decreasing functions on $\mathbf{R}$ and $\mathcal{S}^*(\mathbf{R})$ is the dual space of $\mathcal{S}(\mathbf{R})$.

Let $\{|\cdot|_p; p \in \mathbf{Z}, p \geq 0\}$ be a countable system of consistent Hilbertian norms generating the nuclear topology of $\mathcal{S}$, where $|\cdot|_0$ is the norm of $L^2(\mathbf{R})$. We denote the completion of $\mathcal{S}$ with respect to the norm $|\cdot|_p$ by $\mathcal{S}_p$. Moreover, we assume $|\cdot|_p, p \in \mathbf{Z}, p \geq 0$, satisfy the following conditions:

(i) There exists a constant $\rho, 0 < \rho < 1$, such that $|\cdot|_p \leq \rho |\cdot|_{p+1}$ for all $p$.

(ii) For all $p$, the injection $\mathcal{S}_{p+1} \to \mathcal{S}_p$, is of Hilbert-Schmidt type.

We denote the dual space of $\mathcal{S}_p$ by $\mathcal{S}_p^*$. This space is a Hilbert space with the dual norm of $|\cdot|_p$.

(B) Let $\mu$ be a probability measure on $\mathcal{S}^*$ with the characteristic functional given by

$$C(\xi) \equiv \int_{\mathcal{S}^*} \exp\{i\langle x, \xi\rangle\}\, d\mu(x) = \exp\left[-\frac{1}{2}|\xi|_0^2\right], \quad \xi \in \mathcal{S}.$$

Let $(L^2) = L^2(\mathcal{S}^*, \mu)$ be the space of complex-valued square integrable functionals defined on $\mathcal{S}^*$ and define the *S-transform* by

$$S\varphi(\xi) = C(\xi) \int_{\mathcal{S}^*} \exp\{\langle x, \xi\rangle\}\varphi(x)\, d\mu(x), \quad \varphi \in (L^2).$$

This Hilbert space admits the well-known Wiener-Itô decomposition:

$$(L^2) = \oplus_{n=0}^{\infty} H_n,$$

where $H_n$ is the space of multiple Wiener integrals of order $n \in \mathbf{N}$ and $H_0 = \mathbf{C}$. From this decomposition theorem, each $\varphi \in (L^2)$ is uniquely represented as

$$\varphi = \sum_{n=0}^{\infty} \mathbf{I}_n(f_n), \quad f_n \in L_{\mathbf{C}}^2(\mathbf{R})^{\hat{\otimes}n},$$

where $\mathbf{I}_n(f_n) \in H_n$ and $L^2_{\mathbf{C}}(\mathbf{R})^{\hat{\otimes}n}$ denotes the n-th symmetric tensor product of the complexification of $L^2(\mathbf{R})$.

For each $p \in \mathbf{Z}, p \geq 0$, we define the norm $\|\varphi\|_p$ of $\varphi = \sum_{n=0}^{\infty} \mathbf{I}_n(f_n)$, by

$$\|\varphi\|_p = \left( \sum_{n=0}^{\infty} n! |f_n|_{p,n} \right)^{1/2},$$

where $|\cdot|_{p,n}$ is the norm of $\mathcal{S}_{\mathbf{C},p}^{\hat{\otimes}n}$ (the n-th symmetric tensor product of the complexification of $\mathcal{S}_p$). The norm $\|\cdot\|_0$ is nothing but the $(L^2)$-norm. We put

$$(\mathcal{S})_p = \{\varphi \in (L^2); \|\varphi\|_p < \infty\}$$

for $p \in \mathbf{Z}, p \geq 0$. Let $(\mathcal{S})_p^*$ be the dual space of $(\mathcal{S})_p$. Then $(\mathcal{S})_p$ and $(\mathcal{S})_p^*$ are Hilbert spaces with the norm $\|\cdot\|_p$ and the dual norm of $\|\cdot\|_p$, respectively.

Denote the projective limit space of the $(\mathcal{S})_p, p \in \mathbf{Z}, p \geq 0$, and the inductive limit space of the $(\mathcal{S})_p^*, p \in \mathbf{Z}, p \geq 0$, by $(\mathcal{S})$ and $(\mathcal{S})^*$, respectively. Then $(\mathcal{S})$ is a nuclear space and $(\mathcal{S})^*$ is nothing but the dual space of $(\mathcal{S})$. The space $(\mathcal{S})^*$ is called the space of *Hida distributions* ( or *generalized white noise functionals* ).

Since $\exp\langle \cdot, \xi \rangle \in (\mathcal{S})$, the $S$-transform is extended to an operator $U$ defined on $(\mathcal{S})^*$ :

$$U\Phi(\xi) = C(\xi)\langle\!\langle \Phi, \exp\langle \cdot, \xi \rangle \rangle\!\rangle, \quad \xi \in \mathcal{S},$$

where $\langle\!\langle \cdot, \cdot \rangle\!\rangle$ is the canonical pairing of $(\mathcal{S})$ and $(\mathcal{S})^*$. We call $U\Phi$ the *U-functional* of $\Phi$.

(C) We next introduce the definition of the Lévy Laplacian following Kuo [21] (see also [10]). Let $U$ be a *Fréchet differentiable* function defined on $\mathcal{S}$, i.e., we assume that there exists a map $U'$ from $\mathcal{S}$ to $\mathcal{S}^*$ such that

$$U(\xi + \eta) = U(\xi) + U'(\xi)(\eta) + o(\eta), \quad \eta \in \mathcal{S},$$

where $o(\eta)$ means that there exists a nonnegative integer $p$ depending on $\xi$ such that $o(\eta)/|\eta|_p \to 0$ as $|\eta|_p \to 0$. Then the first variation

$$\delta U(\xi; \eta) = \left. \frac{dU(\xi + \lambda\eta)}{d\lambda} \right|_{\lambda=0}$$

is expressed in the form

$$\delta U(\xi; \eta) = \int_{\mathbf{R}} U'(\xi; u)\eta(u) \, du, \quad \eta \in \mathcal{S}$$

276

by using the generalized function $U'(\xi; \cdot)$. We define the *Hida derivative* $\partial_t \Phi$ of $\Phi$ to be the generalized white noise functional whose $U$-functional is given by $U'(\xi; t)$.

**Definition** A Hida distribution $\Phi$ is called an *L-functional* if for each $\xi \in \mathcal{S}$, there exist $(U\Phi)'(\xi; \cdot) \in L^1_{loc}(\mathbf{R})$, $(U\Phi)''_s(\xi; \cdot) \in L^1_{loc}(\mathbf{R})$ and $(U\Phi)''_r(\xi; \cdot, \cdot) \in L^1_{loc}(\mathbf{R}^2)$ such that the first and second variations are uniquely expressed in the forms:

$$(U\Phi)'(\xi)(\eta) = \int_{\mathbf{R}} (U\Phi)'(\xi; u)\eta(u)\, du,$$

and

$$(U\Phi)''(\xi)(\eta, \zeta) = \int_{\mathbf{R}} (U\Phi)''_s(\xi; u)\eta(u)\zeta(u)\, du$$

$$+ \int_{\mathbf{R}^2} (U\Phi)''_r(\xi; u, v)\eta(u)\zeta(v)\, dudv, \tag{2.1}$$

for $\eta, \zeta \in \mathcal{S}$ such that $\int_T (U\Phi)''_s(\cdot; u)\, du$ is a $U$-functional for any finite interval $T$ on $\mathbf{R}$.

**Definition** Let $D_L$ denote the set of all $L$-functionals. For $\Phi \in D_L$ and any finite interval $T$ on $\mathbf{R}$, the *Lévy Laplacian* $\Delta_L^T$ is defined by

$$\Delta_L^T \Phi = U^{-1}\left[\frac{1}{|T|} \int_T (U\Phi)''_s(\cdot; u)\, du\right].$$

We define the *Lévy Laplacian* $\widetilde{\Delta_L^T}$ acting on the $U$-functionals of $L$-functionals by

$$\widetilde{\Delta_L^T} U[\Phi](\xi) = \frac{1}{|T|} \int_T (U\Phi)''_s(\cdot; u)\, du.$$

**Remark** The explicit conditions for the uniqueness of the above decomposition (2.1) is given in [10, chapter 6].

Let $T$ be a finite interval in $\mathbf{R}$. Take a smooth function $e$ on $\mathbf{R}$ such that $0 \le e(u) \le 1$ for all $u \in \mathbf{R}$, $e(u) = 1$ for $|u| \le 1/2$ and $e(u) = 0$ for $|u| \ge 1$. Let $\rho$ be a smooth function on $\mathbf{R}$ such that $\rho(u) = C \exp(-1/(1 - |u|^2))$ for $|u| < 1$, $\rho(u) = 0$ for $|u| \ge 1$ and $\int_{\mathbf{R}} \rho(u)du = 1$. Put $e_n(u) = e(u/n), \rho_n(u) = n\rho(nu)$ and $\theta_n^T = |\rho_n|_0^{-1}|T|^{-1/2}$, $n = 1, 2, \ldots$. We define an operator $\Delta$ acting on a functional $F(\xi)$, $\xi \in \mathcal{S}$, by

$$\Delta F(\xi) = \lim_{n \to \infty} \int_{\mathcal{S}^*} F''(\xi)(\theta_n^T e_n(x * \rho_n), \theta_n^T e_n(x * \rho_n))\, d\mu(x)$$

if the limit exists, where $x * \rho_n$ is the regularization of $x$.

From now on, we use $j_n(x)$ to denote $e_n(x * \rho_n)$.

Let $D_L^T$ denote the set of all $L$-functionals $\Phi$ satisfying $U\Phi(\eta) = 0$ for $\eta$ with $\operatorname{supp}(\eta) \subset T^c$. We have the following theorem.

**Theorem 2.1** *Let $T$ be a finite interval in $\mathbf{R}$ and let $F$ be a $U$-functional in $U[D_L^T]$. Then we have $\Delta F = \widetilde{\Delta_L^T F}$.*

**Proof** Since $F \in U[D_L^T]$, for any $\xi \in \mathcal{S}$, there exist functions $F_s''(\xi; \cdot) \in L^1(T)$ and $F_r''(\xi; \cdot, \cdot) \in L^1(T^2)$ such that

$$F''(\xi)(\eta, \zeta) = \int_T F_s''(\xi; u)\eta(u)\zeta(u)\, du + \int_{T^2} F_r''(\xi; u, v)\eta(u)\zeta(v)\, dudv,$$

for $\eta, \zeta \in \mathcal{S}$. We can use the Schwarz inequality to get

$$(\theta_n^T)^2 \int_{\mathcal{S}^*} |x * \rho_n(u)||x * \rho_n(v)|\, d\mu(x) \le \frac{1}{|T|}.$$

Hence, by the Fubini theorem and the dominated convergence theorem, we get

$$\int_{\mathcal{S}^*} \int_T F_s''(\xi; u)(\theta_n^T j_n(x)(u))^2\, dud\mu(x)$$
$$= \frac{1}{|T|} \int_T F_s''(\xi; u)e_n(u)^2\, du \longrightarrow \widetilde{\Delta_L^T F}(\xi), \quad \text{as } n \to \infty.$$

On the other hand, by the Fubini theorem again, we have

$$(\theta_n^T)^2 \int_{\mathcal{S}^*} \int_{T^2} F_r''(\xi; u, v)j_n(x)(u)j_n(x)(v)\, dudvd\mu(x) = \int_{T^2} K_n(\xi; u, v)\, dudv, \quad (2.2)$$

where $K_n(\xi; u, v)$ is given by

$$K_n(\xi; u, v) = F_r''(\xi; u, v)(\theta_n^T)^2 \int_{\mathcal{S}^*} j_n(x)(u)j_n(x)(v)\, d\mu(x).$$

Now, let

$$E_0 = \{(u, v) \in T^2; u = v\}, \qquad E_n = \{(u, v) \in T^2; |u - v| > \frac{2}{n}\}.$$

Then the integral (2.2) is equal to

$$\int_{E_n} + \int_{E_n^c} K_n(\xi; u, v)\, dudv.$$

278

Since $\{E_n^c\}$ is decreasing and $\cap E_n^c = E_n$,

$$\int_{E_n^c} |F_r''(\xi; u, v)|\, du dv \longrightarrow \int_{E_0} |F_r''(\xi; u, v)|\, du dv = 0.$$

But

$$\left| \int_{E_n^c} K_n(\xi; u, v)\, du dv \right| \le \frac{1}{|T|} \int_{E_n^c} |F_r''(\xi; u, v)|\, du dv.$$

Hence

$$\lim_{n \to \infty} \int_{E_n^c} K_n(\xi; u, v)\, du dv = 0.$$

For $(u, v) \in E_n$, it is clear that

$$\int_{\mathcal{S}^*} j_n(x)(u) j_n(x)(v)\, d\mu(x) = 0.$$

Consequently, $\Delta F = \widetilde{\Delta_L^T} F$ is obtained. $\qquad\square$

## 3. The Lévy Laplacian as an Infinitesimal Generator

A generalized functional $\Phi$ is called a *normal functional* if its $U$-functional $U\Phi$ is given by a finite linear combination of

$$\int_{A^k} f(u_1, \ldots, u_k) \xi(u_1)^{p_1} \cdots \xi(u_k)^{p_k}\, du_1 \cdots du_k, \tag{3.1}$$

where $f \in L^1(A^k), p_1, \ldots, p_k \in \mathbf{N} \cup \{0\}, k \in \mathbf{N}$, and $A$ is a finite interval on $\mathbf{R}$. Such a functional $\Phi$ is in $D_L$. Let $\mathcal{N}_T$ denote the set of all normal functionals in $D_L^T$. Define an operator $g_t$, $t \ge 0$, acting on a functional $F(\xi)$, $\xi \in \mathcal{S}$, by

$$g_t F(\xi) \equiv \lim_{n \to \infty} \int_{\mathcal{S}^*} F(\xi + \sqrt{2t}\, \theta_n^T j_n(x))\, d\mu(x),$$

if the limit exists. For a normal functional $\Phi$ with $U\Phi$ given by (3.1) and $A \subset T$, we can use the same argument as in the proof of Theorem 2.1 to check that

$$g_t[U\Phi](\xi) = \sum_{\nu_1=0}^{[p_1/2]} \cdots \sum_{\nu_k=0}^{[p_k/2]} \frac{p_1! \cdots p_k!}{(2\nu_1)!!(p_1 - 2\nu_1)! \cdots (2\nu_k)!!(p_k - 2\nu_k)!} \cdot$$

$$\left(\frac{2t}{|T|}\right)^{\nu_1 + \cdots + \nu_k} \int_{A^k} f(u_1, \ldots, u_k) \xi(u_1)^{p_1 - 2\nu_1} \cdots \xi(u_k)^{p_k - 2\nu_k}\, du_1 \cdots du_k.$$

Therefore, $g_t$ is an operator from $U[\mathcal{N}_T]$ into itself. An element of $U[\mathcal{N}_T]$ can be expressed as a finite sum of

$$F(\xi) = \int_{A^k} f(\mathbf{u}) P_k(\xi(\mathbf{u})) d\mathbf{u},$$

where $f \in L^1(A^k)$, $P_k$ is a monomial of $k$ variables and $\xi(\mathbf{u}) = (\xi(u_1), \ldots, \xi(u_k))$. By direct calculations, $g_t F(\xi)$ is given by

$$\int_{A^k} f(\mathbf{u}) \int_{\mathbf{R}^k} P_k\left( \xi(\mathbf{u}) + \left( \frac{2t}{|T|} \right)^{1/2} \mathbf{x} \right) G(x_1, 1) \cdots G(x_k, 1) d\mathbf{x} d\mathbf{u}, \qquad (3.2)$$

where $G(x, t) = (2\pi t)^{-1/2} \exp(-x^2/2t)$.

**Lemma 3.1** *Let $F_i$, $1 \le i \le n$, be functionals in $U[\mathcal{N}_T]$. Then*

$$g_t(F_1 F_2 \cdots F_n) = (g_t F_1)(g_t F_2) \cdots (g_t F_n).$$

**Proof** It is sufficient to prove the case for $n = 2$. Let

$$F_i(\xi) = \int_{A_i^{k_i}} f_i(\mathbf{u}) \xi(u_1)^{p_{1,i}} \cdots \xi(u_{k_i})^{p_{k_i,i}} d\mathbf{u}, \quad i = 1, 2.$$

By direct calculations,

$$g_t(F_1 F_2)(\xi) = \int_{A_1^{k_1} \times A_2^{k_2}} f_1(\mathbf{u}) f_2(\mathbf{v}) \int_{\mathbf{R}^{k_1 + k_2}} P_{k_1}\left( \xi(\mathbf{u}) + \left( \frac{2t}{|T|} \right)^{1/2} \mathbf{x} \right) \cdot$$

$$P_{k_2}\left( \xi(\mathbf{v}) + \left( \frac{2t}{|T|} \right)^{1/2} \mathbf{y} \right) G(x_1, 1) G(x_{k_1}, 1) G(y_1, 1) \cdots G(y_{k_2}, 1) d\mathbf{x} d\mathbf{y} d\mathbf{u} d\mathbf{v}.$$

Therefore, by (3.2), the assertion follows. $\qquad \square$

**Lemma 3.2** *For $F \in U[\mathcal{N}_T]$ and $s, t \ge 0$, we have $g_s(g_t F) = g_{s+t} F$.*

**Proof** By (3.2), $g_s(g_t F)(\xi)$ is given by

$$\int_{A^k \times \mathbf{R}^{2k}} f(\mathbf{u}) P_k(\xi(\mathbf{u}) + \mathbf{z}) G(z_1 - x_1, \frac{2s}{|T|}) G(x_1, \frac{2t}{|T|}) \cdots$$

$$\cdots G(z_k - x_k, \frac{2s}{|T|}) G(x_k, \frac{2t}{|T|}) d\mathbf{x} d\mathbf{z} d\mathbf{u}.$$

Since

$$\int_{\mathbf{R}} G(z - x, s) G(x, t)\, dx = G(z, s + t),$$

we obtain $g_s(g_t F) = g_{s+t} F$. $\qquad\qquad\qquad\qquad\qquad\qquad\qquad\qquad\qquad\qquad\qquad$ $\square$

**Lemma 3.3** *For any $F \in U[\mathcal{N}_T]$, the following equality holds*

$$\lim_{t \to 0} \frac{g_t F(\xi) - F(\xi)}{t} = \widetilde{\Delta_L^T} F(\xi), \quad \xi \in \mathcal{S}.$$

**Proof** For any $F \in U[\mathcal{N}_T]$, there exists $N \in \mathbf{N}$ such that for $\xi$ and $\eta$ in $\mathcal{S}$,

$$F(\xi + \eta) = \sum_{k=0}^{N} \frac{1}{k!} F^{(k)}(\xi)(\eta, \ldots, \eta),$$

By (3.2), we have

$$g_t F(\xi) = \lim_{n \to \infty} \sum_{k=0}^{N} \frac{(2t)^{k/2}}{k!} \int_{\mathcal{S}^*} F^{(k)}(\xi)(\theta_n^T j_n(x), \ldots, \theta_n^T j_n(x)) d\mu(x)$$

for any $t \geq 0$ and $\xi \in \mathcal{S}$. It is easily checked that

$$\lim_{n \to \infty} \int_{\mathcal{S}^*} F'(\xi)(\theta_n^T j_n(x)) d\mu(x) = 0,$$

and

$$\lim_{n \to \infty} \int_{\mathcal{S}^*} F^{(k)}(\xi)(\theta_n^T j_n(x), \ldots, \theta_n^T j_n(x)) d\mu(x)$$

exists and is in $U[\mathcal{N}_T]$ by direct calculations. Hence we get

$$\lim_{t \to 0} \frac{g_t F(\xi) - F(\xi)}{t} = \Delta F(\xi), \quad \xi \in \mathcal{S}.$$

Consequently, Theorem 2.1 implies the assertion of Lemma 3.3. $\qquad\qquad$ $\square$

Now, we introduce two notations to be used later. For $n \in \mathbf{N}$, let $\mathbf{D}_n$ denote the collection of Hida distributions $\Phi$ whose $U$-functionals are of the form $U\Phi = \psi \circ (F_1, F_2, \ldots, F_n)$ with $\psi$ and $F_i's$ satisfying the following conditions:

    (i) $\psi$ is an entire function on $\mathbf{C}^n$.

    (ii) $F_1, F_2, \ldots, F_n \in U[\mathcal{N}_T]$.

    (iii) $\psi \circ (g_t F_1, g_t F_2, \ldots, g_t F_n)$ is a $U$-functional for any $t > 0$.

Let $\mathbf{D}$ denote the linear span of $\{\mathbf{D}_n; n \in \mathbf{N}\}$. Note that if $\Phi \in \mathbf{D}$, then $\Phi \in D_L^T$ for any finite interval.

By using the operator $g_t$, we define an operator $G_t, t \geq 0$, from $\mathbf{D}$ into itself by

$$G_t \Phi = U^{-1}[\psi \circ (g_t F_1, \ldots, g_t F_n)], \quad t \geq 0.$$

If $\psi$ is a polynomial, then $G_t\Phi = U^{-1}[g_t(\psi \circ (F_1,\ldots,F_n))]$ by Lemma 3.1.

**Remark** Suppose $\Phi = U^{-1}[\psi \circ (F_1,\ldots,F_n)] \in \mathbf{D}$ satisfies the following condition

$$\sum_{k_1,\ldots,k_n=0}^{\infty} \frac{1}{k_1!\cdots k_n!}\left|\partial_{z_1}^{k_1}\cdots\partial_{z_n}^{k_n}\psi(0,\ldots,0)\right|\cdot$$

$$\sup_N \int_{\mathcal{S}^*}\left|F_1(\xi + \sqrt{2t}\,\theta_N^T j_N(x))^{k_1}\cdots F_n(\xi + \sqrt{2t}\,\theta_N^T j_N(x))^{k_n}\right| d\mu(x) < \infty.$$

Then by Lemma 3.1 $G_t\Phi$ is given by

$$G_t\Phi = U^{-1}\left[\lim_{n\to\infty}\int_{\mathcal{S}^*} U\Phi(\xi + \sqrt{2t}\,\theta_n^T j_n(x))d\mu(x)\right].$$

**Theorem 3.1** $\{G_t\}_{t\geq 0}$ *has the following properties:*
(a) *For any* $s,t \geq 0$,
$$G_{s+t} = G_s \circ G_t. \tag{3.3}$$

(b) *For any* $\Phi \in \mathbf{D}$ *and* $\xi \in \mathcal{S}$,

$$\lim_{t\to t_0} U[G_t\Phi](\xi) = U[G_{t_0}\Phi](\xi). \tag{3.4}$$

(c) *For any* $\Phi \in \mathbf{D}$ *and* $\xi \in \mathcal{S}$,

$$\lim_{t\to 0} U\left[\frac{G_t - I}{t}\Phi\right](\xi) = U[\Delta_L^T\Phi](\xi). \tag{3.5}$$

**Proof** Since $G_0 = I$, (3.3) is obvious when $s = 0$ or $t = 0$. Let $s$ and $t$ be positive. Suppose $\Phi = U^{-1}[\psi(F_1(\xi),\ldots,F_n(\xi))] \in \mathbf{D}$. By the definition, we have

$$U[G_s(G_t\Phi)](\xi) = \psi(g_s(g_t F_1)(\xi),\ldots,g_s(g_t F_n)(\xi)).$$

Hence, by Lemma 3.2, we get (3.3). The continuities of $\psi$ and $U[G_{(\cdot)}\Phi](\xi)$ for each $\xi \in \mathcal{S}$ imply (3.4). From Lemma 3.3 and the differentiability of $\psi$, it holds that

$$\lim_{t\to 0} U\left[\frac{G_t - I}{t}\Phi\right](\xi) = \frac{d}{dt}\psi(g_t F_1(\xi),\ldots,g_t F_n(\xi))\Big|_{t=0}$$

$$= \sum_{k=1}^{n} \frac{\partial\psi}{\partial z_k}(F_1(\xi),\ldots,F_n(\xi))\widetilde{\Delta_L^T F_k}(\xi)$$

for every $\xi \in \mathcal{S}$. On the other hand, since the second variation of $U\Phi$ is given by

$$U\Phi''(\xi)(\eta,\zeta) = \sum_{k,\ell=0}^{n} \frac{\partial^2 \psi}{\partial z_k \partial z_\ell}(F_1(\xi),\ldots,F_n(\zeta))F_k'(\xi)(\eta)F_\ell'(\xi)(\zeta) +$$

$$\sum_{k=1}^{n} \frac{\partial \psi}{\partial z_k}(F_1(\xi),\ldots,F_n(\xi))F_k''(\xi)(\eta,\zeta),$$

we see that the $U$-functional of the Lévy Laplacian $\Delta_L^T \Phi$ is equal to

$$\sum_{k=1}^{n} \frac{\partial \psi}{\partial z_k}(F_1(\xi),\ldots,F_n(\xi))\widetilde{\Delta_L^T F_k}(\xi).$$

Consequently, we obtain (3.5). $\qquad\qquad\qquad\qquad\qquad\qquad\qquad\qquad\square$

## 4. A Relationship Between the Adjoint Operator of Kuo's Fourier-Mehler Transform and a Group Generated by $\Delta_L^T$

A characterization of Hida distributions was obtained by J. Potthoff and L. Streit [28]. From [28], we see that for any $U$-functional $F$, and $\xi$, $\eta$ in $\mathcal{S}$, the function $F(\lambda\xi+\eta)$, $\lambda \in \mathbf{R}$, extends to an entire function $F(z\xi+\eta)$, $z \in \mathbf{C}$. Hence from (3.2), we can define an operator $g_{it}$, $t \in \mathbf{R}$, by

$$g_{it}F(\xi) = \lim_{n\to\infty} \int_{S^*} F(\xi + \sqrt{2it}\,\theta_n^T j_n(x))d\mu(x).$$

Since $\mu$ is symmetric, the integral is independent of choices of the branch for $\sqrt{2it}$. This fact will be understood without being mentioned again. Since Lemma 3.1 is valid for $g_{it}$ also, we can naturally define $G_{it}$, $t \in \mathbf{R}$, by

$$G_{it}\Phi = U^{-1}[\psi(g_{it}F_1,\ldots,g_{it}F_n)]$$

for $\Phi = U^{-1}[\psi(F_1,\ldots,F_n)] \in \mathbf{D}$. Theorem 3.1 implies the following.

**Theorem 4.1** $\{G_{it}\}_{t\in\mathbf{R}}$ *has the following properties:*

$$G_{i(s+t)} = G_{is} \circ G_{it}, \quad s,t \in \mathbf{R};$$

$$\lim_{t\to t_0} U[G_{it}\Phi](\xi) = U[G_{it_0}\Phi](\xi), \quad \xi \in \mathcal{S};$$

$$\lim_{t\to 0} U\left[\frac{G_{it}-I}{t}\Phi\right](\xi) = U[i\Delta_L^T\Phi](\xi), \quad \xi \in \mathcal{S}.$$

An infinite dimensional Fourier-Mehler transform $\mathbf{F}_\theta$, $\theta \in \mathbf{R}$, on $(\mathcal{S})^*$ was defined by H. -H. Kuo [19] as follows. The transform $\mathbf{F}_\theta\Phi$, $\theta \in \mathbf{R}$, of $\Phi \in (\mathcal{S})^*$ is defined to be the unique generalized white noise functional with the $U$-functional

$$U[\mathbf{F}_\theta\Phi](\xi) = U\Phi(e^{i\theta}\xi)\exp\left[\frac{i}{2}e^{i\theta}\sin\theta|\xi|_0^2\right], \quad \xi \in \mathcal{S}.$$

Moreover, the adjoint operator $\mathbf{F}_\theta^*$ of $\mathbf{F}_\theta$ is given by

$$\mathbf{F}_\theta^*\Phi = \sum_{n=0}^{\infty}\mathbf{I}_n(h_n(\Phi;\theta)) \quad \text{for } \Phi = \sum_{n=0}^{\infty}\mathbf{I}_n(f_n) \in (\mathcal{S}),$$

where

$$h_n(\Phi;\theta) = \sum_{m=0}^{\infty}\frac{(n+2m)!}{n!m!}(\frac{i}{2}\sin\theta)^m e^{i(m+n)\theta}\tau^{\otimes m} * f_{n+2m},$$

$$\tau^{\otimes m} = \int_{\mathbf{R}^m}\delta_{t_1}\otimes\delta_{t_1}\otimes\cdots\otimes\delta_{t_m}\otimes\delta_{t_m}\,dt_1\cdots dt_m.$$

This operator $\mathbf{F}_\theta^*$ is a continuous linear operator on $(\mathcal{S})$. For details, see [9, 19]. The Gross Laplacian $\Delta_G$ (see [2, 3]) and the number operator $N$ are defined on $(\mathcal{S})$ by:

$$\Delta_G\Phi = \int_{\mathbf{R}}\partial_t^2\Phi\,dt, \qquad N\Phi = \int_{\mathbf{R}}\partial_t^*\partial_t\Phi\,dt.$$

See, e.g., [17]. The operator $e^{i\theta N}$ is called the *Fourier-Wiener transform* (see [9]). Now, we introduce an operator $e^{\frac{i}{2}\theta\Delta_G}$ from $(\mathcal{S})$ into itself given by

$$e^{\frac{i}{2}\theta\Delta_G}\Phi = \sum_{n=0}^{\infty}\mathbf{I}_n(\ell_n(\Phi;\theta)), \; \ell_n(\Phi;\theta) = \sum_{m=0}^{\infty}\frac{(n+2m)!}{n!m!}(\frac{i}{2}\theta)^m\tau^{\otimes m} * f_{n+2m}, \quad (4.1)$$

for $\Phi = \sum_{n=0}^{\infty}\mathbf{I}_n(f_n) \in (\mathcal{S})$. Then we have the following lemmas.

**Lemma 4.1** $\mathbf{F}_\theta^* = e^{i\theta N} \circ e^{\frac{i}{2}(e^{i\theta}\sin\theta)\Delta_G}$.

**Proof** Take $\Phi = \sum_{n=0}^{\infty}\mathbf{I}_n(f_n) \in (\mathcal{S})$. From (4.1), we see that

$$e^{\frac{i}{2}(e^{i\theta}\sin\theta)\Delta_G}\Phi = \sum_{n=0}^{\infty}\mathbf{I}_n(\ell_n(\Phi;e^{i\theta}\sin\theta)).$$

Hence,

$$e^{i\theta N}(e^{\frac{i}{2}(e^{i\theta}\sin\theta)\Delta_G}\Phi) = \sum_{n=0}^{\infty}\mathbf{I}_n(e^{in\theta}\ell_n(\Phi;e^{i\theta}\sin\theta)).$$

Since $e^{in\theta}\ell_n(\Phi; e^{i\theta}\sin\theta) = h_n(\Phi;\theta)$, we obtain the assertion. $\qquad\square$

**Lemma 4.2** *For any $\Phi \in (\mathcal{S})$, we have*

$$U[e^{\frac{i}{2}\theta\Delta_G}\Phi](\xi) = \int_{\mathcal{S}^*} U\Phi(\xi + \sqrt{i\theta}\,y)\,d\mu(y). \tag{4.2}$$

**Remark** For any $\Phi \in (\mathcal{S})$, $\xi \in \mathcal{S}$ and $z_1, z_2 \in \mathbf{C}$, the functional $U\Phi(z_1\xi + z_2\eta)$, $\eta \in \mathcal{S}$, can be extended to a functional $\widetilde{U\Phi}(z_1\xi + z_2 y)$, $y \in \mathcal{S}^*$, which is in $(\mathcal{S})$ (cf. [14]). We denote this functional by the same symbol $U\Phi(z_1\xi + z_2 y)$.

**Proof** For $\Phi = \sum_{n=0}^{\infty} \mathbf{I}_n(f_n) \in (\mathcal{S})$, the right-hand side of (4.2) has the following expansion:

$$\sum_{n=0}^{\infty} \int_{\mathbf{R}^n} f_n(\mathbf{u}) \int_{\mathcal{S}^*} \{\xi(u_1) + \sqrt{i\theta}\,x(u_1)\} \cdots \{\xi(u_n) + \sqrt{i\theta}\,x(u_n)\}\,d\mu(x)d\mathbf{u}$$

$$= \sum_{n=0}^{\infty} \sum_{\nu=0}^{[n/2]} \frac{n!}{(2\nu)!!(n-2\nu)!}(i\theta)^{\nu}\langle\xi^{\otimes(n-2\nu)}, \tau^{\nu} * f_n\rangle = \sum_{m=0}^{\infty}\langle\xi^{\otimes m}, \ell_m(\Phi;\theta)\rangle.$$

From (4.1), we see that the last series is equal to $U[e^{\frac{i}{2}\theta\Delta_G}\Phi](\xi)$. $\qquad\square$

Define an operator $J_n$ by

$$U[J_n\Phi](\xi) = U\Phi \circ j_n(\xi), \quad \Phi \in \mathbf{D}, \quad \xi \in \mathcal{S}.$$

For all $n \in \mathbf{N}$ and $\Phi \in \mathbf{D}$, we can easily check $J_n\Phi \in (\mathcal{S})$. Then we have the following.

**Theorem 4.2** *Let $\Phi = U^{-1}[\psi(F_1,\ldots,F_n)] \in \mathbf{D}$. Assume that for all $t > 0$ and $\xi \in \mathcal{S}$,*

$$\sum_{k_1,\ldots,k_n=0}^{\infty} \frac{1}{k_1!\cdots k_n!}\left|\partial_{u_1}^{k_1}\cdots\partial_{u_n}^{k_n}\psi(0,\ldots,0)\right| \cdot$$

$$\sup_N \int_{\mathcal{S}^*}\left|((F_1 \circ j_N)^{k_1}\cdots(F_n \circ j_N)^{k_n})(e^{i\alpha_N(t)}\xi + \sqrt{ie^{i\alpha_N(t)}\sin\alpha_N(t)}\,x)\right|d\mu(x) < \infty,$$

*where $\alpha_N(t) = 2t(\theta_N^T)^2$. Then*

$$\lim_{N\to\infty} U[\mathbf{F}_{\alpha_N(t)}^* J_N\Phi](\xi) = U[G_{it}\Phi](\xi), \quad \xi \in \mathcal{S}. \tag{4.3}$$

**Proof** From Lemma 4.2, we have

$$U[e^{\frac{i}{2}e^{i\alpha_N(t)}\sin\alpha_N(t)\Delta_G}J_N\Phi](\xi) = \int_{S^*} U[J_N\Phi](\xi + \sqrt{ie^{i\alpha_N(t)}\sin\alpha_N(t)}\, y)\, d\mu(y).$$

This functional can be expressed in the form $\sum_{\ell=0}^{\infty}\langle\xi^{\otimes\ell}, f_{N,\ell}\rangle$ with $f_{N,\ell} \in S_{\mathbf{C}}^{\hat{\otimes}\ell}$. Hence, from Lemma 4.1, we get

$$U[\mathbf{F}^*_{\alpha_N(t)}J_N\Phi](\xi) = \sum_{\ell=0}^{\infty} e^{i\alpha_N(t)\ell}\langle\xi^{\otimes\ell}, f_{N,\ell}\rangle.$$

In view of the assumption, we can apply the Lebesgue convergence theorem to get

$$\lim_{N\to\infty} U[\mathbf{F}^*_{\alpha_N(t)}J_N\Phi](\xi) = \lim_{N\to\infty} U[e^{\frac{i}{2}e^{i\alpha_N(t)}\sin\alpha_N(t)\Delta_G}J_N\Phi](e^{i\alpha_N(t)}\xi)$$

$$= \lim_{N\to\infty}\int_{S^*} U[J_N\Phi](e^{i\alpha_N(t)}\xi + \sqrt{ie^{i\alpha_N(t)}\sin\alpha_N(t)}\, y)\, d\mu(y)$$

$$= \sum_{k_1,\ldots,k_n=0}^{\infty} \frac{1}{k_1!\cdots k_n!}\partial_{u_1}^{k_1}\cdots\partial_{u_n}^{k_n}\psi(0,\ldots,0)\cdot$$

$$\lim_{N\to\infty}\int_{S^*}((F_1\circ j_N)^{k_1}\cdots(F_n\circ j_N)^{k_n})(e^{i\alpha_N(t)}\xi + \sqrt{ie^{i\alpha_N(t)}\sin\alpha_N(t)}\, x)\, d\mu(x).$$

Using the similar argument as in the proof of those lemmas in Sections 3, we get

$$\lim_{N\to\infty}\int_{S^*}((F_1\circ j_N)^{k_1}\cdots(F_n\circ j_N)^{k_n})(e^{i\alpha_N(t)}\xi + \sqrt{ie^{i\alpha_N(t)}\sin\alpha_N(t)}\, x)\, d\mu(x)$$

$$= g_{it}(F_1^{k_1}\cdots F_n^{k_n})(\xi) = g_{it}F_1(\xi)^{k_1}\cdots g_{it}F_n(\xi)^{k_n}.$$

Consequently, we obtain (4.3). $\qquad\square$

**Acknowledgements** This research work was partially done during my visit to Louisiana State University in 1992-93. I would like to give my appreciation to LSU Mathematics Department for the warm hospitality. I would like also to express my gratitude to Professors T. Hida, H. Kunita, and H. -H. Kuo for their kind advice and encouragement.

## References

[1] Feller, M. N.: Infinite-dimensional elliptic equations and operators of Lévy type; *Russian Math. Surveys* **41:4** (1986) 119–170

[2] Gross, L.: Abstract Wiener spaces; in:*Proc. 5th Berkeley Symp. Math. Stat. Probab.* **2** 31–42, Univ. of Calif., Berkeley (1965)

[3] Gross, L.: Potential theory on Hilbert space; *J. Func. Anal.* **1** (1967) 123–181

[4] Hida, T.: *Analysis of Brownian Functionals.* Carleton Math. Lecture Notes, No.13, Carleton University, Ottawa, 1975

[5] Hida, T.: *Brownian Motion.* Application of Math., 11, Springer-Verlag, 1980

[6] Hida, T.: A role of the Lévy Laplacian in the causal calculus of generalized white noise functionals; in: *Stochastic Processes, A Festschrift in Honour of Gopinath Kallianpur*, S. Cambanis et al. (eds.) (1993) 131–139

[7] Hida, T.: The impact of classical functional analysis on white noise calculus; *Centro Vito Volterra, Università Degli Studi Di Roma II* (1992)

[8] Hida, T.: Random Fields as Generalized White Noise Functionals; *Proc. IFIP Enschede* (1992)

[9] Hida, T., Kuo, H. -H., and Obata, N.: Transformations for white noise functionals; *J. Funct. Anal.* **111** (1993) 259–277

[10] Hida, T., Kuo, H. -H., Potthoff, J., and Streit, L.: *White Noise: An Infinite Dimensional Calculus.* Kluwer Academic, 1993

[11] Hida, T. and Saitô, K.: White noise analysis and the Lévy Laplacian; in:*Stochastic Processes in Physics and Engineering*, S. Albeverio et al. (eds.) (1988) 177–184

[12] Hida, T., Obata, N., and Saitô, K.: Infinite dimensional rotations and Laplacian in terms of white noise calculus; *Nagoya Math. J.* **128** (1992) 65–93

[13] Itô, K.: Stochastic analysis in infinite dimensions; *Proc. International conference on stochastic analysis*, Evanston, Academic Press (1978) 187–197

[14] Kubo, I. and Takenaka, S.: Calculus on Gaussian white noise I, II, III and IV; *Proc. Japan Acad.* **56A** (1980) 376–380, **56A** (1980) 411–416, **57A** (1981) 433–436, **58A** (1982) 186–189

[15] Kuo, H. -H.: *Gaussian Measures in Banach Spaces.* Lecture Notes in Math. **463**, Springer-Verlag, 1975

[16] Kuo, H. -H.: Fourier-Mehler transforms of generalized Brownian functionals; *Proc. Japan. Acad.* **59A** (1983) 312–314

[17] Kuo, H. -H.: On Laplacian operators of generalized Brownian functionals; *Lecture Notes in Math.* **1203**, Springer-Verlag (1986) 119–128

[18] Kuo, H. -H.: The Fourier transform in white noise calculus; *J. Multi. Anal.* **31** (1989) 311–327

[19] Kuo, H. -H.: Fourier-Mehler Transforms in white noise analysis; in: *Gaussian Random Fields, the Third Nagoya Lévy Seminar* K. Itô and T. Hida (eds.), World Scientific (1991) 257–271

[20] Kuo, H. -H.: Convolution and Fourier transform of Hida distributions; *Lecture Notes in Control and Information Sciences* **176** (1992) 165–176, Springer-Verlag

[21] Kuo, H. -H.; An infinite dimensional Fourier transform; *Aportaciones Matemát. Notas de Investigación* **7** (1992) 1–12

[22] Kuo, H. -H., Obata, N. and Saitô, K.: Lévy Laplacian of generalized functions on a nuclear space; *J. Funct. Anal.* **94** (1990) 74–92

[23] Lévy, P.: *Lecons d'analyse Fonctionnelle.* Gauthier-Villars, Paris, 1922

[24] Lévy, P.: *Problèmes concrets d'analyse Fonctionnelle.* Gauthier-Villars, Paris, 1951

[25] Obata, N.: A characterization of the Lévy Laplacian in terms of infinite dimensional rotation groups; *Nagoya Math. J.* **118** (1990) 111–132

[26] Obata, N.: *Elements of White Noise Calculus.* Lecture Notes, Universität Tübingen, 1992

[27] Polishchuck, E. M.: *Continual Means and Boundary Value Problems in Function Spaces.* Birkhäuther, 1988

[28] Potthoff, J. and Streit, L.: A characterization of Hida distributions; *J. Funct. Anal.* **101** (1991) 212–229

[29] Saitô, K.: Itô's formula and Lévy's Laplacian I and II; *Nagoya Math. J.* **108** (1987) 67–76, **123** (1991) 153–169

[30] Saitô, K.: Lévy's Laplacian in the infinitesimal generator; *Report of the Faculty of Science and Technology, Meijo University* **30** (1990) 13–18

[31] Saitô, K.: On a construction of a space of generalized functionals; *Proc. Pre-seminar for Int. Conf. on Gaussian Random Fields*, part **2** (1991) 20–26

[32] Saitô, K.: The Lévy Laplacian in white noise analysis; *Preprint* (1992)

[33] Yosida, K.: *Functional Analysis.* 3rd edition, 1971, Springer-Verlag

[34] Zhang, Y.: The Lévy Laplacian and Brownian particles in Hilbert spaces; *Preprint* (1993)

Kimiaki Saitô
Department of Mathematics
Meijo University
Nagoya 468, JAPAN
*E-mail*: f43927g@nucc.cc.nagoya-u.ac.jp

H. SATO

# Infinite product and infinite sum

## 1. Introduction

In the natural science, sometimes from a simple fact, a wide and deep theory grows up. In this paper we should begin with a simple problem as follows:

*Let $\{x_k\}$ be a sequence of real numbers such that $x_k + 1 > 0$, $\forall k \in \mathbf{N}$. Then what are the relations between the convergence of the sum $s_n \equiv \sum_{k=1}^n x_k$ and that of the product $m_n \equiv \prod_{k=1}^n (1 + x_k)$ to a positive limit ?*

The answer is that, unless all of entries $x_k$'s have a definite sign, there are no implications between them. But stochastically we have beautiful relations.

**Theorem 1.1** [15] *Let $(X_k, \mathcal{F}_k)$ be a martingale difference sequence such that $X_k + 1 > 0$ a.s., $k \in \mathbf{N}$. Then the sum $S_n \equiv \sum_{k=1}^n X_k$ converges almost surely if and only if the product $M_n \equiv \prod_{k=1}^n (1 + X_k)$ converges to a positive limit almost surely.*

The aim of this paper is to show what we have proved starting from the above simple fact. First when $\{X_k\}$ is an independent random sequence, Theorem 1.1 is strengthened to Theorem 2.1 which shows, for example, that the almost sure convergence of $\sum_k X_k$ implies $\mathbf{E}[\sup_n | \sum_{k=1}^n X_k|] < +\infty$ and $\mathbf{E}[\sup_n \prod_{k=1}^n (1+X_k)] < +\infty$. Then these are extended to the integrable Lévy process $\{X_t\}$ with $\triangle X_t > -1$.

The most important application of Theorem 2.1 is to establish new criteria for the equivalence (mutual absolute continuity) of infinite product measures. In the sequel denote the equivalence of two probabilities $\mu$ and $\nu$ by $\mu \sim \nu$ and the singularity by $\mu \perp \nu$. Let $\mathbf{X} = \{X_k\}$ be an IID, $\mathbf{Y} = \{Y_k\}$ be an independent random sequence, and assume $\mathbf{X}$ and $\mathbf{Y}$ are also independent. Then $\mathbf{X}$ and $\mathbf{X} + \mathbf{Y}$ induce product measures $\mu_{\mathbf{X}} = \prod_k \mu_{X_k}$, and $\mu_{\mathbf{X}+\mathbf{Y}} = \prod_k \mu_{X_k+Y_k}$, respectively. Starting from Theorem 2.1, we gave several necessary, or sufficient conditions for $\mu_{\mathbf{X}+\mathbf{Y}} \sim \mu_{\mathbf{X}}$. In particular we extended the well-known Shepp's theorem on $\mu_{\mathbf{X}+\mathbf{Y}} \sim \mu_{\mathbf{X}}$ where $\mathbf{Y}$ is a deterministic sequence to the case where $\mathbf{Y}$ is random.

When $\mathbf{X}$ is a standard Gaussian sequence we gave several criteria for $\mu_{\mathbf{X}+\mathbf{Y}} \sim \mu_{\mathbf{X}}$ based on Theorem 2.1. But recently Fernique gave a necessary and sufficient condition based on other principles. On the other hand, the Gaussian case provides a typical example of *the multiplicative chaos* defined by Kahane [5] and our discussions gave a negative answer to one of his conjectures.

## 2. Uniform Integrabilities of an Additive Martingale and its Exponential Martingale

A sequence $\{X_k\}$ of independent random variables such that $\mathbf{E}[X_k] = 0, \ k \in \mathbf{N}$ is naturally a martingale difference sequence and Theorem 1.1 is extended to the following theorem.

**Theorem 2.1** [7, 14, 16]   *Let $\{X_k\}$ be a sequence of independent random variables such that $\mathbf{E}[X_k] = 0$, and $X_k + 1 > 0$ a.s. $k \in \mathbf{N}$. Then all of the following are equivalent :*

(A) $\mathbf{E}[\sup_n |\sum_{k=1}^n X_k|] < +\infty.$

(B) $S_n \equiv \sum_{k=1}^n X_k \quad$ *converges in* $\mathbf{L}^1.$

(C) $S_n \equiv \sum_{k=1}^n X_k \quad$ *converges almost surely.*

(D) $\sum_{k=1}^\infty \mathbf{E}[X_k : X_k \geq 1] < +\infty$ *and* $\sum_{k=1}^\infty \mathbf{E}[(X_k)^2 : X_k < 1] < +\infty.$

(E) $M_n \equiv \prod_{k=1}^n (1 + X_k)$ *converges to a positive limit almost surely.*

(F) $M_n$ *converges in* $\mathbf{L}^1.$

(G) $\mathbf{E}[\sup_n M_n] < +\infty.$

(H) $\sum_{k=1}^\infty (1 - \mathbf{E}[\sqrt{1 + X_k}]) < +\infty.$

(I) $\sum_{k=1}^\infty \mathbf{E}[(1 - \sqrt{1 + X_k})^2] < +\infty.$

On the other hand, let $\mathcal{X} = (X_t, \mathcal{F}_t), \quad t \in T \equiv [0, +\infty)$ be a *qàdlàg* (right continuous and left limit path) martingale such that $X_0 = 0$ a.s. and define the *exponential martingale* of $\mathcal{X}$ by

$$\mathcal{E}(\mathcal{X})_t \equiv \exp[X_t - \frac{1}{2}\langle X^c \rangle_t] \prod_{s \leq t}(1 + \triangle X_s) \exp[-\triangle X_s], \quad t \in T,$$

where $X_t^c$ denotes the continuous part of $\mathcal{X}$, $\langle X^c \rangle_t$ denotes the quadratic variation of $X^c$, and $\triangle X_s \equiv X_s - X_{s-}$ [10]. Then Theorem 2.1 is extended to the Lévy process as follows.

**Theorem 2.2** [16]   *Let $\{X_t\}_{t \in T}$ be an integrable Lévy process such that $X_0 = 0$ a.s., $\mathbf{E}[X_t] = 0, \ t \in T$ and*

$$\triangle X_t + 1 > 0 \quad for \ every \ t \in T \ a.s.$$

*Then all of the following statements are equivalent :*

(A) $\mathbf{E}[\sup_t |X_t|] < +\infty$.

(B) $\{X_t : t \in T\}$ *are uniformly integrable.*

(C) $X_\infty \equiv \lim_{t\to+\infty} X_t$ *exists almost surely.*

(E) $\mathcal{E}(\mathcal{X})_\infty \equiv \lim_{t\to+\infty} \mathcal{E}(\mathcal{X})_t > 0$ *a.s.*

(F) $\{\mathcal{E}(\mathcal{X})_t : t \in T\}$ *is a uniformly integrable martingale.*

(G) $\mathbf{E}[\sup_t \mathcal{E}(\mathcal{X})_t] < +\infty$.

## 3. Random Translations

Let $\mathbf{X} = \{X_k(\omega)\}$ be an IID on a probability space $(\Omega, \mathcal{F}, \mathbf{P})$ and $\mathbf{Y} = \{Y_k(t)\}$ be a sequence of independent random variables on another probability space $(T, \Sigma, \sigma)$. Then $\mathbf{X} + \mathbf{Y} = \{X_k(\omega) + Y_k(t)\}$ is defined as a sequence of independent random variables on the probability space $(\Omega \times T, \mathcal{F} \times \Sigma, \mathbf{P} \times \sigma)$ and $\mathbf{X}$ and $\mathbf{Y}$ are independent. Denote the probability measures on the sequence space $\mathbf{R}^\infty$ induced from $\mathbf{X}$, $\mathbf{Y}$ and $\mathbf{X} + \mathbf{Y}$ by $\mu_{\mathbf{X}}$, $\mu_{\mathbf{Y}}$ and $\mu_{\mathbf{X}+\mathbf{Y}}$, respectively. Then they are product measures

$$\mu_{\mathbf{X}} = \prod_k \mu_{X_k}, \quad \mu_{\mathbf{Y}} = \prod_k \mu_{Y_k} \quad \text{and} \quad \mu_{\mathbf{X}+\mathbf{Y}} = \prod_k \mu_{X_k+Y_k}$$

where $\mu_{X_k}$, $\mu_{Y_k}$ and $\mu_{X_k+Y_k}$ are the corresponding marginal distributions. In the remaining part of this paper, we always use these notations and assume the mutually absolute continuity

$$\mu_{X_k+Y_k} \sim \mu_{X_k} \quad for \ every \quad k \in \mathbf{N}.$$

Then we have

**Fact 3.1** [7] $\quad \mu_{\mathbf{X}+\mathbf{Y}} \sim \mu_{\mathbf{X}} \quad or \quad \mu_{\mathbf{X}+\mathbf{Y}} \perp \mu_{\mathbf{X}}$.

Our first problem is to characterize $\mu_{\mathbf{X}+\mathbf{Y}} \sim \mu_{\mathbf{X}}$ *in terms of the distributions of* $\mathbf{Y}$. When $\mathbf{Y}$ is a deterministic sequence or both $\mathbf{X}$ and $\mathbf{Y}$ are Gaussian, the following results are known.

**Fact 3.2** [20] $\quad$ *Let* $\mathbf{y} = \{y_k\}$ *be a real sequence.*

(Sh.1) $\mu_{\mathbf{X}+\mathbf{y}} \sim \mu_{\mathbf{X}} \quad \Rightarrow \quad \mathbf{y} \in \ell^2$.

(Sh.2) $\mu_{\mathbf{X}+\mathbf{y}} \sim \mu_{\mathbf{X}}$ *for every* $\mathbf{y} \in \ell^2$ *if and only if* $\mu_{X_1}$ *has a positive density* $f(x) \equiv \dfrac{d\mu_{X_1}}{dx}(x)$ *which is an absolutely continuous function and satisfies*

$$\mathbf{I}_1(f) \equiv \int_{-\infty}^{+\infty} \frac{f'(x)^2}{f(x)} \, dx \ < +\infty.$$

**Fact 3.3** [13]   *If $X_1$ obeys to a standard Gaussian law $N(0,1)$, and $Y_k$ obeys to a Gaussian law $N(0,v_k)$, $k \in \mathbf{N}$, then $\mu_{\mathbf{X}+\mathbf{Y}} \sim \mu_{\mathbf{X}}$ if and only if $\sum_k (v_k)^2 < +\infty$.*

The last condition implies $\mathbf{E}[\sum_k (Y_k)^4] = 3\sum_k (v_k)^2 < +\infty$, so that $\mathbf{Y} \in \ell^4$ a.s. Conversely, since $\mathbf{Y}$ is Gaussian, $\mathbf{Y} \in \ell^4$ a.s. implies $\sum_k (v_k)^2 < +\infty$ [2] so that $\mu_{\mathbf{X}+\mathbf{Y}} \sim \mu_{\mathbf{X}}$.

There are several contributions when $\mathbf{Y}$ is deterministic (for example [1]), but when $\mathbf{Y}$ is random, Kitada and Sato [8] are the first who analyse the problem systematically. Our starting point is the following theorem which is an immediate corollary of Theorem 2.1.

**Theorem 3.4** [8]   *Let*

$$Z_k(x) \equiv \frac{d\mu_{X_k+Y_k}}{d\mu_{X_k}}(x) - 1 \; , \quad k \in \mathbf{N}.$$

*Then all of the following are equivalent :*
- (C) $\sum_k Z_k(X_k)$ *converges almost surely.*
- (D) $\sum_k \mathbf{E}[Z_k(X_k) : Z_k(X_k) \geq 1] < +\infty$ *and* $\sum_k \mathbf{E}[Z_k(X_k)^2 : Z_k(X_k) < 1] < +\infty$.
- (H) $\sum_k (1 - \mathbf{E}[\sqrt{1 + Z_k(X_k)}]) < +\infty$.
- (J) $\mu_{\mathbf{X}+\mathbf{Y}} \sim \mu_{\mathbf{X}}$.

**Necessary Conditions**   There have been several contributions for necessary conditions [8, 12, 18, 20], but very recently Hino unified and improved them to the following theorem.

**Theorem 3.5** [4]   *If $\mu_{\mathbf{X}+\mathbf{Y}} \sim \mu_{\mathbf{X}}$, then for any $\varepsilon > 0$ we have*
- ($L_\varepsilon^2$) $\sum_k \mathbf{P}(|Y_k| > \varepsilon)^2 < +\infty$.
- ($M_\varepsilon$) $\sum_k \mathbf{E}[Y_k : |Y_k| \leq \varepsilon]^2 < +\infty$.
- ($M_\varepsilon^2$) $\sum_k \mathbf{E}[(Y_k)^2 : |Y_k| \leq \varepsilon]^2 < +\infty$.

**Corollary 3.6** [12]   *Let $\varepsilon = \{\varepsilon_k\}$ be a Rademacher sequence, $\mathbf{a} = \{a_k\}$ be a real sequence and define $\mathbf{a} \cdot \varepsilon \equiv \{a_k \varepsilon_k\}$. Then $\mu_{\mathbf{X}+\mathbf{a}\cdot\varepsilon} \sim \mu_{\mathbf{X}}$ implies $\mathbf{a} \in \ell^4$.*

**Theorem 3.7** [11]   *Assume $\mu_{\mathbf{X}+\mathbf{a}\cdot\varepsilon} \sim \mu_{\mathbf{X}}$ for every $\mathbf{a} \in \ell^4$. Then $\mu_{X_1}$ has a positive density $f(x) \equiv \dfrac{d\mu_{X_1}}{dx}(x)$ which is continuously differentiable, the derivative is an absolutely continuous function and satisfies*

$$\mathbf{I}_2(f) \equiv \int_{-\infty}^{+\infty} \frac{f''(x)^2}{f(x)} \, dx \; < +\infty.$$

## Sufficient Conditions

**Theorem 3.8** [8]  *If $\mu_{X_1}$ has a positive density $f(x) \equiv \dfrac{d\mu_{X_1}}{dx}(x)$ which is continuously differentiable, the derivative is an absolutely continuous function and satisfies*

$$\int_{-\infty}^{+\infty} \sup_{|t|<\varepsilon} \frac{f''(x+t)^2}{f(x)} \, dx \; < +\infty$$

*for some $\varepsilon > 0$ and $\mathbf{Y}$ is symmetric, then both*

$(L_\varepsilon)$ $\sum_k \mathbf{P}(|Y_k| > \varepsilon) < +\infty$

*and*

$(M_\varepsilon^2)$ $\sum_k \mathbf{E}[(Y_k)^2 : |Y_k| \le \varepsilon]^2 < +\infty$

*imply $\mu_{\mathbf{X}+\mathbf{Y}} \sim \mu_{\mathbf{X}}$.*

**Theorem 3.9** [19]  (1) *If $\mu_{X_1}$ has a positive density $f(x) \equiv \dfrac{d\mu_{X_1}}{dx}(x)$ which is continuously differentiable, the derivative is an absolutely continuous function and satisfies $\mathbf{I}_2(f) < +\infty$ and $\mathbf{Y}$ is symmetric, then $\mathbf{Y} \in \ell^4$ a.s. implies $\mu_{\mathbf{X}+\mathbf{Y}} \sim \mu_{\mathbf{X}}$.*
(2) $\mathbf{I}_1(f) \le \frac{3}{2}\sqrt{\mathbf{I}_2(f)}$ *for every probability density $f$ for which $\mathbf{I}_2(f)$ is defined..*

## 4. Gaussian Case

In this section we assume that $\mathbf{X} = \{X_k\}$ is a standard Gaussian sequence, that is, every $X_k$ obeys to $N(0,1)$. Then the previous discussions imply the following results.

**Fact 4.1** [7]  *If $\mathbf{y} = \{y_k\}$ is a deterministic sequence, then $\mu_{\mathbf{X}+\mathbf{Y}} \sim \mu_{\mathbf{X}}$ if and only if $\mathbf{y} \in \ell^2$.*

As an immediate corollary of the above fact we have

**Fact 4.2** ($\Leftarrow$ Fact 4.1)  $\mathbf{Y} \in \ell^2$ *a.s.* $\implies$ $\mu_{\mathbf{X}+\mathbf{Y}} \sim \mu_{\mathbf{X}}$

The above fact is true even if $\mathbf{Y}$ is not independent of $\mathbf{X}$. As a special case of Theorem 3.8, we have

**Fact 4.3** ($\Leftarrow$ Theorem 3.8)  *If $\mathbf{Y}$ is symmetric, then both*

$(L_\varepsilon)$ $\sum_k \mathbf{P}(|Y_k| > \varepsilon) < +\infty$

*and*

$(M_\varepsilon^2)$ $\sum_k \mathbf{E}[(Y_k)^2 : |Y_k| \le \varepsilon]^2 < +\infty$

*imply $\mu_{\mathbf{X}+\mathbf{Y}} \sim \mu_{\mathbf{X}}$.*

In particular we have

**Fact 4.4** ($\Leftarrow$ Theorem 3.9)  *If $\mathbf{Y}$ is symmetric, then $\mathbf{Y} \in \ell^4$ a.s. implies $\mu_{\mathbf{X}+\mathbf{Y}} \sim \mu_{\mathbf{X}}$.*

Recently, based on the 0-1 law of a Gaussian measure, Fernique gave a necessary and sufficient condition as follows.

**Theorem 4.5** [3]   *Let $\mathbf{X}$ be a standard Gaussian sequence. Then $\mu_{\mathbf{X}+\mathbf{Y}} \sim \mu_{\mathbf{X}}$ if and only if there exists a real sequence $\{\alpha_k\}$ such that*

(F.1)  *The series $\sum_k \mathbf{P}(|X_k| > \alpha_k)$,   $\sum_k (\mathbf{P} \times \sigma)(|X_k| + |Y_k| > \alpha_k)$ and $\sum_k (\mathbf{P} \times \sigma \times \sigma)(|X_k| + |Y_k + Y_k'| > \alpha_k)$ are convergent, where $\mathbf{Y}' = \{Y_k'\}$ is an independent copy of $\mathbf{Y}$ which is also independent of $\mathbf{X}$.*

(F.2)  $\sum_k \mathbf{E}_{\sigma \times \sigma}[(\exp[Y_k Y_k'] - 1)\mathbf{P}(|X_k + Y_k + Y_k'| \leq \alpha_k)]$ *is convergent.*

Furthermore he proved an interesting theorem.

**Theorem 4.6** [3]   *Let $\mathbf{X}$ be a standard Gaussian sequence. Then all of the following are equivalent:*

(G.1)  $\mu_{\mathbf{X}+\theta\mathbf{Y}} \sim \mu_{\mathbf{X}}$ *for every $\theta \in \mathbf{R}$, where $\theta\mathbf{Y} \equiv \{\theta Y_k\}$.*

(G.2)  *There exists a positive, increasing and divergent sequence $\mathbf{a} = \{a_k\}$ such that $\mu_{\mathbf{X}+\mathbf{a}\cdot\mathbf{Y}} \sim \mu_{\mathbf{X}}$, where $\mathbf{a} \cdot \mathbf{Y} \equiv \{a_k Y_k\}$.*

(G.3)  *There exists an independent random sequence $\mathbf{U} = \{U_k\}$ such that $\mu_{\mathbf{U}}$ is non-singular to $\mu_{\mathbf{Y}}$ and $\sum_k \mathbf{E}[\exp[\theta^2 U_k U_k'] - 1]$ is convergent for every $\theta \in \mathbf{R}$, where $\mathbf{U}' = \{U_k'\}$ is an independent copy of $\mathbf{U}$.*

In any cases $\mu_{\mathbf{X}+\mathbf{Y}} \sim \mu_{\mathbf{X}}$ is equivalent to the uniform integrability of the local densities

$$Q_n(\omega) \equiv \int_T \prod_{k=1}^n \exp[Y_k(t)X_k(\omega) - \frac{1}{2}Y_k(t)^2] \, d\sigma(t)$$

$$= \int_T \prod_{k=1}^n \exp\left(Y_k(t)X_k(\omega) - \frac{1}{2}\mathbf{E}^{\mathbf{X}}[(Y_k(t)X_k(\omega))^2]\right) d\sigma(t) , \quad n \in \mathbf{N}.$$

## 5. Multiplicative Chaos

Let $\{X_k(t,\omega)\}_{t\in T}$, $k \in \mathbf{N}$ be an independent family of centered Gaussian processes. Then a finite measure $\lambda$ on $(T, \Sigma)$ is transformed into a random measure defined by

$$\lambda(A,\omega) \equiv \lim_n \int_A \prod_{k=1}^n \exp\left[X_k(t,\omega) - \frac{1}{2}\mathbf{E}[X_k(t,\omega)^2]\right] d\lambda(t), \quad A \in \Sigma$$

and Kahane [5] called this transformation a *multiplicative chaos*. The reader recognizes that $\{Q_n\}$ above provides a typical example of the multiplicative chaos. Then the problem is to characterize the uniform integrability of

$$Q_n^\lambda(\omega) \equiv \int_T \prod_{k=1}^n \exp\left[X_k(t,\omega) - \frac{1}{2}\mathbf{E}[X_k(t,\omega)^2]\right] \, d\lambda(t), \quad n \in \mathbf{N}.$$

Kahane proved the following theorem which is important since it gives a final answer to the Høegh-Krohn model problem [9] in the quantum field theory.

**Theorem 5.1** [5] *Let $T \equiv [0,1]^d$ and $d\lambda(t) \equiv dt$, and assume*

$$c_k(s,t) \equiv \mathbf{E}[X_k(s,\omega)X_k(t,\omega)] \geq 0, \quad s,t \in T, \ k \in \mathbf{N}$$

*and that there exists $a > 0$ such that*

$$\sum_k c_k(s,t) = a\log^+ \frac{1}{\|s-t\|} + O(1), \quad s,t \in T.$$

*Then $\{Q_n^\lambda\}$ are uniformly integrable if*

$$J(\lambda) \equiv \int_T \int_T \exp[\frac{1}{2}\sum_k c_k(s,t)] \, d\lambda(s)d\lambda(t) < +\infty \quad (i.e.\ a < 2d)$$

*and $\lim_n Q_n^\lambda = 0$ a.s. if $J(\lambda) = +\infty$ (i.e., $a \geq 2d$).*

Then he gave the following conjecture.

**Conjecture** [6] *A multiplicative chaos $\{Q_n^\lambda\}$ is uniformly integrable if and only if there exists a sequence of finite measures $\{\lambda_m\}$ such that $J(\lambda_m) < +\infty$ for every $m \in \mathbf{N}$ and $\lambda = \sum_m \lambda_m$ (convergence in total variation).*

Sato and Tamashiro [18] gave a negative answer to the above conjecture by giving examples of $\lambda$ such that

$\{Q_n^\lambda\}$ *is uniformly integrable but* $J(\lambda) = +\infty$,

and that

$J(\lambda) < +\infty$ *but* $\{Q_n^\lambda\}$ *is not uniformly integrable.*

## References


[1] Chatterji, S. D. and Mandrekar, V.: Quasi-invariance of measures under translation; *Math. Zeitschrift* **154** (1977) 19–29

[2] Fernique, X.: Intégrabilité des vecteurs gaussiennes; *C.R. Acad. Sci. Paris* **270** (1970) 1698–1699

[3] Fernique, X.: Sur l'équivalence de certaines mesures produit; *Probab. Th. Rel. Fields* **98** (1994) 77–90

[4] Hino, M.: On equivalence of product measures by random translation; *J. Math. Kyoto Univ.* (1994) (to appear)

[5] Kahane, J. -P.: Sur le chaos multiplicatif; *Ann. Sci. Math. Québec* **9** (1985) 105–150

[6] Kahane, J. -P.: From Riesz products to random sets; *Harmonic Analysis*, Igari, S. (ed.) *Proceedings of a conference held in Sendai, Japan, August 1990* (1990) 125–139, Springer (Tokyo, Berlin, Heidelberg)

[7] Kakutani, S.: On the equivalence of infinite product measures; *Ann. Math.* **49** (1948) 214–224

[8] Kitada, K. and Sato, H.: On the absolute continuity of infinite product measure and its convolution; *Prob. Th. Rel. Fields* **81** (1989) 609–627

[9] Kusuoka, S.: Høegh-Krohn's model of quantum fields and the absolute continuity of measures; in: *Ideas and Methods in Quantum and Statistical Physics*, Albeverio, S., Fenstad, J.E., Holden, H., and Lindstrøm, T. (eds.), (1992) Cambridge University Press, London

[10] Meyer, P. A.: Un cours sur les integrals stochastiques; *Lecture Notes in Math.* **511** (1976) 245–400, Springer-Verlag, Berlin, Heidelberg, New York

[11] Okazaki, Y.: On equivalence of product measure by symmetric random $\ell_4$-translation; *J. Func. Anal.* **115** (1993) 100-103

[12] Okazaki, Y. and Sato, H.: Distinguishing a random sequence from a random translate of itself; *Ann. Probab.* (1993) (to appear)

[13] Rozanov, Yu. A.: On the density of one Gaussian measure with respect to another; *Th. Probab. its Appl.* **7** (1962) 82–87

[14] Sato, H.: On the convergence of the product of independent random variables; *J. Math. Kyoto Univ.* **27** (1987) 381–385

[15] Sato, H.: Convergence of sum and product of a martingale difference sequence; *Hiroshima Math. J.* **18** (1988) 69–72

[16] Sato, H.: Uniform integrabilities of an additive martingale and its exponential; *Stochastics and Stochastics Reports* **30** (1990) 163–169

[17] Sato, H. and Tamashiro, M.: Absolute continuity of one-sided random translations; *Stochastic Processes and Their Applications* (1992) (to appear)

[18] Sato, H. and Tamashiro, M.: Multiplicative chaos and random translation; *Ann. l'Inst. H. Poincaré* (1993) (to appear)

[19] Sato, H. and Watari, C.: Some integral inequalities and absolute continuity of a symmetric random translation; *J. Func. Anal.* **114** (1993) 257–266

[20] Shepp, L. A.: Distinguishing a sequence of random variables from a translate of itself; *Ann. Math. Stat.* **36** (1965) 1107–1112

Hiroshi Sato
Department of Mathematics
Faculty of Science
Kyushu University
Fukuoka 812, JAPAN
*E-mail*: sato@math.kyushu-u.ac.jp

A. SENGUPTA

# A limiting measure in Yang-Mills theory

## 1. Introduction

In this lecture we describe a result from quantum Yang-Mills theory: a certain family of measures $\mu_T$ ($T$ a positive parameter) converges, as $T \downarrow 0$, to a measure $\mu_0$. We will explain the main ideas involved in this result, present a sketch of the proof (details being in [9]), and discuss some related results and questions. The measure $\mu_T$ is a probability measure which underlies the quantum theory of gauge fields in a certain setting, and is sometimes called the Yang-Mills measure; $\mu_T$ is defined on an infinite dimensional non-linear space $\mathcal{C}$ and is given by the heuristic expression $d\mu_T(\omega) = N_T^{-1} e^{-S_{YM}(\omega)/T}$. The notation is explained in more detail later, but for now we note that $S_{YM}$ (the "Yang-Mills functional"), is a function on the space $\mathcal{C}$. It is a reasonable guess that $\mu_T$ might converge, as $T \downarrow 0$, to a probability measure $\mu_0$ on the space $\mathcal{C}_0$ of minima of $S_{YM}$. In section 3 we describe a precise formulation of this in a certain special setting, and sketch a proof of the result (details and related results are in [9]). In quantum physics, the measure $\mu_T$ would form the basis of a quantum theory of gauge fields in four dimensions if it could be defined rigorously in that setting. The construction of $\mu_T$ in four dimensions, unlike the two-dimensional theory which we work with here, is an unsolved problem of enormous difficulty. Our interest is, however, in a purely mathematical setting; physically this would correspond to gauge theory on a two dimensional torus. In this setting, $\mu_T$ is a rigorously defined object (the construction is carried out in [8]). In the physical context, the limit $T \downarrow 0$ may be interpreted in a variety of ways. If one views $T$ as corresponding to Planck's constant, then the limit $T \downarrow 0$ provides a comparison with classical physics, and is sometimes called the classical or semiclassical limit. In the language of statistical physics, one may view the limit $\lim_{T \downarrow 0} \mu_T$ as a "low-temperature limit".

In section 2, we set up the notation and summarize basic results from quantum gauge theory relevant to us here. In section 3, we outline the argument proving the convergence of $\mu_T$ to a probability measure $\mu_0$, for $SU(2)$ gauge theory over the torus. We conclude with a brief survey of related results and questions.

## 2. Gauge Theory over the Torus

**2.1 The surface** : $\Sigma$ will denote a torus equipped with a Riemannian metric. For convenience, we will take the metric to be scaled so that the total area of $\Sigma$ is 1. We will also work with some chosen point $m \in \Sigma$.

**2.2 The group** : We will work with the group $G = SU(2)$, consisting of all $2 \times 2$ complex matrices $A$, with determinant 1, for which $A^*A = I$ (wherein $A^*$ is the conjugate transpose of $A$, and $I$ is the identity matrix). Thus :

$$SU(2) = \left\{ \begin{pmatrix} a & b \\ -\bar{b} & \bar{a} \end{pmatrix} : a, b \in \mathbf{C}, |a|^2 + |b|^2 = 1 \right\}.$$

Taking $a = x_1 + ix_2$ and $b = x_3 + ix_4$, we see that $SU(2)$ can be identified with the three-dimensional unit sphere $S^3$ in $\mathbf{R}^4$.

**2.3 The Lie algebra** : The Lie algebra of $SU(2)$ is the tangent space to this sphere at the point $I \in SU(2)$, where $I = \begin{pmatrix} 1 & 0 \\ 0 & 1 \end{pmatrix}$. It is more useful to keep working in terms of matrices, and take this tangent space to be the space :

$$\underline{g} \stackrel{\text{def}}{=} \{\text{all } 2 \times 2 \text{ complex matrices } A, \text{ with } A^* = -A \text{ and } \text{tr}(A)=0\}$$
$$= \left\{ \begin{pmatrix} ia_1 & a_2 + ia_3 \\ -a_2 + ia_3 & -ia_1 \end{pmatrix} : a_1, a_2, a_3 \in \mathbf{R}^3 \right\}.$$

Thus $\underline{g}$ is a three-dimensional real vector space and is a Lie algebra (i.e. if $A, B \in \underline{g}$ then $[A, B] \stackrel{\text{def}}{=} AB - BA$ is also in $\underline{g}$). The matrix

$$\begin{pmatrix} ia_1 & a_2 + ia_3 \\ -a_2 + ia_3 & -ia_1 \end{pmatrix} \in \underline{g}$$

corresponds to the vector $(0, a_1, a_2, a_3)$ tangent to the unit sphere $S^3$ at the point $(1, 0, 0, 0)$ (which corresponds to $I \in SU(2)$).

There is a useful inner-product $\langle \cdot, \cdot \rangle_{\underline{g}}$ on $\underline{g}$ :

$$\langle A, B \rangle_{\underline{g}} = \text{tr}(A^*B) = -\text{tr}(AB).$$

Thus if

$$A = \begin{pmatrix} ia_1 & a_2 + ia_3 \\ -a_2 + ia_3 & -ia_1 \end{pmatrix} \quad \text{and} \quad B = \begin{pmatrix} ib_1 & b_2 + ib_3 \\ -b_2 + ib_3 & -ib_1 \end{pmatrix},$$

then we have $\langle A, B \rangle_{\underline{g}} = 2(a_1 b_1 + a_2 b_2 + a_3 b_3)$. Thus $\langle \cdot, \cdot \rangle_{\underline{g}}$ is, up to the factor 2, just the usual inner-product on $\{0\} \times \mathbf{R}^3 \subset \mathbf{R}^4$, the tangent space to $S^3$ at $(1,0,0,0)$. The reason why this inner product on $\underline{g}$ is useful is that it is $Ad-$ invariant, i.e $\langle xAx^{-1}, xBx^{-1} \rangle_{\underline{g}} = \langle A, B \rangle_{\underline{g}}$ for every $A, B \in \underline{g}$ and $x \in G$.

**2.4 Connections** : A *connection* $\omega$ is a smooth $\underline{g}-$valued $1-$form on $\Sigma$. Thus, for each vector $X \in T_p\Sigma$ (the tangent space to $\Sigma$ at $p \in \Sigma$), $\omega(X)$ is a matrix in $\underline{g}$. The mapping $T_p\Sigma \to \underline{g} : X \mapsto \omega(X)$ is linear, and if $Y$ is a smooth vector field then $p \mapsto \omega(Y_p)$ is smooth. The set $\mathcal{A}$ of all connections is an infinite dimensional linear space.

**2.5 Curvature** : If $\omega$ is a connection then its *curvature* $\Omega^\omega$ is the smooth $\underline{g}-$valued $2-$form defined as follows. For any $p \in \Sigma$ and any $X, Y \in T_p\Sigma$,

$$\Omega^\omega(X, Y) = d\omega(X, Y) + [\omega(X), \omega(Y)].$$

(Explanation of notation : choosing a basis $\{E_1, E_2, E_3\}$ of $\underline{g}$, and local coordinates $(x^1, x^2)$ on $\Sigma$, we can write $\omega = \sum_{r,j} \omega_j^r E_r dx^j$; then $d\omega = \sum_{r,i,j} \frac{\partial \omega_j^r}{\partial x^i} E_r dx^i \wedge dx^j$.)

**2.6 Gauge transformations** : A *gauge transformation* is a smooth map $\phi : \Sigma \to G$ satisfying $\phi(m) = I$ (a convenient normalization). The set of $\mathcal{G}$ all gauge transformations is a group under pointwise multiplication.

**2.7 Action of $\mathcal{G}$ on $\mathcal{A}$** : The group $\mathcal{G}$ acts on $\mathcal{A}$ by : $(\omega, \phi) \mapsto \omega^\phi$, where $\omega^\phi$ is the $1-$form given by $\omega^\phi(X) = \phi(p)^{-1} \omega(X) \phi(p) + \phi(p)^{-1} d\phi(X)$, for every $X \in T_p\Sigma$ and $p \in \Sigma$.

**2.8 The quotient space $\mathcal{C}$** : We denote by $\mathcal{C}$ the quotient of $\mathcal{A}$ under the action of $\mathcal{G}$ :

$$\mathcal{C} = \mathcal{A}/\mathcal{G}.$$

Thus $\mathcal{C}$ is, in a reasonable sense, an infinite dimensional non-linear space.

**2.9 The Yang-Mills action functional $S_{YM}$** : For $\omega \in \mathcal{A}$, $S_{YM}(\omega)$ is defined as follows. Choose at each point $p \in \Sigma$, an orthonormal basis $\big(e_1(p), e_2(p)\big)$ of $T_p\Sigma$. Denote by $d\sigma$ the area-measure on $\Sigma$ corresponding to the inner-product on $\Sigma$. Then:

$$S_{YM}(\omega) = \int_\Sigma \|\Omega^\omega\big(e_1(p), e_2(p)\big)\|_{\underline{g}}^2 \, d\sigma(p).$$

This definition makes sense because (by the $Ad-$invariance of the metric $\langle \cdot, \cdot \rangle_{\underline{g}}$) the number $\|\Omega^\omega(e_1, e_2)\|_{\underline{g}}$ is independent of the specific choice of the orthonormal basis $(e_1, e_2)$ in $T_p\Sigma$, and depends smoothly on $p$.

**2.10** $S_{YM}$ **on** $\mathcal{C}$ : It may be verified that $S_{YM}(\omega)$ retains the same value if $\omega$ is replaced by $\omega^{\phi}$ for any $\phi \in \mathcal{G}$. Thus there is induced a well-defined function on the quotient space $\mathcal{C}$, and we denote this function again by $S_{YM}$.

**2.11 The Yang-Mills measure** $\mu_T$ **on** $\mathcal{C}$ : This is a probability measure on $\mathcal{C}$, informally specified by :

$$d\mu_T(\omega) = \frac{1}{N_T}e^{-S_{YM}(\omega)/T}\,[\mathcal{D}\omega].$$

**2.12 Comparison with Wiener measure** : Recall that Wiener measure $\nu_T$ on the space of paths $x : [0,1] \to \mathbf{R}^n$ is given by the formal expression :

$$d\nu_T(x) = \frac{1}{N_T'}e^{-\frac{1}{T}\int_0^1 ||x'(t)||^2\,dt}\,[\mathcal{D}x].$$

This looks similar to the expression for $d\mu_T(\omega)$. Note, however, that $\mathcal{C}$ is a non-linear space, and that the Yang-Mills action $S_{YM}(\omega)$ has a quartic dependence on $\omega$, as opposed to the quadratic nature of the action $\int_0^1 ||x'(t)||^2\,dt$ for paths.

**2.13 The rigorous** $\mu_T$ : There is a well-defined probability measure $\mu_T$, living on a certain 'closure' of $\mathcal{C}$, which is the rigorous version of the informal measure described above. We will denote this closure also by $\mathcal{C}$. The construction of $\mu_T$ is carried out in [8]. The strategy used in the construction is to view the torus $\Sigma$ as the quotient of the unit disk with certain opposite arcs glued together. The Yang-Mills measure for the disk is Gaussian (it can be shown that $\mathcal{C}$, for the disk, is a linear space on which $S_{YM}$ is quadratic), and the topological procedure of gluing can be implemented probabilistically by means of a certain conditioning. The conditioning involves the use of a stochastic differential equation with values in $G$.

We shall not need the details of the construction. What will be of use to us is the "loop expectation value formula" described below.

**2.14 Overall goal** : The main goal is to determine a probability measure $\mu_0$ on the set $\mathcal{C}_0$ of minima of $S_{YM}$ such that

$$\lim_{T\downarrow 0} \int_{\mathcal{C}} f\,d\mu_T = \int_{\mathcal{C}_0} f\,d\mu_0$$

for 'interesting' functions $f$. In order to understand what these 'interesting' functions are we need the concept of 'holonomy'.

**2.15 Holonomy** : Let $[0,1] \to \Sigma : t \mapsto C(t)$ be a piecewise smooth loop, with $C(0) = C(1) = m$ (recall that $m$ is a chosen point on $\Sigma$). Let $\omega$ be a connection. Then there is a piecewise smooth map $[0,1] \to G : t \mapsto g_t$ such that $g_0 = I = \begin{pmatrix} 1 & 0 \\ 0 & 1 \end{pmatrix}$, and :

$$\frac{dg_t}{dt} g_t^{-1} = -\omega\big(C'(t)\big) \qquad (*)$$

The path $t \mapsto g_t$ is said to describe *parallel transport* by $\omega$ along the path $C$ on $\Sigma$. The *holonomy* of $\omega$ around $C$ is defined to be :

$$g(C;\omega) = g_1 \in G.$$

An easily verifiable fact is that $g(C;\omega) = g(C;\omega^\phi)$, for every gauge transformation $\phi$ (note : this would not be true without the normalization $\phi(m) = I$).

The notion of holonomy described above makes sense for smooth connections. However, the measure $\mu_T$ is not supported by smooth connections, or even pointwise defined ones. Thus to define holonomy in a sense which can be used in conjunction with $\mu_T$, it is necessary to extend the notion of holonomy to :

**2.16 Stochastic holonomy** : By viewing equation $(*)$, in a certain way, as a Fisk-Stratonovich stochastic differential equation, one has a $\mu_T$−almost-everywhere defined random variable

$$\mathcal{C} \to G : \omega \mapsto g(C;\omega).$$

This idea of viewing $(*)$ as a stochastic differential equation in a manner relevant to our use is due to L. Gross ([5] and [3] develop quantum gauge theory over the plane $\mathbf{R}^2$). A similar idea, in closely related but independent usage, has also been developed in several works of Albeverio et al. (for instance [1]). The precise definition of stochastic holonomy in the context of compact surfaces (in our case, the torus) appears in [8] : this definition requires that $C$ have certain properties ($C$ must be of the type described in the theorem below).

**2.17 Interesting functions on $\mathcal{C}$** : If $C_1,...,C_k$ are well-behaved closed curves on $\Sigma$, all based at $m$, and $f$ is a measurable function on $G$ then the function on $\mathcal{C}$ specified by $\omega \mapsto f\big(g(C_1;\omega),...,g(C_k;\omega)\big)$ is of interest.

The following result describes the joint distribution of stochastic holonomies for certain families of loops. Recall our earlier remark that the construction of $\mu_T$ involves a quotient map $q : D \to \Sigma$, where $D$ is the closed unit disk in $\mathbf{R}^2$ and $q$

pastes together certain arcs on the boundary $\partial D$. Consider a triangulation of $D$ whose bonds are either radial or are 'cross-radial' in the sense that a bond meets each radial line at most once. By adding a few bonds if necessary, we can ensure that the projection by $q$ of this triangulation yields a triangulation of $\Sigma$. We require also that $q(O) = m$, and that $m$ is a vertex of the triangulation. The triangulation of $\Sigma$ used in the statement of the theorem below must be of this type.

**Theorem** (Loop expectation formula) [8]  *Let $\mathcal{T} = \{\Delta_1, ..., \Delta_n\}$ be a triangulation of $\Sigma$, with bonds $\{e_1, \bar{e}_1, ..., e_M, \bar{e}_M\}$. Let $C_1, ..., C_k$ be closed curves, based at $m$, made up of bonds. Then for any bounded measurable function $f$ on $G^k$ :*

$$\int_C f\big(g(C_1;\omega), ..., g(C_k;\omega)\big)\, d\mu_T(\omega) =$$

$$\frac{1}{Z_T} \int_{G^M} f\big(x(C_1), ..., x(C_k)\big) \prod_{i=1}^{n} Q_{T|\Delta_i|}\big(x(\partial\Delta_i)\big)\, dx_{e_1}...dx_{e_M}$$

*wherein $Z_T$ is the normalizing constant obtained by setting $f = 1$ (or see the Corollary below for an explicit formula), $b \mapsto x_b \in G$ satisfies $x_{\bar{b}} = x_b^{-1}$ (here $b$ runs over the bonds, and $\bar{b}$ denotes the orientation-reverse of the bond $b$), $x(C) = x_{b_r}...x_{b_1}$ if the loop $C$ is the composite $b_r....b_1$ (the $b_i$ being bonds), $Q_t(x)$ is the density function at time $t$ of Brownian motion on $G$ with respect to Haar measure on $G$ of total mass $1$, and $|\Delta_i|$ denotes the area enclosed by the $2-$simplex $\Delta_i$.*

Our discussion of the limiting measure for the $\mu_T$ will focus on two loops $S_1$ and $S_2$ , based at $m$, which are the standard generators of the homotopy classes of based loops on $\Sigma$. For these two loops, the above theorem simplifies to :

**Corollary** (Loop expectation for basic loops)  *For any bounded measurable function $f$ on $G^2$,*

$$\int_C f\big(g(S_1;\omega), g(S_2;\omega)\big) d\mu_T(\omega) = \frac{1}{Z_T} \int_{G^2} f(a, b) Q_T(b^{-1}a^{-1}ba)\, dadb,$$

*wherein*

$$Z_T = \int_{G^2} Q_T(b^{-1}a^{-1}ba)\, dadb$$

*with $da$ and $db$ being the Haar measure on $G$ of total mass $1$.*

Henceforth, we will *assume* that the function $f$ is *invariant under conjugation,* i.e. that

$$f(x^{-1}ax, x^{-1}bx) = f(a, b) \qquad \text{for every} \quad x, a, b \in G.$$

This is essentially a harmless assumption since it is readily seen that the right side of the loop expectation value formula, for say the basic loops, is invariant under the replacement of $f(\cdot,\cdot)$ by the conjugate $f\big(x(\cdot)x^{-1}, x(\cdot)x^{-1}\big)$. Such invariant functions are more significant for other reasons as well.

## 3. The Limiting Expectation Values

Recall that our goal is to determine a probability measure $\mu_0$ on the set $C_0$ of minima of $S$ such that

$$\lim_{T\downarrow 0} \int_C f \, d\mu_T = \int_{C_0} f \, d\mu_0$$

for interesting functions $f$. More precisely, we would like to calculate the limit of $\int_C f\big(g(C_1;\omega), ..., g(C_k;\omega)\big) \, d\mu_T(\omega)$, as $T \downarrow 0$, where $C_1, ..., C_n$ are loops on $\Sigma$ as discussed earlier. By expressing each loop $C_i$ as a composite of contractible loops and the basic loops $S_1$ and $S_2$, it is possible to reduce the problem to calculating the limit of $\int_C f\big(g(S_1;\omega), g(S_2;\omega)\big) \, d\mu_T(\omega)$, as $T \downarrow 0$. Roughly speaking, the reason is that the stochastic holonomy around a contractible loop is $I$ in the limit $T \downarrow 0$ (for example, for the loop $\partial\Delta_i$, a factor $Q_{T|\Delta_i|}(z)$ appears in the expectation value formula, and as $T \downarrow 0$, this becomes a 'delta function' at $I$). See [9] for a detailed argument. So we focus on calculating $\lim_{T\downarrow 0} \int_C f\big(g(S_1;\omega), g(S_2;\omega)\big) \, d\mu_T(\omega)$.

**Theorem** (The limiting formula)  *If $f$ is a continuous function on $G^2$, then*

$$\lim_{T\downarrow 0} \int_C f\big(g(S_1;\omega), g(S_2;\omega)\big) \, d\mu_T(\omega) = \int_{K\times K} f(k, k') \, dk \, dk'$$

*wherein*

$$K = \left\{ \begin{pmatrix} e^{it} & 0 \\ 0 & e^{-it} \end{pmatrix} : t \in \mathbf{R} \right\} \subset G,$$

*and $dk$ and $dk'$ are the unit mass Haar measure on $K$.*

**The minima of $S_{YM}$, and $\mu_0$:** If $\omega$ is a minimum of $S_{YM}$ then the joint conjugacy-class of the holonomies $g(S_1;\omega)$ and $g(S_2;\omega)$ specifies $\omega$ uniquely up to gauge transformations (the minimum value of $S_{YM}$ is 0, corresponding to 'flat' connections, i.e. connections with zero curvature). More precisely, the map

$$C_0 \to G^2 : \omega \mapsto \big(g(S_1;\omega), g(S_2;\omega)\big)$$

is a bijection onto the set

$$\mathcal{F} = \{(a, b) \in G^2 : aba^{-1}b^{-1} = I\}.$$

If $g \in G$ and $(a, b) \in \mathcal{F}$ then $(gag^{-1}, gbg^{-1}) \in \mathcal{F}$. This conjugacy action transfers to a corresponding action of $G$ on $\mathcal{C}_0$, and the bijection between $\mathcal{C}_0$ and $\mathcal{F}$ induces a bijection between the quotient spaces $\mathcal{C}_0/G$ and $\mathcal{F}/G$. Let $W$ be the two-element group of transformations on $K \times K$ consisting of the identity map, and the map $(k_1, k_2) \mapsto (k_1^{-1}, k_2^{-1})$. Then the map $K^2/W \to \mathcal{F}/G : [a, b]_W \mapsto [a, b]_{\mathcal{F}}$ (in the obvious notation) is a bijection. The bijection between $\mathcal{C}_0/G$ and $\mathcal{F}/G$ then produces a bijection between $\mathcal{C}_0/G$ and $K^2/W$. The normalized Haar measure $dk dk'$ on $K^2$ projects to a probability measure on $K^2/W$, and transferring this measure to $\mathcal{C}_0/G$ we obtain the sought for measure $\mu_0$ (which may be lifted up to $\mathcal{C}_0$).

**Sketch Proof of the Limiting Formula** : We will describe here the main ideas used; details are in [9]. The first strategy one might try is to use the map $C : (a, b) \mapsto b^{-1}a^{-1}ba$ as a change of variables and then use the fact that $\lim_{T \downarrow 0} \int_G f(z)Q_T(z) \, dz = f(I)$ for nice enough $f$. However, it turns out that every point on $C^{-1}(I)$ is a critical point of $C$ and therefore $C$ cannot be used to change variables. (This problem is not as acute for surfaces of genus $> 1$, a case which presents other problems; see [4, 7, 10]). Our strategy is to expand $Q_t(x)$ in a 'Fourier series' on $G$ :

$$Q_t(x) = \sum_{n=1}^{\infty} e^{-tC_n/2} n \chi_n(x)$$

wherein $C_n = (n^2 - 1)/\kappa^2$ ($\kappa$ is a numerical constant determined by the metric on $\underline{g}$), $\chi_n$ is the character of the $n$–dimensional irreducible representation of $SU(2)$, and the convergence of the above series is pointwise and absolute.

We use the following coordinates on $SU(2)$ : every $g \in SU(2)$ can be expressed as a product :

$$g = k_\phi a_\theta k_\psi \qquad (\phi, \theta, \psi) \in [0, 2\pi] \times [0, \pi] \times [0, 2\pi]$$

wherein $k_\phi = \begin{pmatrix} e^{i\phi} & 0 \\ 0 & e^{-i\phi} \end{pmatrix}$ and $a_\theta = \begin{pmatrix} \cos\frac{\theta}{2} & i\sin\frac{\theta}{2} \\ i\sin\frac{\theta}{2} & \cos\frac{\theta}{2} \end{pmatrix}$.

Then the normalized Haar measure on $SU(2)$ can be expressed as :

$$dg = \frac{\sin\theta}{8\pi^2} \, d\phi d\theta d\psi.$$

This is the usual "surface-area" measure on the 'unit-sphere' $S^3 \subset \mathbf{R}^4$, normalized to have total mass 1. Actually, $(\phi, \theta, \psi)$ is not a genuine coordinate system on $SU(2)$, but it is good enough; in detail, the map $(0, 2\pi) \times (0, \pi) \times (0, 2\pi) \to SU(2) : (\phi, \theta, \psi) \mapsto k_\phi a_\theta k_\psi$ is a two-to-one diffeomorphism onto a dense open subset of $SU(2)$, and the expression for $dg$ is obtained by pulling back the unit-mass Haar measure to $[0, 2\pi] \times [0, \pi] \times [0, 2\pi]$.

Then we have :

$$\lim_{T \downarrow 0} \frac{\int_{G^2} f(a, b) Q_T(b^{-1}a^{-1}ba)\, dadb}{\int_{G^2} Q_T(b^{-1}a^{-1}ba)\, dadb} =$$

$$\lim_{T \downarrow 0} \frac{\sum_{n=1}^{\infty} e^{-TC_n/2} n \int_{G^2} f(a, b) \chi_n(b^{-1}a^{-1}ba)\, dadb}{\sum_{n=1}^{\infty} e^{-TC_n/2} n \int_{G^2} \chi_n(b^{-1}a^{-1}ba)\, dadb}.$$

A standard calculation (see [2] Exercise 2.4.17.3) shows that :

$$\int_{G^2} \chi_n(b^{-1}a^{-1}ba)\, dadb = \frac{1}{n}.$$

This implies that $\int_{G^2} Q_T(b^{-1}a^{-1}ba)\, dadb = \sum_{n=1}^{\infty} e^{-TC_n/2} \to \infty$, as $T \downarrow 0$. We can therefore apply a L'Hôpital argument to deduce that :

$$\lim_{T \downarrow 0} \frac{\int_{G^2} f(a, b) Q_T(b^{-1}a^{-1}ba)\, dadb}{\int_{G^2} Q_T(b^{-1}a^{-1}ba)\, dadb} = \lim_{n \to \infty} \frac{n \int_{G^2} f(a, b) \chi_n(b^{-1}a^{-1}ba)\, dadb}{n \int_{G^2} \chi_n(b^{-1}a^{-1}ba)\, dadb}$$

$$= \lim_{n \to \infty} n \int_{G^2} f(a, b) \chi_n(b^{-1}a^{-1}ba)\, dadb,$$

provided the latter limit exists as a finite number. Now we argue that this limit does exist (and is finite) and we calculate its value. First we write the integral using the coordinates on $SU(2)$:

$$n \int_{G^2} f(a, b) \chi_n(b^{-1}a^{-1}ba)\, dadb =$$

$$\frac{2n}{\pi} \int_0^\pi dt \left[ \sin^2 t \int_{[0, 2\pi] \times [0, \pi] \times [0, 2\pi]} \frac{d\phi\, \sin(\theta) d\theta\, d\psi}{8\pi^2} f(k_\phi a_\theta k_\psi, k_t) \chi_n(k_t a_\theta k_{-t} a_{-\theta}) \right],$$

where we have used the conjugation invariance of $f$ to simplify the $db$−integral to an integral over $K$ (see Proposition 2.5.2 of [2]).

At this stage, we use the explicit formula (see section 2.5 in [2]):

$$\chi_n(g) = \frac{\sin\left[n \cos^{-1}\{\frac{1}{2}Tr(g)\}\right]}{\sin\left[\cos^{-1}\{\frac{1}{2}Tr(g)\}\right]} \quad \text{for every } g \in SU(2) \setminus \{\pm I\}.$$

From this we notice, after some calculation, a fortunate identity :

$$n \sin^2 t \sin \theta \, \chi_n(k_t a_\theta k_{-t} a_{-\theta}) = \frac{\partial}{\partial \theta} \left\{ - \cos \left[ n \cos^{-1} A(\theta, t) \right] \right\},$$

wherein

$$A(\theta, t) \stackrel{\text{def}}{=} \frac{1}{2} Tr(k_t a_\theta k_{-t} a_{-\theta}) = 1 - 2 \left( \sin^2 \frac{\theta}{2} \right) \sin^2 t.$$

In view of this, we can integrate by parts (at this stage we need $f$ to be smooth, so we should actually go back and observe that it suffices to prove the original result for smooth $f$). Thus :

$$n \int_{G^2} f(a, b) \chi_n(b^{-1} a^{-1} ba) \, da db =$$

$$B.T. + \frac{2}{\pi} \int_0^\pi dt \int_{[0,2\pi] \times [0,\pi] \times [0,2\pi]} \frac{d\phi \, d\theta \, d\psi}{8\pi^2} \frac{\partial f(k_\phi a_\theta k_\psi, k_t)}{\partial \theta} \cos \left[ n \cos^{-1} A(\theta, t) \right].$$

The second term on the right is oscillatory and, after some work, the Riemann-Lebesgue Lemma (RLL) implies that it goes to 0 as $n \to \infty$. The boundary term $B.T.$ is :

$$B.T. = -\frac{2}{\pi} \int_0^\pi dt \int_{[0,2\pi]^2} \frac{d\phi.d\psi}{8\pi^2} \left\{ f(k_\phi a_\pi k_\psi, k_t) \cos(2nt) - f(k_{\phi+\psi}, k_t) \right\}$$

$$\to \frac{2}{\pi} \int_0^\pi dt \int_{[0,2\pi]^2} \frac{d\phi.d\psi}{8\pi^2} f(k_{\phi+\psi}, k_t), \quad \text{as } n \to \infty, \text{ (by RLL)}$$

$$= \int_K dk \int_0^\pi \frac{dt}{\pi} f(k, k_t)$$

$$= \int_{K \times K} f(k, k') \, dk \, dk',$$

where, in the last step, we used the conjugation-invariance property of $f$, and $a_\pi k_t a_\pi^{-1} = k_{-t}$.

## 4. Other Results, Other Questions

The result whose proof we have described is for $SU(2)-$gauge theory over the torus. For compact oriented surfaces of genus $> 1$, and for a general class of gauge groups, the limit has been studied in different ways by Forman [4], by Witten [10], and by King and Sengupta [7]. Determination of the limiting measure for gauge theory over the torus in the case of general compact groups $G$ has not been carried out yet (for $G = SO(3)$ this is done in [9]). The measure $\mu_0$, in the cases where it is known, can be identified with a volume measure associated to a natural symplectic structure on the moduli space (this is the quotient by the group of all gauge transformations not necessarily satisfying the restriction $\phi(m) = I$) of flat connections (more information on this may be found in [4, 6, 10]).

A natural question is whether there is a large deviation principle for the family $\{\mu_T\}_{T>0}$. With $\mu_T$ taken as a measure on the infinite dimensional space $\mathcal{C}$, one might suspect that the Yang-Mills functional $S_{YM}$ plays the role of the large deviation rate functional. However, it seems more sensible and convenient to address the problem restricting our attention to the loop expectation value formula for $\mu_T$, for a given set of loops (in this form, this question was raised by Funaki at the end of the talk). This is then a finite dimensional problem since, with a fixed choice of the loops $C_i$, we may view $\mu_T$ as a measure on $G^k$ (for some positive integer $k$).

# References


[1] Albeverio, S., Høegh-Krohn, R., and Holden, H.: Representation and Construction of Multiplicative Noise; *J. Funct. Anal.* **78** (1988)154–184

[2] Brocker, T. and tom Dieck, T.: *Representations of Compact Lie Groups.* Springer-Verlag (1985)

[3] Driver, B. K.: $YM_2$ : Continuum Expectations, Lattice Convergence, and Lassos; *Commun. Math. Phys.* **123** (1989) 575–616

[4] Forman, R.: Small volume limits of 2-d Yang-Mills; *Commun. Math. Phys.* **151** (1993) 39–52

[5] Gross, L., King, C., and Sengupta, A.: Two Dimensional Yang-Mills Theory via Stochastic Differential Equations; *Ann. Phys.* **194** (1989) 65–112

[6] King, C. and Sengupta, A.: An Explicit Description of the Symplectic Structure of Moduli Spaces of Flat Connections; *Preprint* (1994)

[7] King, C. and Sengupta, A.: The Semiclassical Limit of the Two Dimensional Quantum Yang-Mills Model; *Preprint* (1994)

[8] Sengupta, A.: Quantum Gauge Theory on Compact Surfaces; *Ann. Phys.* **221** (1993) 17–52

[9] Sengupta, A.: The Semiclassical Limit for $SU(2)$ and $SO(3)$ Gauge Theory on the Torus; *Preprint* (1994)

[10] Witten, E.: On Quantum Gauge Theories in Two Dimensions; *Commun. Math. Phys.* **141** (1991) 153–209



Ambar Sengupta
Department of Mathematics
Louisiana State University
Baton Rouge, LA 70803-4918, U.S.A.
*E-mail*: sengupta@marais.math.lsu.edu

S. WATANABE

# Some refinements of Donsker's delta functions

## 1. Introduction

Let $W = W_0(\mathbf{R}^d)$ be the Banach space of $d$-dimensional continuous paths $w :$ $[0,1] \to \mathbf{R}^d$ such that $w(0) = 0$ endowed with the supremum norm and $P$ be the ($d$-dimensional) standard Wiener measure on $W$ so that $(W, P)$ be the $d$-*dimensional Wiener space*. A *Donsker's delta function* is a formal composite $\delta_x(w(1))$ where $\delta_x(\cdot)$ is the Dirac delta function at $x \in \mathbf{R}^d$. This may be considered as a formal or fictitious density of the ($d$-dimensional) pinned Brownian measure (Brownian bridge measure) $P_{0,0}^{1;x}(\cdot) = P(\cdot|w(1) = x)$ with respect to the Wiener measure $P$. Of course, the pinned Brownian measure is singular to the Wiener measure and yet, it may be useful to consider such a formal density on the Wiener space similarly as the Dirac delta functions are useful in the operational calculus on the Euclidean space. A Donsker's delta function was first introduced by H. -H. Kuo [8] in the frame of white noise analysis. In particular, Kuo obtained the Wiener chaos expansion (i.e., Itô's multiple Wiener integral expansion) of Donsker's delta function so that we can determine the class of $L^2$-Wiener functionals which can be coupled with Donsker's delta function. Inspired by Kuo's work, we studied Donsker's delta functions in the frame of the Malliavin calculus [20]. Then we were able to determine more general class of Wiener functionals, not necessarily in $L^2$ nor with easily computable Wiener chaos expansions, which can be coupled with Donsker's delta functions. Note that, since the pinned Brownian measure is a nice Borel measure on $W$, any Borel function on $W$ can be coupled with the Donsker's delta function as far as it is integrable with respect to the pinned measure; Wiener functionals are however not Borel functions but *equivalence classes* of $P$-measurable functions. So our first refinement of Donsker's delta functions is concerned with determining as precisely as possible the class of Wiener functionals which can be coupled with Donsker's delta functions. This can be studied as a problem of determining all *Sobolev spaces* in the Malliavin calculus to which Donsker delta functions can belong. We will review the results in section 2 under a more general setting of Donsker's delta functions as explained in the next paragraph.

The second refinement or generalization is concerned with defining Donsker's delta functions $\delta_x(F)$ for $d$-dim.Wiener functionals $F$ more general than linear and nondegenerate Wiener functionals (i.e., more general than those in the first order Wiener chaos with nondegenerate covariance, typically the functional $w(1)$ ). Then the method based on the Wiener chaos expansions or $S$-transforms are no longer available. However, the method in the Malliavin calculus based on the integration by parts on the Wiener space can be used effectively to define a Donsker's delta function $\delta_x(F)$ for a $d$-dimensional functional $F$, which is nondegenerate and smooth, as an element in a Sobolev space with negative differentiability index. More generally, the composite or pull-back $T(F)$ can be defined for any Schwartz distribution $T$ on $\mathbf{R}^d$. The fact that $T(F)$ can be defined for any Schwartz distribution is important; we can deduce the differentiability in $x$ of Banach space valued map $x \to \delta_x(F)$ from the fact that $T(F)$ can be defined for $T$ being the derivatives in $x$ of the Dirac delta function $\delta_x$ . We can study the regularity of Donsker's delta functions in this way and thereby deduce the regularity of the probability density of $F$ and the conditional expectations given $F$. Also, a regular conditional probability distribution given $F$ (and thereby a nice disintegration formula) can be obtained (cf. Airault-Malliavin [2], Lescot [10]) from Donsker's delta functions by appealing to a theorem of Sugita [18] and Airault-Malliavin [2] that a positive element in Sobolev spaces defines a Borel measure, actually a regular measure in the sense of Itô [7], on the Wiener space.

The regularity assumption on $F$ can be weakened considerably and we can determine the class of Sobolev spaces containing $[(1 - \Delta)^{\beta/2}\delta_x](F)$, where $\beta \geq 0$ and $\Delta$ the Laplacian on $\mathbf{R}^d$, rather precisely with a help of the interpolation theory [22, 23, 24]. The results will be reviewed in section 2. Also, the interpolation theory helps us to find many non-differentiable Wiener functionals which can be coupled with Donsker's delta functions. This will be reviewed in section 3.

As a final refinement and generalization of Donsker's delta functions by Kuo, we discuss the Wiener chaos expansion of $\delta_x(F)$ in the case of *Itô functionals or Itô maps* $F$ defined by stochastic differential equations (SDE's) with nondegenerate diffusion coefficients. We can obtain the chaos expansion under the assumption of the Hölder continuity of coefficients by extending the method of *Levi's sums* for heat kernels [25]. The $L^2$-chaos kernels are shown to coincide with those given by the Krylov-Veretennikov-Zbonkin formula [19, 26]. A formal Wiener chaos thus obtained is usually not found in a Sobolev space in the Malliavin calculus but we can show that it can actually be found there under a little bit more stringent regularity assumption on the coefficients of SDE. The results will be reviewed in section 4.

## 2. Donsker's Delta Functions in the Malliavin Calculus

Malliavin calculus is, in short, a study of Wiener functionals under the *differentiable structure* on Wiener space [11]. Let $(B, H, \mu)$ be an abstract Wiener space in a usual standard notation; in particular, $H$ is the Cameron-Martin space. Basic notions are those of differential operators like the gradient operator (or the Gross-Shigekawa $H$-derivative) $D$, its dual divergence operator (or the Skorohod operator) $D^*$, the Ornstein-Uhlenbeck operator $L = -D^*D$ and those of Sobolev spaces $\mathcal{D}_p^\alpha$, $1 < p < \infty$, $\alpha \in \mathbf{R}$, of real Wiener and generalized Wiener functionals. Roughly,

$$\mathcal{D}_p^\alpha = (I - L)^{-\alpha/2}(\mathcal{L}_p)$$

with the norm

$$\|F\|_{p,\alpha} = \|(I - L)^{\alpha/2} F\|_{\mathcal{L}_p},$$

where $\mathcal{L}_p$ is the usual $L_p$ space. In particular,

$$\mathcal{D}_p^0 = \mathcal{L}_p, \quad \mathcal{D}_{p'}^{\alpha'} \subset \mathcal{D}_p^\alpha \quad \text{if} \quad p < p' \quad \text{and} \quad \alpha < \alpha', \quad (\mathcal{D}_p^\alpha)' = \mathcal{D}_{p/(p-1)}^{-\alpha}.$$

So, roughly, $F \in \mathcal{D}_p^\alpha$ if and only if the derivatives of $F$ up to the $\alpha$-th order belong to $\mathcal{L}_p$ and therefore, $p$ and $\alpha$ are called the *integrability index* and the *differentiability index* of the Sobolev space, respectively. Similarly as in the Schwartz theory of distributions, we define the space of test Wiener functionals by

$$\mathcal{D}_{\infty-}^\infty = \bigcap_{\alpha > 0} \bigcap_{1 < p < \infty} \mathcal{D}_p^\alpha$$

and then its dual, the space of generalized Wiener functionals, is given by

$$\mathcal{D}_{1+}^{-\infty} = \bigcup_{\alpha > 0} \bigcup_{1 < p < \infty} \mathcal{D}_p^{-\alpha}.$$

$\mathcal{D}_{\infty-}^\infty$ is an algebra and $\mathcal{D}_{1+}^{-\infty}$ is a $\mathcal{D}_{\infty-}^\infty$-module. The *generalized expectation* $E(f \cdot \Phi)$ for $f \in \mathcal{D}_p^\alpha$ and $\Phi \in \mathcal{D}_{p/(p-1)}^{-\alpha}$ is by definition the natural coupling of $f$ and $\Phi$, in particular, the generalized expectation $E(\Phi)$ of $\Phi \in \mathcal{D}_{1+}^{-\infty}$ is the natural coupling

310

of $\Phi$ with $\mathbf{1}$, the test Wiener functional identically equal to 1. $E(\Phi)$ coincides with the usual expectation $\int_B \Phi(w)\mu(dw)$ when $\Phi$ is defined by an integrable Wiener functional. Furthermore, we introduce the following notations

$$\mathcal{D}^{\alpha}_{\infty-} := \bigcap_{1<p<\infty} \mathcal{D}^{\alpha}_p \quad \text{and} \quad \mathcal{D}^{\alpha}_{1+} := \bigcup_{1<p<\infty} \mathcal{D}^{a}_p$$

so that

$$\mathcal{D}^0_{\infty-} = \mathcal{L}_{\infty-} := \bigcap_{1<p<\infty} \mathcal{L}_p \quad \text{and} \quad \mathcal{D}^0_{1+} = \mathcal{L}_{1+} := \bigcup_{1<p<\infty} \mathcal{L}_p.$$

$\mathcal{D}^{\alpha}_{\infty-}$ is a Fréchet algebra for each $\alpha \geq 0$ (cf. Kusuoka [9]). If we consider, more generally, $E$-valued Wiener functionals, $E$ being a real separable Hilbert space, the corresponding spaces are denoted by $\mathcal{L}_p[E]$, $\mathcal{D}^{\alpha}_p[E]$, $\mathcal{D}^{\infty}_{\infty-}[E]$, etc. We refer to e.g., [5, 11, 21] for details of the above notions and their basic facts like Meyer's equivalence of norms.

Let $\mathcal{S}(\mathbf{R}^d)$ be the real Schwartz space of rapidly decreasing $C^{\infty}$-functions on $\mathbf{R}^d$ and $\mathcal{S}'(\mathbf{R}^d)$ be the real space of Schwartz tempered distributions on $\mathbf{R}^d$. Choose a system of increasing and compatible norms $|\ \ |_{S_n}$, $n \in \mathbf{Z}$, on $\mathcal{S}(\mathbf{R}^d)$ which are complete in $\mathcal{S}'(\mathbf{R}^d)$ such that $\mathcal{S}(\mathbf{R}^d) = \bigcap_{n=1}^{\infty} S_n$ and $\mathcal{S}'(\mathbf{R}^d) = \bigcup_{n=1}^{\infty} S_{-n}$, $S_n$ being the closure in $\mathcal{S}'(\mathbf{R}^d)$ of $\mathcal{S}(\mathbf{R}^d)$ w.r.t. $|\ \ |_{S_n}$, (cf. [5, 21] for an example of such norms; a particular choice is irrelevant, however, in future discussions).

Let $F : B \to \mathbf{R}^d$ be a $d$-dimensional Wiener functional or map, i.e., an equivalence class of $\mu$-measurable maps coinciding each other $\mu$-almost surely. For given $T \in \mathcal{S}'(\mathbf{R}^d)$, we want to define the composite $T \circ F = T(F(w))$ as an element in $\mathcal{D}^{-\infty}_{1+}$. If $T = \varphi \in \mathcal{S}(\mathbf{R}^d)$, then $\varphi \circ F$ is obviously a Wiener functional such that $\varphi \circ F \in \bigcap_{1<p<\infty} \mathcal{D}^0_p (= \mathcal{L}_{\infty-})$.

**Definition 1** We say that $T \circ F$ is defined in $\mathcal{D}^{-\infty}_{1+}$ and $T \circ F = \Phi$ if $\Phi \in \mathcal{D}^{-\infty}_{1+}$ exists and, in addition, there exist $1 < p < \infty$, $\alpha \geq 0$ and $n \in \mathbf{Z}^+$ such that $T \in \mathcal{S}_{-n}$, $\Phi \in \mathcal{D}^{-\alpha}_p$ and the following holds: For any sequence $\{\varphi_k\}$ in $\mathcal{S}(\mathbf{R}^d)$ such that $|\varphi_k - T|_{S_{-n}} \to 0$ as $k \to \infty$, it holds that $\|\varphi_k \circ F - \Phi\|_{p,-\alpha} \to 0$ as $k \to \infty$.

It is obvious that $\Phi$ is uniquely determined whenever $T \circ F$ is defined in the sense of Definition 1. In the following, we are primarily interested in the case of $T \in \mathcal{S}'(\mathbf{R}^d)$ given by $T = (1 - \Delta)^{\beta/2}\delta_x$, $\beta \geq 0$, $\delta_x$ being the Dirac delta function at $x \in \mathbf{R}^d$ and $\Delta$ the Laplacian. $\delta_x \circ F$, when it is defined, is called *Donsker's delta function*. This notion is important for the study of the density $p_F$ of the law of $F$

and the conditional expectations given $F = x$. For example, if $\delta_x(F)$ can be defined in $\mathcal{D}_{1+}^{-\infty}$ in such a way that, for some $\alpha > 0$ and $1 < p < \infty$, the map

$$\mathbf{R}^d \ni x \to \delta_x(F) \in \mathcal{D}_p^{-\alpha}$$

is weakly continuous, then $p_F(x) = E(\delta_x(F))$ is continuous in $x$ and for every $G \in \mathcal{D}_{p/(p-1)}^\alpha$, $E(G \cdot \delta_x(F)) = E(G|F = x)p_F(x)$ is also continuous in $x$.

**Definition 2**  A $d$-dimensional Wiener map $F$ is called *regular in the sense of Malliavin* if

(i)  $F \in \mathcal{D}_{\infty-}^\infty[\mathbf{R}^d]$, i.e., $F = (F^1, \ldots, F^d)$ with $F^i \in \mathcal{D}_{\infty-}^\infty$ and

(ii)  the Malliavin covariance $\sigma_F$ of $F$ (defined by $\sigma_F^{ij}(w) =< DF^i, DF^j >_{H^*}$ ) is nondegenerate in the sense that $(\det \sigma_F)^{-1} \in \mathcal{L}_{\infty-}$.

It is well-known [5, 21] that, if $F$ is regular in the sense of Malliavin then $T \circ F$ is defined in $\mathcal{D}_{1+}^{-\infty}$ for every $T \in \mathcal{S}'(\mathbf{R}^d)$ and $T \circ F \in \bigcup_{\alpha>0} \bigcap_{1<p<\infty} \mathcal{D}_p^{-\alpha}$. In particular, for every $m = 0, 1, \ldots$ and $x \in \mathbf{R}^d$,

$$[(1 - \Delta)^m \delta_x] \circ F \in \bigcap_{1<p<\infty} \mathcal{D}_p^{-2k} \quad \text{provided} \quad k > m + d/2.$$

Next, we introduce several notations for spaces of functions defined on $\mathbf{R}^d$, [1, 15]. Let $L_p(\mathbf{R}^d)$ be the usual $L_p$-space with respect to the Lebesgue measure with the norm $|\ \ |_p$ . Define the family of Bessel potential spaces by

$$L_p^\alpha(\mathbf{R}^d) = \{f \mid (1 - \Delta)^{\alpha/2} f \in L_p(\mathbf{R}^d)\}, \quad \alpha \in \mathbf{R}, \quad 1 < p < \infty,$$

with the norm $|f|_{p,\alpha} = |(1 - \Delta)^{\alpha/2} f|_p$ so that

$$\mathcal{S}(\mathbf{R}^d) \subset L_p^\alpha(\mathbf{R}^d) \subset L_p(\mathbf{R}^d) \subset L_p^{-\alpha}(\mathbf{R}^d) \subset \mathcal{S}'(\mathbf{R}^d), \quad \alpha \geq 0.$$

Let $C^\nu(\mathbf{R}^d)$, $\nu \geq 0$, be the Banach space of $[\nu]$-times continuously differentiable functions on $\mathbf{R}^d$ whose $[\nu]$-th derivatives are uniformly Hölder continuous with exponent $\nu - [\nu]$ with the norm

$$|f|_{C^\nu} = \sum_{\mathbf{n}; |\mathbf{n}| \leq [\nu]} |\partial^{\mathbf{n}} f|_\infty + \sum_{\mathbf{n}; |\mathbf{n}| = [\nu]} \sup_{x \neq y} |\partial^{\mathbf{n}} f(x) - \partial^{\mathbf{n}} f(y)|/|x - y|^{\nu - [\nu]}$$

with the notations $\mathbf{n} = (n_1, \ldots, n_d)$, $|\mathbf{n}| = \sum_{i=1}^d n_i$ and $\partial^{\mathbf{n}} = \partial_1^{n_1} \cdots \partial_d^{n_d}$, $(\partial_i = \partial/\partial x_i)$. Note that if $0 \leq \nu < \nu'$, then $C^{\nu'}(\mathbf{R}^d) \subset C^\nu(\mathbf{R}^d)$ with the compact inclusion and furthermore, $(1 - \Delta)^{-\nu/2}(C^0(\mathbf{R}^d)) \subset C^\nu(\mathbf{R}^d)$.

Let $F : B \to \mathbf{R}^d$ be a $d$-dimensional Wiener functional.

**Definition 3** For $\delta > 0$, we say that $F$ is *in the class* $\mathbf{D}(\delta)$ if $F$ satisfies the following conditions:

(i) $F \in \mathcal{D}_{\infty-}^{1+\delta}[\mathbf{R}^d]$.

(ii) $(\det \sigma_F)^{-1} \in \mathcal{L}_{\infty-}$.

(iii) The density $p_F$ of the law of $F$ is essentially bounded, i.e., $|p_F|_\infty$ is finite.

**Remark 1** It is well-known that the assumptions (i) and (ii) in Definition 3 with $\delta$ large enough implies (iii); actually it is enough to assume that $1 + \delta \geq 2l \geq d$ for some integer $l$. In applications, however, there are some cases in which we can verify the assumption (iii) even for small $\delta$ from other sources. If $F$ is regular, then clearly it is in the class $\mathbf{D}(\delta)$ for every $\delta > 0$.

**Theorem 1** [24] *If* $F \in \mathbf{D}(\delta)$ *for some* $\delta > 0$, *then* $(1 - \Delta)^{\beta/2} \delta_x \circ F$ *can be defined in* $\mathcal{D}_{1+}^{-\infty}$ *for every* $0 \leq \beta < \delta$. *To be more precise,*

$$(1 - \Delta)^{\beta/2} \delta_x \circ F \in \mathcal{D}_p^{-\alpha} \tag{1}$$

*and, furthermore, the map*

$$x \in \mathbf{R}^d \to (1 - \Delta)^{\beta/2} \delta_x \circ F \in \mathcal{D}_p^{-\alpha}$$

*is bounded and continuous provided that* $0 \leq \beta < \alpha$ *and* $1 < p < \infty$ *satisfy*

$$0 \leq \beta < \alpha \wedge \delta - d/q, \quad 1/p + 1/q = 1. \tag{2}$$

*If* $F$ *is regular, then it is clearly in the class* $\mathbf{D}(\delta)$ *for every* $\delta > 0$. *Hence* (1) *holds with bounded and continuous dependence in* $x$ *if* $0 \leq \beta < \alpha$ *and* $1 < p < \infty$ *satisfy*

$$0 \leq \beta < \alpha - d/q, \quad 1/p + 1/q = 1. \tag{3}$$

Since $(1 - \Delta)^{-\beta/2}(C^0(\mathbf{R}^d)) \subset C^\beta(\mathbf{R}^d)$, we have

**Corollary 1** *Let* $F \in \mathbf{D}(\delta)$ *and* $0 \leq \beta < \alpha$ *and* $1 < p < \infty$ *satisfy* (2). *Then* $u \in C^\beta(\mathbf{R}^d)$ *for every* $G \in \mathcal{D}_q^\alpha$, *where* $u$ *is defined by*

$$u(x) := E[G \cdot \delta_x \circ F] = E[G | F = x] p_F(x).$$

**Remark 2** In order that (1) holds for every regular $F$ and $x \in \mathbf{R}^d$, the condition (3) is also necessary. To see this, choose $d$ orthonormal elements $h_1, \ldots, h_d$ in $H$

and define $F = (F^1, \ldots, F^d)$ by $F^i = D^*h_i$, $i = 1, \ldots, d$, where $D^*h$ is the first order chaos associated to $h \in H$. Clearly, $F$ is regular, standard $d$-dim. Gaussian distributed and it is easy to see [13, 22] that

$$(1 - L)^{-\alpha/2}[\delta_0(F)] =$$
$$\frac{1}{\Gamma(\alpha/2)(2\pi)^{d/2}} \int_0^\infty e^{-t} t^{\alpha/2-1}(1 - e^{-2t})^{-d/2} \exp\left[-\frac{e^{-2t}|F|^2}{2(1 - e^{-2t})}\right] dt.$$

From this, it is easy to deduce that $\delta_0 \circ F \in \mathcal{D}_p^{-\alpha}$ if and only if $\alpha > 0$ and $1 < p < \infty$ satisfy

$$d/q < \alpha, \quad 1/p + 1/q = 1.$$

This proves our assertion when $\beta = 0$. The case of general $\beta$ can be discussed similarly.

It has been obtained by Airault-Malliavin [2] and Sugita [18] that every positive element in $\mathcal{D}_{1+}^{-\infty}$ defines a finite Borel measure on $B$ which does not charge on slim sets (a regular measure in the sense of Itô [7]). Let $F : B \to \mathbf{R}^d$ be in the class $\mathbf{D}(\delta)$ for some $\delta$ and, for each $x \in \mathbf{R}^d$, let $\mu_x(dw)$ be the Borel measure corresponding to $\delta_x \circ F$ which is clearly positive element in $\mathcal{D}_{1+}^{-\infty}$. Then we have a family of uniformly bounded Borel measures $\{\mu_x\}_{x \in \mathbf{R}^d}$ continuous in $x$ with respect to the narrow topology on the space of bounded Borel measures on $B$. This gives a version of regular conditional probability distribution given $F$: Namely we have the following *disintegration formula*: For every bounded Borel function $G(w)$ on $B$ and every bounded Borel function $f$ on $\mathbf{R}^d$,

$$E[G(w)f(F(w))] = \int_{\mathbf{R}^d} f(x)\left\{\int_B G(w)\mu_x(dw)\right\} dx. \tag{4}$$

The class of $G$ can be extended to that of Wiener functionals $G$ in certain Sobolev spaces with positive differentiability index by replacing $G$ in the right-hand side of (4) by its quasi-continuous modification. The total measure $\mu_x(B)$ coincides with the density $p_F(x) = E[\delta_x(F)]$ of the law of $F$. Another refinement of the disintegration formula was obtained by Lescot [10].

Finally we give typical examples. We take as the basic space $(B, \mu)$ the classical $r$-dimensional Wiener space $(W_0^r, P)$ where, for an arbitrary but fixed $T > 0$, $W_0^r = \{w \in \mathcal{C}([0, T] \to \mathbf{R}^r) \mid w(0) = 0\}$ and $P$ is the $r$-dimensional Wiener measure.

Let $\sigma(t, x) : [0, T] \times \mathbf{R}^d \to \mathbf{R}^d \otimes \mathbf{R}^r$ and $b(t, x) : [0, T] \times \mathbf{R}^d \to \mathbf{R}^d$ be continuous functions, uniformly Lipschitz continuous in $x$ and uniformly Hölder continuous in

$t$, i.e., there exist constants $K > 0$ and $0 < \mu \leq 1$ such that

$$|\sigma(t, x) - \sigma(s, y)| + |b(t, x) - b(s, y)| \leq K\{|x - y| + |t - s|^\mu\} \tag{5}$$

for all $x, y \in \mathbf{R}^d$, $0 \leq s \leq t \leq T$. For each $0 \leq s < T$ and $x \in \mathbf{R}^d$, we denote by $X(s, x; t)[w] = X(s, x; t)$ the unique solution to the following Itô SDE on $\mathbf{R}^d$:

$$dX(t) = \sigma(t, X(t))dw(t) + b(t, X(t))dt, \quad s \leq t \leq T \tag{6}$$

$$X(s) = x \in \mathbf{R}^d.$$

Then (cf. [3]), for fixed $s \leq t$ and $x$,

$$X(s, x; t) \in \mathcal{D}^1_{\infty-}[\mathbf{R}^d].$$

In order to have

$$X(s, x; t) \in \mathbf{D}(\delta),$$

we assume further that

(A.I) $\sigma$ is bounded and uniformly nondegenerate, i.e., there exists a constant $c > 0$ such that $\sigma \cdot \sigma * (s, x) \geq c \cdot I$, where $[\sigma \cdot \sigma*]^{ij}(s, x) = \sum_{k=1}^r \sigma_k^i(s, x)\sigma_k^j(s, x)$, $i, j = 1, \ldots, d$.

Then, for fixed $x \in \mathbf{R}^d$ and $0 \leq s < t \leq T$, $d$-dimensional Wiener functional $F = X(s, x; t)$ satisfies (ii) of Definition 3 and by a well-known Gaussian estimate of the fundamental solution $p(s, x; t, y)$ to the backward heat equation

$$\frac{\partial u}{\partial s} + L_x^s u = 0, \tag{7}$$

where

$$L_x^s = \frac{1}{2}[\sigma \cdot \sigma*]^{ij}(s, x)\frac{\partial}{\partial x_i}\frac{\partial}{\partial x_j} + b^i(s, x)\frac{\partial}{\partial x_i}.$$

The condition (iii) is also satisfied [17].

Let $\mu \geq 0$ and $1 < p < \infty$ be given.

**Definition 4** We say that a function $f(t, x)$ on $[0, T] \times \mathbf{R}^d$ *satisfies* $(A_{\mu,p})$ if

(i) $f(t, x)$ is $C^{[\mu]}$ in $x$ with derivatives up to $[\mu]$-th order bounded on $[0, T] \times \mathbf{R}^d$,

(ii) each $[\mu]$-th derivative in $x$ can be expressed as a sum $f_1(t, x) + f_2(t, x)$ where $f_1(t, \cdot) \in L_p^{\mu - [\mu]}(\mathbf{R}^d)$ with $\sup_{t \in [0, T]} |f_1(t, \cdot)|_{p, \mu - [\mu]} < \infty$ and $f_2(t, \cdot)$ is $C^1$ in $x$ with first order derivatives bounded on $[0, T] \times \mathbf{R}^d$.

We assume further that,

(A.II) for some $\delta > 0$,(each component) of $\sigma$ satisfies the condition $(A_{1+\delta,p})$ with $p > d \vee 2$ and $b$ satisfies $(A_{1+\delta,p})$ with $p > d/2 \vee 1$.

Then, it was proved in [24] that

$$X(s,x;t) \in \mathcal{D}^{1+\delta}_{\infty-}[\mathbf{R}^d]$$

and hence if the cocfficients $\sigma$ and $b$ satisfy (5), (A.I) and (A.II), then, for fixed $x \in \mathbf{R}^d$ and $0 \le s < t \le T$,

$$X(s,x;t) \in \mathbf{D}(\delta).$$

Now we can conclude that $(1-\Delta)^{\beta/2}\delta_y(X(s,x;t))$ can be defined in $\mathcal{D}^{-\infty}_{1+}$ for every $0 \le \beta < \delta$. To be more precise, under the assumptions (5), (A.I) and (A.II) of coefficients, $(1-\Delta)^{\beta/2}\delta_y(X(s,x;t))$ belongs to $\mathcal{D}^{-\alpha}_p$ if $0 \le \beta < \alpha$ and $1 < p < \infty$ satisfy (2). The generalized expectation $E[\delta_y(X(s,x;t))]$ coincides with the fundamental solution $p(s,x;t,y)$ to the heat equation (7).

## 3. A Class of Wiener Functionals that can be Coupled with Donsker's Delta Functions

In the previous section, we have studied the Sobolev spaces to which Donsker's delta functions can belong. Any Wiener functional in the Sobolev space, dual to which Donsker's delta functions can belong, can be coupled with Donsker's delta functions and the space dual to $\mathcal{D}^{-\alpha}_p$ is $\mathcal{D}^{\alpha}_q$, $1/p+1/q = 1$. Thus, for example, we can conclude the following from Cor. 1:

Let $\delta > 0$ and $F \in \mathbf{D}(\delta)$. Then $\delta_x \circ F$ can be coupled with any $G \in \mathcal{D}^{\alpha}_{\infty-}$ for every $\alpha > 0$. Furthermore, $u$ defined by $u(x) = E[G \cdot \delta_x \circ F]$ has the following regularity property: $u \in C^{\beta}(\mathbf{R}^d)$ for every $0 \le \beta < \alpha \wedge \delta$.

So it is important to give many examples of Wiener functionals belonging to fractional order Sobolev spaces with positive differentiability indices.

Consider the classical $r$-dimensional Wiener space $(W = W^r_0, P)$, where $W^r_0 = \{w \in C([0,T] \to \mathbf{R}^r) \mid w(0) = 0\}$ endowed with the standard $r$-dimensional Wiener measure $P$. Here, $T > 0$ be arbitrary but fixed.

**Example 1** Let $T > 0$ be arbitrary but fixed and consider the following four types of Wiener functionals of $w = (w^1(t),\ldots,w^r(t)) \in W$ :

(i) $G^{(1)}_f(w) = \int_0^T f(w(t))dt$.

(ii) $G_{f,i}^{(2)}(w) = \int_0^T f(w(t))dw^i(t), \quad i = 1, \ldots, r.$

(iii) $G_f^{(3)}(w) = \int_0^T dt \int_0^t f(w(t) - w(s))ds.$

(iv) $G_{f;i,j}^{(4)}(w) = \int_0^T dw^j(t) \int_0^t f(w(t) - w(s))d\widetilde{w}^i(s), \quad i,j = 1, \ldots, r,$

where $\sim$ is the backward Itô integral.

We say $f \in \tilde{\mathbf{L}}_p^\alpha(\mathbf{R}^r)$, $\alpha > 0, 1 < p < \infty$, if $f$ can be written as $f = f_1 + f_2$, where $f_1 \in \mathbf{L}_p^\alpha(\mathbf{R}^r) := (1 - \Delta)^{-\alpha/2}(\mathbf{L}_p(\mathbf{R}^r))$ and $f_2 \in C^{[\alpha]+1}(\mathbf{R}^r)$. Then we have the following:

(a) Let $\alpha \geq 0$ and $p > (r/2) \vee 1$. Then $G_f^{(1)}$, $G_f^{(3)} \in \mathcal{D}_{\infty-}^\alpha$ if $f \in \tilde{\mathbf{L}}_p^\alpha(\mathbf{R}^r)$.

(b) Let $\alpha \geq 0$ and $p > r \vee 2$. Then $G_{f,i}^{(2)}$, $G_{f;i,j}^{(4)} \in \mathcal{D}_{\infty-}^\alpha$ for $i,j = 1, \ldots, r$ if $f \in \tilde{\mathbf{L}}_p^\alpha(\mathbf{R}^r)$.

As an application, we can deduce the following from (b) combined with the Tanaka formula and the fact that the Heaviside function $Y(x) = 1_{[0,\infty)}(x)$ is in $\tilde{\mathbf{L}}_p^\alpha(\mathbf{R}^1)$ if and only if $\alpha p < 1$:

Let $r = 1$ and $l(t, x)$ be the local time of $\{w(t)\}$ at $x$. Then for fixed $t > 0$ and $x \in \mathbf{R}$,

$$l(t, x) \in \mathcal{D}_{\infty-}^\alpha \quad \text{for every } \alpha < 1/2.$$

This is a little bit improvement of the result in [13].

**Example 2** Suppose we are given a scalar function $U(t, x)$ and a vector function $\mathbf{V}(t, x) = \{V_k(t, x)\}_{k=1}^r$ defined on $[0, T] \times \mathbf{R}^d$. Let $\alpha > 0$ and assume that $U$ satisfies $(A_{\alpha,p})$ (cf. Definition 4 above) with $p > d/2 \vee 1$ and $V_k$ satisfies $(A_{\alpha,p})$ with $p > d \vee 2$, $k = 1, \ldots, r$. Let $X(s, x; t)$ be the solution of SDE (6) in the previous section with coefficients satisfying the conditions (5), (A.I) and (A.II), and define, for fixed $0 \leq s < t \leq T$, a Wiener functional $A_s^t(X) = A_s^t(X)[w]$ by

$$A_s^t(X) = \int_s^t U(u, X(s, x; u))du + \int_s^t V_k(u, X(s, x; u))dw^k(u)$$
$$- \frac{1}{2} \int_s^t |\mathbf{V}(u, X(s, x; u))|^2 du.$$

Then we can show that $A_s^t(X) \in \mathcal{D}_{\infty-}^\alpha$. If we assume further that $U$ and $\mathbf{V}$ are of linear growth in $x$ uniformly in $t$, then we can show that $G(w) = \exp[A_s^t(X)] \in \mathcal{D}_{\infty-}^{\alpha'}$ for every $\alpha' < \alpha$, by the following general fact [24, Theorem 1.4]:

If $F \in \mathcal{D}_{\infty-}^\alpha$, $\alpha > 0$, and $\exp F \in \mathcal{L}_{\infty-}$, then $\exp F \in \mathcal{D}_{\infty-}^{\alpha'}$ for every $\alpha' < \alpha$.

As an example of applications, let $A(t,x)$ be a smooth function in $(t,x)$ with bounded derivatives on $[0,T] \times \mathbf{R}_+^d$ where $\mathbf{R}_+^d = \{x \in \mathbf{R}^d | x_1 \geq 0\}$. Set

$$V(t,x) = A(t,x), \quad x \in \mathbf{R}_+^d$$

$$= -A(t,\check{x}), \quad x \in \mathbf{R}_-^d = \{x \in \mathbf{R}^d | x_1 < 0\},$$

where $\check{x} = (-x_1, x_2, \ldots, x_d)$ for $x = (x_1, x_2, \ldots, x_d) \in \mathbf{R}^d$. Then we can conclude the following regularity for a Girsanov density:

$$\exp[\int_s^t V(u, X(s,x;u))dw^1(u) - \frac{1}{2}\int_s^t V(u, X(s,x;u))^2 du] \in \mathcal{D}_{\infty-}^\alpha$$

for every $\alpha < \frac{1}{d\sqrt{2}}$.

The above results in this section are based on the complex interpolation theory [16]. Details will be discussed in elsewhere.

## 4. Wiener Chaos Expansions of Donsker's Delta Functions Associated With Itô Functionals

Throughout this section, let $T > 0$ be arbitrary but fixed and take, as our basic space $(B,\mu)$, the $d$-dimensional Wiener space $(W,P)$ with $W = \{w \in C([0,T] \to \mathbf{R}^d) | w(0) = 0\}$. It is well-known that $\mathcal{L}_2 = \mathcal{D}_2^0$ is decomposed into the sum of mutually orthogonal subspaces called *Wiener's homogeneous chaos*:

$$\mathcal{L}_2 = \mathcal{C}^{(0)} \oplus \mathcal{C}^{(1)} \oplus \cdots \oplus \mathcal{C}^{(n)} \oplus \cdots.$$

We denote by $J_n$, $n = 0, 1, \ldots$, the orthogonal projection onto $\mathcal{C}^{(n)}$ so that $F = \sum_n J_n F$ for every $F \in \mathcal{L}_2$. For fixed $0 \leq s < t \leq T$, we denote by $\mathcal{L}_{2,[s,t]}$, $(\mathcal{C}_{[s,t]}^{(n)})$, the closed subspace consisting of all $F \in \mathcal{L}_2$, (resp. $F \in \mathcal{C}^{(n)}$), which are $\mathbf{B}[s,t]$-measurable, where $\mathbf{B}[s,t]$ is the $\sigma$-field on $W$ generated by $\{w(u) - w(v), s \leq v < u \leq t\}$. Let

$$D_{[s,t]}^{(n)} = \{(t_1, \ldots, t_n) \in [s,t]^n \mid s < t_1 < \cdots < t_n < t\}.$$

It is well-known (Itô [6]) that every $F \in \mathcal{C}_{[s,t]}^{(n)}$ can be represented by a system

$$(F^{(k_1,\ldots,k_n)}(t_1,\ldots,t_n))$$

of functions in $L_2(D_{[s,t]}^{(n)})$ in the form of iterated Itô integral:

$$F = \sum_{(k_1,\ldots,k_n)} \int \cdots \int_{D_{[s,t]}^{(n)}} F^{(k_1,\ldots,k_n)}(t_1,\ldots,t_n)dw^{k_1}(t_1) \cdots dw^{k_n}(t_n),$$

where the indices $(k_1, \ldots, k_n)$ range over all $n$-tuples of $1, 2, \ldots, d$, i.e., $(k_1, \ldots, k_n) \in \{1, 2, \ldots, d\}^n$. Furthermore, we have

$$\| F \|_{\mathcal{L}_2}^2 = \sum_{(k_1, \ldots, k_n)} \int \cdots \int_{D_{[s,t]}^{(n)}} |F^{(k_1, \ldots, k_n)}(t_1, \ldots, t_n)|^2 \, dt_1 \cdots dt_n.$$

It is well-known that $\mathcal{C}^{(n)} \subset \mathcal{D}_{\infty-}^\infty$ for each $n = 0, 1, \ldots$ and the projection $J_n : \mathcal{L}_2 \to \mathcal{C}^{(n)}$ can be uniquely extended by continuity as an operator from $\mathcal{D}_{1+}^{-\infty}$ into $\mathcal{C}^{(n)}$, which is denoted again by $J_n$. It is not easy to obtain in general an explicit form of $J_n[\delta_x \circ F]$ when a Donsker's delta function $\delta_x \circ F$ can be defined. Kuo [8] obtained this in the case of $F(w) = w(1)$. Here, we discuss the case of Itô functional $F(w) = X(s, x; t)[w]$ obtained as the solution to SDE (6).

Let $\sigma(t, x)$ and $b(t, x)$ be given as in section 2 satisfying (5) and the nondegeneracy condition (A.I). *We assume that $r = d$ in this section.* Let $X(s, x; t)$ be the solution to SDE (6). A Krylov-Zbonkin-Veretennikov formula [19, 26] states that the kernel

$$\Pi_n^{(k_1, \ldots, k_n)}(s, x; t, y)[t_1, \ldots, t_n] \in L^2(D_{[s,t]}^{(n)})$$

representing $J_n[\delta_y(X(s, x; t))]$ (which is clearly $\mathbf{B}[s,t]$-measurable) should be given, for each $0 \le s < t \le T$ and $x, y \in \mathbf{R}^d$, by

$$\Pi_n^{(k_1, \ldots, k_n)}(s, x; t, y)[t_1, \ldots, t_n] = \int_{\mathbf{R}^d} \cdots \int_{\mathbf{R}^d} p(s, x; t_1, y_1) \cdot$$
$$\cdot A_{k_1, y_1}^{t_1}[p(t_1, y_1; t_2, y_2)] \cdots A_{k_n, y_n}^{t_n}[p(t_n, y_n; t, y)] dy_1 \cdots dy_n, \qquad (8)$$

where $p(s, x; t, y) = \Pi_0(s, x; t, y)$ is the fundamental solution to the heat equation (7) and the first order differential operators $A_{k,x}^s$ are defined by

$$A_{k,x}^s = \sigma_k^i(s, x) \frac{\partial}{\partial x^i}, \quad k = 1, \ldots, d. \qquad (9)$$

This formula, when $\sigma(s, x) \equiv \sigma$ (a constant non-singular matrix) and $b(s, x) \equiv 0$, reduces to

$$p(s, x; t, y) =$$
$$[2\pi(t - s)]^{-\frac{d}{2}} \{\det[\sigma \cdot \sigma^*]\}^{-\frac{1}{2}} \exp\left[ -\frac{1}{2} < \sigma^{-1} \frac{(y - x)}{(t - s)^{\frac{1}{2}}}, \sigma^{-1} \frac{(y - x)}{(t - s)^{\frac{1}{2}}} > \right] \qquad (10)$$

and for $n = 0, 1, \ldots,$

$$\Pi_n^{(k_1,\ldots,k_n)}(s, x; t, y)[t_1, \ldots, t_n] =$$
$$p(s, x; t, y)(t - s)^{-n/2}\mathbf{m}! \cdot \prod_{k=1}^{d} H_{m_k}\left(\left[\sigma^{-1}\frac{y - x}{(t - s)^{1/2}}\right]^k\right), \tag{11}$$

where $\mathbf{m} = (m_1, \ldots, m_d)$ is determined from $(k_1, \ldots, k_n)$ by

$$m_k = \sharp\{l | k_l = k\}, \quad k = 1, \ldots, d,$$

so that $|\mathbf{m}| = \sum_{k=1}^{d} m_k = n$. Also, $\mathbf{m}! = m_1! \cdots m_d!$ and for $m = 0, 1, \ldots,$

$$H_m(x) = \frac{(-1)^m}{m!} e^{x^2/2} \frac{d^m}{dx^m}(e^{-x^2/2}), \quad x \in \mathbf{R}^1.$$

Hence, for $n = 0, 1, \ldots,$

$$J_n[\delta_y(x + \sigma[w(t) - w(s)])] = p(s, x; t, y) \sum_{\mathbf{m}; |\mathbf{m}| = n} \mathbf{m}! \cdot$$
$$\cdot \prod_{k=1}^{d} H_{m_k}\left(\left[\sigma^{-1}\frac{y - x}{(t - s)^{1/2}}\right]^k\right) \cdot \prod_{k=1}^{d} H_{m_k}\left(\frac{w^k(t) - w^k(s)}{(t - s)^{1/2}}\right)$$

and this is the formula by Kuo [8].

In the general case of coefficients, however, it is by no means obvious that

$$\Pi_n^{(k_1,\ldots,k_n)}(s, x; t, y)[t_1, \ldots, t_n] \in L^2(D_{[s,t]}^{(n)}).$$

A standard estimate of the heat kernel

$$\left|\frac{\partial}{\partial x_i}p(s, x; t, y)\right| \leq C_1(t - s)^{-d/2-1/2} \exp\left[-C_2\frac{|y - x|^2}{t - s}\right]$$

helps us to prove that $\Pi_n^{(k_1,\ldots,k_n)}(s, x; t, y)$ is in $L^1(D_{[s,t]}^{(n)})$ but not in $L^2(D_{[s,t]}^{(n)})$. A proof given by Hu-Meyer [4, p. 66] is based on an integration by parts so that a more stringent regularity assumption is required on the coefficients than (5). In [25], this problem has been solved by extending the classical method of *Levi's sum* for constructing heat kernels.

Let $\alpha \in \mathbf{R}$ be given. By a *chaos kernel of order* $\alpha$, we mean a sequence

$$E(s, x; t, y) = \{E_n^{(k_1,\ldots,k_n)}(s, x; t, y)[t_1, \ldots, t_n] \mid 1 \leq k_1, \ldots, k_n \leq d\}_{n=0}^{\infty}$$

of $L^2(D_{[s,t]}^{(n)})$-functions depending on $(s, x, t, y)$, $0 \leq s < t \leq T$, $x, y \in \mathbf{R}^d$ such that the dependence in $(s, x, t, y)$ is measurable in the obvious sense and

$$\|E_n^{(k_1,\ldots,k_n)}(s, x; t, y)\|_{L^2(D_{[s,t]}^{(n)})} \leq C_1(t-s)^{\alpha - d/2 - 1/2} \exp\left[ -C_2 \frac{|y-x|^2}{t-s} \right]$$

for every $k_1, \ldots, k_n$, $n = 0, 1, \ldots$, where constants $C_1, C_2 > 0$ may depend on $n$. The totality of chaos kernels of order $\alpha$ is denoted by $\mathcal{E}_\alpha$. Given two chaos kernels $E(s, x; t, y)$ and $F(s, x; t, y)$, we define new chaos kernels $[E \natural F](s, x; t, y)$ and $[E \flat_k F](s, x; t, y)$, $k = 1, 2, \ldots, d$, as follows:

$$[E \natural F]_n^{(k_1,\ldots,k_n)}(s, x; t, y)[t_1, \ldots, t_n] \equiv$$

$$\sum_{l+m=n, l \geq 0, m \geq 0} \int_{t_l}^{t_{l+1}} d\tau \int_{\mathbf{R}^d} dz\, E_l^{(k_1,\ldots,k_l)}(s, x; \tau, z)[t_1, \ldots, t_l]$$

$$\times F_m^{(k_{l+1},\ldots,k_n)}(\tau, z; t, y)[t_{l+1}, \ldots, t_n],$$

where $n \geq 0$, $(t_1, \ldots, t_n) \in D_{[s,t]}^{(n)}$, and $t_0 = s, t_{n+1} = t$. On the other hand, for $n \geq 1$, $(t_1, \ldots, t_n) \in D_{[s,t]}^{(n)}$,

$$[E \flat_k F]_n^{(k_1,\ldots,k_n)}(s, x; t, y)[t_1, \ldots, t_n] \equiv$$

$$\sum_{l+m=n-1, l \geq 0, m \geq 0} \delta_{k_{l+1}, k} \int_{\mathbf{R}^d} dz\, E_l^{(k_1,\ldots,k_l)}(s, x; t_{l+1}, z)[t_1, \ldots, t_l]$$

$$\times F_m^{(k_{l+2},\ldots,k_n)}(t_{l+1}, z; t, y)[t_{l+2}, \ldots, t_n]$$

and we define $[E \flat_k F]_n^{(k_1,\ldots,k_n)}(s, x; t, y)[t_1, \ldots, t_n] = 0$ if $n = 0$.

Let $E \in \mathcal{E}_{\alpha_1}$ and $F \in \mathcal{E}_{\alpha_2}$. It was shown in [25, Prop. 3.5] that if $\alpha_1 > -1/2$ and $\alpha_2 > -1/2$ then $E \natural F$ is well-defined with $E \natural F \in \mathcal{E}_{a_1 + a_2 + 1/2}$ and if $\alpha_1 > 0$ and $\alpha_2 > 0$ then $E \flat_k F$ is well-defined with $E \flat_k F \in \mathcal{E}_{a_1 + a_2}$ for $k = 1, \ldots, d$.

Let $\sigma(t, x)$ and $b(t, x)$ be given with assumptions (12) and (A.I), where (12) is the Hölder continuity assumption which is weaker than (5):

$$|\sigma(t, x) - \sigma(t, y)| + |b(t, x) - b(t, y)| \leq K\{|x-y|^\beta + |t-s|^{\beta/2}\} \qquad (12)$$

for all $x, y \in \mathbf{R}^d$ and $0 \leq s \leq t \leq T$, for some $\beta > 0$ and $K > 0$. Let

$$\Delta(x, s; t, y) = \{\Delta_n^{(k_1,\ldots,k_n)}(s, x; t, y)[t_1, \ldots, t_n] \mid 1 \leq k_1, \ldots, k_n \leq d\}_{n=0}^\infty$$

be the chaos kernel given by (10) and (11) with $\sigma = \sigma(t, y)$ (the coefficients *frozen at* $(t, y)$). Then $\Delta \in \mathcal{E}_{1/2}$. Define chaos kernels, $(A_k - \overline{A}_k)\Delta$, $k = 1, \ldots, d$ and $(L - \overline{L})\Delta$ by

$$(A_k - \overline{A}_k)\Delta(s, x; t, y) = \{(A_{k,x}^s - \overline{A}_{k,x}^{t,y})\Delta_n^{(k_1, \ldots, k_n)}(s, x; t, y)[t_1, \ldots, t_n]\}$$

and

$$(L - \overline{L})\Delta(s, x; t, y) = \{(L_x^s - \overline{L}_x^{t,y})\Delta_n^{(k_1, \ldots, k_n)}(s, x; t, y)[t_1, \ldots, t_n]\},$$

where $L_x^s$ and $A_{k,x}^s$ are defined by (7) and (9), respectively,

$$\overline{A}_{k,x}^{t,y} = \sigma_k^i(t, y)\frac{\partial}{\partial x^i}, \quad k = 1, \ldots, d$$

and

$$\overline{L}_x^{t,y} = \frac{1}{2}[\sigma \cdot \sigma*]^{ij}(t, y)\frac{\partial}{\partial x_i}\frac{\partial}{\partial x_j}.$$

Then $(A_k - \overline{A}_k)\Delta \in \mathcal{E}_{\beta/2}$ and $(L - \overline{L})\Delta \in \mathcal{E}_{\beta/2 - 1/2}$, (cf. [25]). Hence, for $\alpha > 0$, we can define an operator

$$\Lambda : \mathcal{E}_\alpha \to \mathcal{E}_{\alpha + \beta/2}$$

by

$$\Lambda(E) = \sum_{k=1}^{d} E\flat_k[(A_k - \overline{A}_k)\Delta] + E\sharp[(L - \overline{L})\Delta].$$

Define $\Pi^{(l)} \in \mathcal{E}_{1/2}$, $l = 0, 1, \ldots$, by

$$\Pi^{(l)} = \Delta + \Lambda(\Delta) + \cdots + \Lambda^l(\Delta).$$

Then we can show that, for each $n = 0, 1, \ldots$ and $(s, x, t, y)$, $(\Pi^{(l)})_n^{(k_1, \ldots, k_n)}(s, x; t, y)$ converges in $L^2(D_{[s,t]}^{(n)})$ as $l \to \infty$. This convergence is compact uniform in $(s, x, t, y)$ and, furthermore, the limit $\Pi_n^{(k_1, \ldots, k_n)}(s, x; t, y)$ can be shown to coincide with that given by (8). This shows, in particular, that $\Pi_n^{(k_1, \ldots, k_n)}(s, x; t, y)$ is in $L^2(D_{[s,t]}^{(n)})$ for every $n$ under the assumption of Hölder continuity only. Note that, when $n = 0$, $(\Pi^{(l)})_0(s, x; t, y)$ converges to the fundamental solution $p(s, x; t, y)$ of the heat equation (7) and this is precisely the classical Levi sum.

We do not know if the sequence $[\Pi_n(s, x; t, y)](w) \in \mathcal{C}^{(n)}$, $n = 0, 1, \ldots$, defined by

$$[\Pi_n(s, x; t, y)](w) =$$

$$\sum_{(k_1, \ldots, k_n)} \int \cdots \int_{D_{[s,t]}^{(n)}} \Pi_n^{(k_1, \ldots, k_n)}(s, x; t, y)[t_1, \ldots, t_n] dw^{k_1}(t_1) \cdots dw^{k_n}(t_n)$$

322

determines an element $\Phi \in \mathcal{D}_{1+}^{-\infty}$, that is, $\Phi \in \mathcal{D}_{1+}^{-\infty}$ exists such that

$$J_n[\Phi] = [\Pi_n(s, x; t, y)], \quad n = 0, 1, \dots.$$

We know, however, that under the assumptions (5), (A.I) and (A.II) of coefficients, $\delta_y(X(s, x; t))$ can be defined in $\mathcal{D}_{1+}^{-\infty}$ and then we can conclude that

$$J_n[\delta_y(X(s, x; t))] = [\Pi_n(s, x; t, y)].$$

Cf. [25] for details.

## References


[1] Adams, R. A.: *Sobolev Spaces*. Academic Press, 1987

[2] Airault, H. and Malliavin, P.: Intégration géométrique sur l'espace de Wiener; *Bull. Sc. Math., 2e série* **112** (1988) 3–52

[3] Bouleau, N. and Hirsch, F.: *Dirichlet Forms and Analysis on Wiener Space*. Walter de Gruyter, 1991

[4] Hu, Y. Z. and Meyer, P. A.: Chaos de Wiener et intégrale de Feynman; in:*Sém. de Prob. XXII* J. Azéma, P. A. Meyer, and M. Yor (eds.) *LNM* **1321** (1988) 51–71, Springer-Verlag

[5] Ikeda, N. and Watanabe, S.: *Stochastic Differential Equations and Diffusion Processes*. Second edition, North-Holland/Kodansha, 1988

[6] Itô, K.: Multiple Wiener integral; *J. Math. Soc. Japan* **3** (1951) 157–169

[7] Itô, K.: An elementary approach to Malliavin fields; in:*Asymptotic problems in probability theory: Wiener functionals and asymptotics*, K. D. Elworthy and N.Ikeda (eds.), *Pitman RNMS* **284** (1993) 35–103, Longman

[8] Kuo, H. -H.: Donsker's delta function as a generalized Brownian functionals and its application; in:*Theory and Applications of Random Fields, Proc. IFIP Conf. Bangalore 1982*, G. Kallianpur (ed.), *LNCIS* **49** (1983) 167–178, Springer-Verlag

[9] Kusuoka, S.: On the foundation of Wiener Riemannian manifolds; in:*Stochastic Analysis, Path Integration and Dynamics*, K. D. Elworthy and J. -C. Zambrini (eds.), *Pitman RNMS* **200**(1989) 103–164, Longman

[10] Lescot, P.: Un théorème de désintégration en analyse quasi-sure; *Séminaire de Prob. XXVII* J. Azéma, P. A. Meyer, and M. Yor (eds.), *LNM* **1557** (1993) 256–275, Springer-Verlag

[11] Malliavin, P.: *Stochastic Analysis*. Springer-Verlag, 1994 (to appear)

[12] Meyer, P. A.: Retour sur la théorie de Littlewood-Paley; in: *Séminaire de Prob. XV, 1987/1980*, J. Azéma and M. Yor (eds.), *LNM* **850** (1981) 151–166, Springer-Verlag

[13] Nualart, D. and Vives, J.: Smoothness of Brownian local times and related functionals; *Potential Analysis* **1** (1992) 257–263

[14] Ren, J. and Watanabe, S.: A convergence theorem for probability densities and conditional expectations of Wiener functionals; in preparation

[15] Stein, E. M.: *Singular Integrals and Differentiability Properties of Functions.* Princeton Univ. Press, 1970

[16] Stein. E. M. and Weiss, G.: *Introduction to Fourier Analysis on Euclidean Spaces.* Princeton Univ. Press, 1971

[17] Stroock, D. W. and Varadhan, S. R. S.: *Multidimensional Diffusion Processes.* Springer-Verlag, 1979

[18] Sugita, H.: Positive generalized Wiener functions and potential theory over abstract Wiener spaces; *Osaka J. Math.* **25** (1988) 665–698

[19] Veretennikov, A. Yu. and Krylov, N. V.: On explicit formulas for solutions of stochastic differential equations; *Math. USSR Sbornik* **29** (1976) 239–256

[20] Watanabe, S.: Malliavin calculus in terms of generalized Wiener functionals; in: *Theory and Applications of Random Fields, Proc. IFIP Conf. Bangalore 1982*, G. Kallianpur (ed.), *LNCIS* **49** (1983) 284–290, Springer-Verlag

[21] Watanabe, S.: *Lectures on Stochastic Differential Equations an Malliavin Calculus.* Tata Institute of Fundamental Research/Springer, 1984

[22] Watanabe, S.: Donsker's $\delta$-functions in the Malliavin calculus; in: *Stochastic Analysis, Liber Amicorum for Moshe Zakai*, E. Mayer-Wolf, E. Merzbach, and A. Shwartz (eds.), (1991), 495–502, Academic Press

[23] Watanabe, S.: A fractional calculus on Wiener space; in: *Stochastic Processes, A Festschrift in Honour of Gopinath Kallianpur*, S. Cambanis, J. K. Ghosh, R. L. Karandikar, and P. K. Sen (eds.), (1993) 341–346, Springer-Verlag

[24] Watanabe, S.: Fractional order Sobolev spaces on Wiener space; *Probab. Theory Relat. Fields* **95** (1993) 175–198

[25] Watanabe, S.: Stochastic Levi sums; *Comm. Pure Appl. Math.* (1994) to appear

[26] Zbonkin, A. K. and Krylov, N. V.: On strong solutions of stochastic differential equations; *Sel. Math. Sov.* **1** (1981) 19–61, Birkhäuser. (Originally published in Russian in: *Proc. School Seminar in Random Processes*, Vilnius, 1975)

Shinzo Watanabe
Department of Mathematics
Faculty of Science
Kyoto University
Kyoto 606, JAPAN
*E-mail*: watanabe@kusm.kyoto-u.ac.jp